Lecture Notes in Morphogenesis

Lecture Notes in Morphogenesis is an interdisciplinary book series which aims to face the questions of emergence, individuation and becoming of forms from several different points of view: those of pure and applied mathematics, of computational algorithms, of biology, of neurophysiology, of cognitive and social structures. The set of questions above concerns all the manifestations of Being, all the manifestations of Life. At the heart of contemporary embryogenesis lies an essential question: How can form emerge from the constant, chaotic flow? How can a sequence of purely informational elements—an a-signifying combination of chemical substances organized in the DNA molecule—evolve into the highly complex and structured forms of the living organism? A similar question can be asked when we deal with the morphogenesis of vision in neural systems and with the creation of evolving synthetic images, since digital technology makes possible the simulation of emergent processes both of living bodies and of visual forms. Finally the very idea that abstract structures of meaning could be captured in terms of morphodynamic evolution opens the door to new models of semiolinguistics, semiotic morphodynamics, and cognitive grammars. An entire heritage of ideas and concepts has to be reconsidered in order to face new and challenging problems: the theoretical framework opened by Goethe with the introduction of the word "Morphogenesis" is developed by D'Arcy Thompson in "On Growth and Form", it is reorganized with new theoretical insights by the classical structuralism of Levi-Strauss and formalized by the dynamical structuralism of René Thom. The introduction of the post-structuralists ideas of individuation (in Gilbert Simondon and Gilles Deleuze) and plasticity of structures builds a bridge to contemporary problems of morphogenesis at a physical, biological, social and transindividual level. The objective of this book series is to provide suitable theoretical and practical tools for describing evolutionary phenomena at the level of Free boundary problems in Mathematics, Embryogenesis, Image Evolution in Visual Perception, Visual Models of Morphogenesis, Neuromathematics, Autonomy and Self-Organization, Morphogenetic Emergence and Individuation, Theoretical Biology, Cognitive Morphodynamics, Cities Evolution, Semiotics, Subjectivation processes, Social movements as well as new frontiers of Aesthetics. To submit a proposal or request further information, please use the PDF Proposal Form or contact directly: *Dr. Thomas Ditzinger (thomas.ditzinger@springer.com).*

Pietro Conte · Anna Caterina Dalmasso ·
Maria Giulia Dondero · Andrea Pinotti
Editors

Algomedia. The Image at the Time of Artificial Intelligence

Springer

Editors
Pietro Conte
Department of Philosophy "Piero Martinetti"
University of Milan
Milan, Italy

Anna Caterina Dalmasso
Department of Philosophy "Piero Martinetti"
University of Milan
Milan, Italy

Maria Giulia Dondero
Département des Romanes
F.R.S.-FNRS/Université de Liège
Liège, Belgium

Andrea Pinotti
Department of Philosophy "Piero Martinetti"
University of Milan
Milan, Italy

ISSN 2195-1934 ISSN 2195-1942 (electronic)
Lecture Notes in Morphogenesis
ISBN 978-3-032-08725-6 ISBN 978-3-032-08726-3 (eBook)
https://doi.org/10.1007/978-3-032-08726-3

This Springer imprint is published by the registered company Springer Nature Switzerland AG
The registered company address is: Gewerbestrasse 11, 6330 Cham, Switzerland

Contents

Editors and Contributors

About the Editors

Pietro Conte is an associate professor in Aesthetics in the Department of Philosophy "Piero Martinetti" at the University of Milan and a member of the research centers "EXT. Extended Realities" and "HCAI. Human-Centered Artificial Intelligence." His research focuses on illusion, hyperrealism, and the multifarious practices of unframing, especially in relation to immersive (virtual) environments. He has explored this thematic cluster in the monographs *Unframing Aesthetics* (2020) and *In carne e cera. Estetica e fenomenologia dell'iperrealismo* (2014).

Anna Caterina Dalmasso is a tenure-track assistant professor in Aesthetics at the Department of Philosophy "Piero Martinetti" at the University of Milan, where she teaches "Media Theory and AI". She is a member of the research centers "EXT. Extended Realities" and "HCAI. Human-Centered Artificial Intelligence." She is also an associate researcher at the Centre Prospéro at UCLouvain—Brussels, where she was a Marie Curie cofund fellow (2017–2019), and a member of the research center *Vivre par(mi) les écrans* at the University "Jean Moulin" Lyon 3. A specialist of Merleau-Ponty's philosophy, her work combines phenomenology, aesthetics, film and media studies, media archaeology, and visual culture. Her current research focuses on the aesthetics of immersive environments and the status of virtual and algorithmic images.

Maria Giulia Dondero is a research director of the National Belgian Fund for Scientific Research (F.R.S.-FNRS) and a professor at the University of Liège. She is the president of the International Association for Visual Semiotics (IAVS/AISV), co-founder and editor-in-chief of the journal *Signata*, and co-director of the collection "Sigilla." Her main research fields are the relation between semiotics and art history, the theory of photography, the diagram in scientific discourse, and generative AI. She is the author of four books: *The Language of Images. The Forms and the Forces* (2020); *The Semiotic Challenge of Scientific Images. A Test Case for Visual Meaning* (with J. Fontanille, 2014); *Sémiotique de la photographie* (with P. Basso, 2011); *Le sacré dans l'image photographique* (2009). She has co-edited 30 collective works and special issues on scientific and artistic images and on the semiotic theory of visual language. She has been Visiting Professor at UNESP, São Paulo; Instituto Nacional de Antropología e Historia (INAH), Mexico City; Paris 2-Panthéon-Assas; CELSA-Sorbonne University; University of Turin, and La Sapienza Rome. She has been invited as a visiting researcher at the University of Southern California (2020) and at Purdue University (2022).

Andrea Pinotti is a full professor in aesthetics in the Department of Philosophy "Piero Martinetti" at the University of Milan, where he coordinates the research centre "EXT. Extended Realities" and has directed the ERC Advanced project *An-Iconology. History, Theory, and Practices of Environmental Images* (2019–2025). His studies focus on image theories and visual culture studies, virtual and augmented reality, empathy theories, memorialisation, and monumentality. He has been a fellow of various international institutions, among which The Italian Academy at Columbia University; The Warburg Institute in London; the ZFL in Berlin; the IEA, the EHESS, and the FMSH in Paris. In 2018, he was awarded the Wissenschaftspreis der Aby-Warburg-Stiftung in Hamburg. In 2023 he was appointed as a Panofsky professor at the Zentralinstitut für Kunstgeschichte in Munich. Among his recent publications is the volume *At the Threshold of the Image. From Narcissus to Virtual Reality* (2021, English ed. 2025).

Contributors

Martina Bacaro University of Bologna, Bologna, Italy

Bruno Bachimont Compiègne University of Technology, Compiègne, France

Marion Colas-Blaise University of Luxembourg, Luxembourg, Luxembourg

Pietro Conte University of Milan, Milan, Italy

Anna Caterina Dalmasso University of Milan, Milan, Italy

Adrien Deliège University of Liège, Liège, Belgium

Maria Giulia Dondero FNRS/Université de Liège, Liège, Belgium

Enzo D'Armenio University of Lorraine, Metz, France

Ruggero Eugeni Università Cattolica del Sacro Cuore, Milan, Italy

Alfio Ferrara University of Milan, Milan, Italy

Remo Gramigna University of Turin, Turin, Italy

Massimo Leone University of Turin, Turin, Italy;
Korea University, Seoul, Korea

Lev Manovich City University of New York, New York, NY, USA

Sergio Picascia University of Milan, Milan, Italy

Andrea Pinotti University of Milan, Milan, Italy

Sofia Pirandello University of Milan, Milan, Italy

Domenico Quaranta Accademia di Belle Arti di Brera, Milan, Italy

Alberto Romele Sorbonne Nouvelle University, Paris, France

Aluminé Rosso University of Liège, Liège, Belgium;
University of Lyon, Lyon, France

Maria Serafini University of Milan, Milan, Italy

Samuel Solé École normale supérieure, Paris, France;
Sorbonne Nouvelle University, Paris, France

Andrea Valle University of Turin, Turin, Italy

Cristina Voto University of Turin, Turin, Italy

Chapter 1
Introduction

Pietro Conte, Anna Caterina Dalmasso, Maria Giulia Dondero, and Andrea Pinotti

In recent years, the pervasive integration of Artificial Intelligence (AI) and algorithmic systems into the analysis, production, circulation, and interpretation of images has profoundly reshaped the visual landscape. This transformation is not merely technological but epistemological, aesthetic, and political. Images are no longer simply created or captured—they are computed, inferred, and generated through probabilistic models trained on vast datasets. As a result, the very ontology of the image is shifting: from representation to prediction, from indexical trace to synthetic projection.

This volume responds to the urgent need for a critical examination of the implications of this shift. It brings together interdisciplinary perspectives to explore how algorithmic systems are redefining the conditions of image-seeing and image-making. Key questions addressed include the operations and politics of machine vision; the semiotics of figure and pattern recognition; originality, creativity, and authorship; the aesthetic and institutional value of AI-generated art; the complex ethical and legal challenges posed by bias and copyright; and the broader cultural imaginaries that emerge from these new visual regimes.

Historically, images have generally been understood as representations—visual artifacts that reflect and imitate (painting) or capture and index (photography and cinema) aspects of the real world. Despite some significant exceptions, such as

P. Conte (✉) · A. C. Dalmasso · A. Pinotti
Department of Philosophy "Piero Martinetti", University of Milan, Milan, Italy
e-mail: pietro.conte@unimi.it

A. C. Dalmasso
e-mail: annacaterina.dalmasso@unimi.it

A. Pinotti
e-mail: andrea.pinotti@unimi.it

M. G. Dondero
F.R.S./Département des Romanes, FNRS/Université de Liège, Liège, Belgium
e-mail: mariagiulia.dondero@uliege.be

P. Conte et al. (eds.), *Algomedia. The Image at the Time of Artificial Intelligence*,
Lecture Notes in Morphogenesis, https://doi.org/10.1007/978-3-032-08726-3_1

first and foremost so-called abstract or non-figurative images, this account holds both for the mainstream theoretical models (e.g. the innumerable variants on the Platonic doctrine of mimesis, Husserl's phenomenology of image consciousness, Panofsky's iconology, the analytic theories of depiction) as well as for ordinary, everyday language.

This representationalist paradigm presupposes a strong relationship between image and world: the image is explicitly or implicitly understood as being linked to an already existing reference. However, algorithmic images—those produced by generative adversarial networks (GANs), diffusion models, and transformer-based architectures—disrupt this traditionally accepted logic, and reframe such images as computational outputs derived from probabilistic inference and massive data processing. They are not captured but synthesised; not recorded but inferred.

These images emerge from the manipulation of latent spaces: mathematical abstractions that encode statistical regularities across massive datasets. The generative process is not anchored in an alleged physical referent but in a probabilistic model of visual possibility. As such, algorithmic images challenge traditional notions of realism and indexicality. They are not representations of what is, but projections of what could or might be, based on patterns learned from what has been.

This computational turn also impacts the way images are received. While the act of viewing has always been culturally mediated, the rise in use of algorithmic recommendation systems, automated content moderation, and personalised feeds has transformed spectatorship into a dynamic interaction between human perceptual faculties and machine-driven curation, mediating what we see and how we see it. Audiences no longer simply choose what to see; increasingly, what they see is chosen for them by opaque algorithmic processes. The only option left to the human actor is to minimally orient the process through selective operations, adjustments in "temperature" (from "low" to "high" randomness) and "weight" (the difference in focus between different elements of the prompt), and fine-tuning the database. The image becomes a node in a feedback loop of data extraction and behavioural prediction. This convergence of production and reception within the algorithmic loop raises profound questions about agency, attention, and the commodification of vision.

The contemporary algorithmic scenario poses a radical challenge to the experience of iconicity not only objectively, but also subjectively. Can we still speak of "images" if the above-mentioned representationalist paradigm is substantially overthrown? Can we still speak of "vision" if no visual or tactile organs are at work during the generation through AI models, but only machinic interactions?

Machine vision—the ability of computational systems to process, analyse and generate visual information—underpins a wide range of technologies, from facial recognition and autonomous vehicles to surveillance systems and content moderation tools. Unlike human vision, which is embodied, situated, and context-sensitive, machine vision operates through mathematical models that detect, classify, and infer patterns in data.

This mode of "seeing" does not simply replicate human perception; it structurally reconfigures it. Machine vision privileges features and correlations that may be invisible to the human eye. It translates the world into data vectors and statistical correlations, thereby reshaping what counts as visible, knowable, and actionable. For example, convolutional neural networks (CNNs) and transformer-based architectures enable the detection of objects, faces, and gestures in vast amounts of image data, automating tasks which traditionally depend on human cognition.

At the heart of machine vision lies the operation of figure recognition: the ability to analyse visual data and identify discernible forms—faces, bodies, gestures, objects—and to attach semantic labels to them. This computational semiotics raises questions about representation, meaning-making, and the status of images as signifying practices. Machine figure recognition is not merely a technical feat but a space where cultural codes are inscribed, contested, and rearticulated.

This process is not neutral: the supposed neutrality of machine vision is a myth; its outputs are conditioned by training data, algorithmic parameters, and the institutional frameworks within which it operates. Consequently, critical scholarship must examine how machine vision reproduces, amplifies, or disrupts existing social hierarchies, taxonomies and classifications that reflect cultural narratives and power relations. The semiotic dimension of machine vision thus intersects with its political and ethical implications. What is recognised, and how, is always a question of meaning and value. And of biases.

Bias is one of the most pressing ethical and political challenges in algorithmic media. Because machine learning models are trained on existing datasets, they inherit and potentially amplify the biases embedded in those datasets. In the realm of image production and recognition, this can lead to the reinforcement of stereotypes, the marginalisation of under-represented communities, and the perpetuation of discriminatory patterns. For example, facial recognition systems have been shown to perform unevenly across different demographic groups, thereby perpetuating the racial and gender biases inherent in the data.

Bias in algorithmic systems is not merely a technical flaw to be fixed; it is a structural issue that reflects broader social hierarchies and exclusions. Addressing it requires more than fairness metrics or dataset diversification. It demands critical inquiry into who controls the data, whose perspectives are encoded or excluded, and how algorithmic classifications perpetuate particular worldviews. Scholars and artists have begun to address these issues by exposing, subverting, or reimagining the datasets and models that shape algorithmic vision. These interventions highlight the need for inclusive design practices, regulatory frameworks, and cultural critiques that foreground questions of justice, representation, and epistemic accountability.

However, to what extent is it possible for a human agent to operate critically both *with* the machine and *against* the machine at the same time? To what extent does the use of algorithmic technologies allow, or prevent, original intervention?

The rise of generative AI has reignited long-standing debates about originality and creativity. In the Western aesthetic tradition, the archetypal creative image-maker, the artist, is often portrayed as a solitary genius, whose unique vision gives rise to distinctive works. In the last century, analyses of film production methods—a collaborative undertaking involving the director, the director of photography and many other actors—certainly promoted greater awareness of the inadequacy of this solipsistic model of creativity and authorship suggesting that pre-cinematic media should also be reconsidered in terms of creative cooperation. Consider, for example, the role of workshops and ateliers in the case of painting and sculpture. Nevertheless, the romantic myth of the solitary genius persists and continues to this day. Algorithmic systems, by contrast, operate through recombination: they generate novel outputs by remixing elements of prior cultural artefacts from their training data, based on vast repositories of pre-existing images and texts. This raises fundamental questions: can such outputs be considered original if they are the result of statistical remixing rather than intentional, conscious invention? Who—or what—is the author of an AI-generated image? To what extent does the human prompter, designer, or trainer retain authorship over the output?

Recent scholarship suggests that creativity in algorithmic contexts should be understood as a collaborative process. Rather than viewing AI as an autonomous creator, it is more productive to understand it as a collaborator or infrastructure which extends creativity. Artists working with AI engage in iterative processes of prompting, curating, fine-tuning, and interpreting. The creative act becomes distributed across human and non-human agents, across code, data, and interface, so as to result in a complex entanglement of multiple agencies:

1. the initial prompt submitted by the user to the model;
2. the prompt as revised and reformulated by the model itself, and accessible only through the paid tier of the service (typically the case with ChatGPT/DALL·E);
3. the annotated database upon which the algorithms operate through abstraction processes, with the aim of producing a statistical average. This average is based on all image annotations in the dataset, specifically those prompted by the query;
4. The operation of the algorithms themselves, which are tasked with ensuring the alignment between the embeddings of the prompt and those of the image annotations;
5. the stochastic or random dimension of computational processing, which can be tentatively evaluated by repetitively submitting the same prompt hundreds of times. It is only through such repetition that we can assess the agency of randomness, by examining the range of visual outputs produced.

This entangled co-authorship challenges romantic notions of authorship and calls for new models that account for procedural, collaborative, and machine-based forms of creativity.

Moreover, the network of co-authorship extends beyond the artist to include developers, dataset curators, platform providers, and even users who contribute prompts and feedback. Each of these actors plays a role in shaping the conditions under which images are generated and interpreted. Authorship, in this context, is not a singular act but a distributed process involving multiple enunciative instances: a veritable choreography of human intention and algorithmic affordance, in which agency is negotiated between human and non-human agents.

In this complex framework, it is worth asking whether the question of artistic value still has any relevance for the production and reception of AI-generated images. Critics often argue that algorithmic art is derivative, stereotypical, or manneristic—constrained by the biases and patterns of its training data. Proponents, however, emphasise the generative potential of these systems: their capacity to uncover latent patterns, facilitate experimentation, and democratise access to creative tools.

The growing institutional recognition of AI-generated art, in the form of exhibitions, auctions, and critical discourse, has further legitimised these practices. Yet this raises new questions: How does the commodification of algorithmic art intersect with platform capitalism? Who profits from the circulation of AI-generated works? How are value and ownership negotiated in a context where authorship is distributed and the boundaries between tool and creator are blurred? These questions are inextricably linked to wider discussions about labour, exploitation, and the monetisation of creativity in the digital economy.

Such issues are not merely economic; they are also aesthetic and philosophical. They compel us to rethink the criteria by which we evaluate artistic practices, the role of institutions in shaping cultural legitimacy, and the evolving relationship between art, technology, and society. And they also raise significant legal implications.

The legal frameworks governing intellectual property were not designed to accommodate non-human agents. As a result, the rise of AI-generated content poses significant challenges to existing notions of copyright, authorship, and liability that remain unresolved in many jurisdictions. Can an AI-generated image be copyrighted? If so, who holds the rights—the developer, the user, the platform, or some combination of these? These questions become even more complex when AI systems generate content that resembles or reproduces existing works. In such cases, issues of plagiarism, copyright infringement, and unauthorised reproduction come to the fore. The probabilistic nature of generative models complicates attribution and enforcement, especially when outputs are derived from vast and opaque training datasets.

Beyond copyright, broader legal concerns include data privacy and biometric rights. The use of personal data, such as facial images, to train or operate machine vision systems raises serious ethical questions, particularly when consent is unclear or absent. These challenges underscore the need for robust regulatory frameworks that can balance innovation with accountability, individual rights with collective interests.

As the contributions in this volume demonstrate, algorithmic media are not neutral tools; they are active agents in the shaping of contemporary visual culture. They influence the way we see, understand, and represent the world. They redistribute agency, transform creative labour, and recalibrate cultural notions of originality and authorship. They embed biases and contentions that reflect broader socio-political dynamics.

A critical approach to algorithmic imagery must therefore be interdisciplinary. It must draw on philosophy, media theory, computer science, semiotics, law, and cultural studies. It must attend to the technical operations of algorithms as well as their aesthetic, ethical, and political ramifications. It must investigate not only how images are made and consumed but also how they make and remake the conditions of cultural intelligibility and political possibility.

This volume invites scholars from various disciplines into a shared dialogue about the futures of visual culture in the age of algorithms. By situating algorithmic images within the broader histories of media, art, and technology, it seeks to illuminate both the continuities and the ruptures that define our evolving iconosphere: a world in which images are no longer merely human-controlled representations but computational projections that shape and pre-empt the realities to come. A world that we are going to see and live *through AI*—as staged by the watershed exhibition inaugurated in April 2025 at the Jeu de Paume in Paris and explicitly devoted to "Exploring latent spaces"—a world no longer structured solely by human intentionality but by the coupling of human beings and *algomedia.*

* * *

Guided by the central tenet of the volume, the first section, dedicated to *The Black Box of Computational Imaging*, is structured at the intersection of philosophy of technology, computer science, and media theory. This section aims to set out the key conceptual foundations to orient the reader through the interdisciplinary contributions that follow.

Bruno Bachimont's contribution addresses fundamental epistemological questions, seeking to shed light on the gap between algorithmic computation and human interpretation in the processing, generation, and understanding of digital content. What are algorithms, and how can these constructs manipulate not only numerical data, but also complex discursive content with semantic structure? How do the cognitive operations performed by humans differ from the machine-driven "interpretation" of the world, which produces plausible outputs without engaging with meaning or context? Which operations are aimed at translating two seemingly irreconcilable forms of language, and how do they impact the content itself? Only by recognising this fundamental divide can we understand the epistemological implications of algorithmic media and their capacity to shape meaning and model the world. Investigating the historical and conceptual roots of formalisation, Bachimont analyses AI models to reframe them as "meta-archives" of knowledge.

The chapter by Sergio Picascia and Alfio Ferrara aims to show how AI enables machines to both analyse and generate visual content, with the aim of making the computational aspects of AI as accessible as possible to readers with a background in the humanities. What does a machine do when it "sees"? How can the visible world be processed by algorithms? How does AI-based image generation work? What operations and computational processes underlie the remarkable ability of algorithms to manipulate visual content? By clarifying key terms such as convolutional neural network, transformer, and latent space, Picascia and Ferrara provide readers with the tools needed to decode the innovations which have, in recent years, revolutionised AI research and applications. The chapter explores in particular the functioning of generative models to explain how AI systems transform random noise or textual prompts into images. It also raises important concerns about the need to address ethical issues surrounding AI-generated content, including privacy, bias, and copyright infringement.

Drawing on his extensive experimentation with AI models such as Midjourney and Stable Diffusion, Lev Manovich focuses on visual generative media to analyse how they are reshaping aesthetics and visual culture. Which operations are specific to the way AI processes and synthesises images, particularly through text-to-image models? In what ways does generative AI engage with and ultimately align with a lineage of artistic strategies that remix, reframe, or recompose existing cultural material? And what are the limitations of a form of creativity rooted in database training, which tends to favour stereotypical outputs and foster forms of mannerism? Combining both theoretical approach and hands-on experimentation, Manovich seeks to pinpoint the profound transformation afoot in traditional image-making practices, namely the shift from a representational paradigm to one grounded in predictive logic, in which AI generates new images through probabilistic inference.

Although algorithms have long played a role in shaping digital media over the past decades, AI systems have now become central to the very functioning, interface design, and logic of almost every technology and device. The section *Algorithmic Mediascapes* explores some of the most significant and disruptive domains in which this convergence can be observed, by combining theoretical and genealogical approaches.

In recent years we have witnessed a process of virtualisation that has normalised various forms of telepresence and access to simulated worlds. AI itself participates in this process as it is enabled by a virtual language system that regulates the relationship between datasets and algorithmic outputs. Yet, the conception of virtuality largely pre-dates the era of digital and algorithmic media: it can be traced back to the notion of *virtus* through Aristotle and Aquinas, as a way to explain the well-known *dynamis/energeia* dyad. But is the philosophical and theological notion of "virtual" still pertinent to its usage in digital media and generative AI models? What happens when we move to artificial virtuality? Andrea Pinotti draws a parallel between the metaphysical conception of "virtual" as a reservoir of latent potentialities and the so-called "latent space" which is actualised through probabilistic processes. This dynamic becomes especially evident in one particular application of generative AI: Neural Style Transfer, which renders the content of an image in a chosen artistic style, thereby inviting discussion of the distinction—and supposed separability—between form and content. In conclusion, the contribution focuses on identifying the analogies and differences between AI and the notion of "virtual" in virtual reality, considering both as technologies which actualise virtualities.

Extending further the focus on immersive media and extended realities, Sofia Pirandello's contribution addresses Augmented Reality (AR) technologies as their functioning relies entirely on AI for bodily tracking and monitoring users' behaviour. She points out that the AI-assisted automation that enables these technologies, especially remote control through virtual gestures, is often perceived by the audience as a form of occult spellbinding magic. Indeed, references to magic also inform the rhetoric used to communicate and market these interfaces. But what role does AI play in AR interfaces in constructing user experience through behavioural data extraction? And what are the implications of magic in technological discourse, particularly in terms of surveillance and control? Drawing on recent literature exploring how the epistemic background of modern media is imbued with spectral and paranormal connotations, Pirandello examines, on the one hand, the technical processes by which AR operates through real-time sensing, recognition, and mutual adaptation between users and AI systems; on the other hand, she seeks to show how this form of "*automagic*" simultaneously conceals both the presence of hardware and the invisible human labour which underpins and sustains these systems.

Indeed, AI has expanded the range of image-making techniques, ushering in a new era of image production which marks a significant rupture with the visual media of both the distant and more recent past. But in what ways does algorithmic imagery depart from traditional forms of imaging? What are the points of divergence and continuity between these regimes? And what are the biopolitical consequences of such a transformation in imaging technologies? Focusing on the case study of the SARS-CoV 2 virus representation through micrographic imaging technologies, Ruggero Eugeni analyses how increasingly sophisticated AI interventions, particularly through deep learning algorithms, process, segment, classify, and enhance raw data, transforming it into visually and scientifically meaningful representations. Building on this analysis, Eugeni conceptualises a transition from modern *techno-images*, created by directly capturing light or other electromagnetic flows to produce visible traces, to contemporary *algo-images*, where visual representations are subordinated to algorithmic processing. This shift, he argues, reflects a more profound transformation in the political economy of light, moving from image production to data extraction, with significant implications for surveillance and control.

The logic of data extraction reverberates across numerous domains within the framework of surveillance capitalism. The most salient example is facial recognition technologies: biometric systems designed to identify individuals based on their unique facial features. These systems raise ethical, legal, and political concerns regarding the governance of biometric data but also impact on personal identity, due to the symbolic and cultural meanings attached to the face. How does facial recognition operate? In what ways do machines transform human bodily features into data vectors? What happens when human faces, instead of being looked at, are processed as massive datasets in machine-to-machine networks? Samuel Solé's contribution explores the technical functioning of facial recognition technologies, examining the different stages of acquisition, detection, normalisation, extraction, and recognition, in order to highlight the underlying processes and operative logics. Moreover, he situates the logic underpinning these technologies within a broader epistemic paradigm that can be traced back to nineteenth-century anthropometry, which established the foundations of modern identification systems. His analysis is then enriched with a genealogy of face-reading practices, showing how the biases embedded in contemporary AI systems inherit and amplify the discriminatory operations intrinsic to these earlier techniques.

Various forms of pre-emption and premediation are embedded within algorithmic media, informing both the technical architecture of AI systems and their wide range of applications, including automated decision-making, recommendation systems, and behavioural modelling. But how is this continuous form of prediction performed by AI across different fields? What distinguishes algorithmic prognostication from earlier forms of forecasting and prophecy? How does this anticipatory activity resonate with the enduring human drive to foresee the future, which has shaped civilisations throughout history? Anna Caterina Dalmasso investigates the predictive logic inherent in algorithmic media both in its intrinsically probabilistic structure and in its broader cultural role, and interprets AI as a contemporary form of divination. Situating AI within a broader historical trajectory of predictive technologies, echoing ancient and modern futurological practices, she examines the logic of forecasting inherent in contemporary mediality showing how an anticipatory attitude shapes both technological infrastructures and affective-cultural orientations. She calls for a reflection on how algorithmic foresight generates self-reinforcing loops and forecloses futurity itself, thereby serving late capitalism's tendency to control and neutralise the unpredictable.

The section on *Algorithms and the Arts* is devoted to investigating how artificial intelligence is dramatically changing the multifaceted realm of artistic practices and all the traditional notions related to it, such as style, creativity, invention, authorship, originality, reproduction, and plagiarism.

Domenico Quaranta explores the relationship between humans and machines in the creative process facilitated by generative models which can produce text, images, sound, and, more recently, video in response to textual prompts. Artists describe these technologies in different ways: some see them as mere tools and executors, while others regard them as collaborators and conversational partners. This dual interpretation raises thorny questions: is machine-generated content created *by* artificial intelligence, or *with* it? Can full authorship be attributed to a non-human agent? And can machine-generated content be granted copyright if it is seemingly produced with minimal direct human involvement? Furthermore, in an era where artistic production is increasingly mediated by AI, the creative act exists at the intersection of multiple interests and sociopolitical and economic influences. We are active users of machine learning technologies, but at the same time, we are also used by them: while exploiting AI's potential, we provide unpaid labour in the form of prompts and feedback, which are collected and classified to feed the AI system in a continuous cycle of refinement. To counter this situation and raise awareness of the dominant role played by AI systems and the companies behind them, artists can oppose the *strategic* use of AI—understood as the uncritical acceptance of the direction of development proposed by machine learning corporations—with a *tactical* use grounded in reflection and techno-political critique.

The interplay between algorithmic economies and the arts is particularly evident in blockchain technology and the growing phenomenon of non-fungible tokens (NFTs). Marion Colas-Blaise's essay focuses precisely on this issue and is structured in three parts. The first section draws on Nelson Goodman's distinction between autographic and allographic art forms to explore how NFTs incorporate mechanisms for verifying authenticity, replication, scarcity, and preservation. The second section focuses on reception practices: What does it mean to transfer a certificate of ownership of a digital artwork? How can NFTs be collected and exhibited? Finally, the third section investigates the question of when an NFT becomes a work of art. To what extent do these new reception practices differ from traditional aesthetic experiences? How are artistic criteria redefined by the rise and spread of NFTs? In order to address these questions, the author traces the origins of cryptoart back to twentieth-century art movements.

The advent of algorithmic media is not only revolutionising the way images are produced, but also how they are consumed. Aluminé Rosso addresses the novelty introduced by contemporary exhibitions of artworks co-produced with artificial intelligence, analysing both their historical evolution and their institutional and aesthetic implications. After drawing a parallel between salient features of AI-generated art and key traits of twentieth Century conceptual art, the author examines emblematic ways in which performative AI installations have been used as opportunities to reshape artistic formats, curatorial practices, and audience experiences. Focusing on notable case studies, the chapter illustrates how performative AI installations render the artwork actively responsive to the audience's presence, often without the audience themselves being aware of their role in shaping it. Thus, AI co-produced art complicates traditional notions of exhibition and audience interaction, envisioning not only new formats, but also new temporalities, forms of participation, and aesthetic regimes. How does the emergence of AI-generated art contribute to dissolving the traditional boundaries between performance, installation, and interactive art? How do AI tools enable the creation of performing images that challenge both aesthetic conventions and institutional roles? Referencing artists such as Harold Cohen, Anicka Yi, and Trevor Paglen, Rosso highlights the profound transformation of museums into dynamic spaces of co-creation and experimentation.

In addition to image creation and consumption, algorithms are having a profound impact on scientific research in disciplines such as art history. In the final chapter of this section, Maria Giulia Dondero and Adrien Deliège consider how applying deep learning-based computational methods to large image databases can help analyse the evolution of styles in the history of painting and establish unexpected genealogies of forms. Combining semiotic and computational approaches, the authors survey and analyse the most significant current projects in digital art history, highlighting both their strengths and their limitations. Then, under the auspices of Henri Focillon and Aby Warburg, a new project is described that considers complex individual and collective poses, incorporating contemporary photography into the database to identify surviving ancient poses. Drawing inspiration from the research of the Paris School of Semiotics, this approach offers a unique perspective on the forms and forces within images, with the ultimate goal of analysing continuous bodily gestures in large collections of paintings and photographs.

The advent of algorithmic media has given rise to ethical, social, and political challenges, profoundly reshaping today's mediascape and contemporary technoculture. Providing a critical counterpoint to many of the themes addressed throughout the volume, the section *Cultural Critique of Algorithmic Media* seeks to map some of the most significant ethical and biopolitical issues arising from the algorithmic revolution, focussing in particular on techno-anaesthetics, identity, decolonial perspectives, and the evolving notion of authorship.

Particularly through the new image-making techniques introduced by generative models, AI is having a significant impact on visual culture and aesthetics. But how is the visuality of AI itself taking shape within the collective imaginary? How can we represent algorithms and their specific modes of operation, which are inherently non-visual? Alberto Romele's reflection takes its cue precisely from the representations and visuals—mostly stock imagery—that increasingly populate our global imagination and visual culture, and yet are rarely scrutinised in the context of AI ethics. His essay critically explores the visual representations of AI through which we depict and personify algorithms, many of which are generated by the very same text-to-image models. Romele shows how these representations shape public perception and ethical discourse, by elaborating visual codes which tend to placate rather than provoke critical thought, side-lining ethical complexities and fostering a depoliticised imaginary. Drawing on Gilbert Simondon's essay *On Techno-Aesthetics*, the author reflects on the technical reconfiguration of human sensibility and considers how the relationship between technology and perception should be understood within today's evolving mediascape, advocating for a repoliticisation of visual culture and a cultural critique of AI.

Video games were among the first virtual and simulated environments to make use of procedural image generation. Since these early experiments, AI systems and deep learning techniques have become increasingly integral to video games, both at a technical-functional level and at the level of gameplay experience construction, as well as at the level of identity and emotional engagement. How can we classify the different modalities through which AI operates across these various levels in video games? How should we conceptualise AI's specific form of agency within video game design and structural functioning? Even in the absence of true cognitive autonomy, can AI be regarded as an enunciating subject within virtual worlds? Enzo D'Armenio investigates these questions from a semiotic perspective: starting from Émile Benveniste's theory of enunciation, he develops a typology of AI enunciations and identifies emerging AI trends in video games. In particular, he examines the key points where AI intervenes as both co-enunciator and enunciator, highlighting its ability to foster empathy by simulating contingency management.

Cristina Voto's contribution focuses on the biases and structural asymmetries embedded in AI-generated imagery and image-processing systems, which stem from the datasets on which models are trained and tend to reflect and reinforce existing inequalities or to result in predatory predictions. But how can AI influence cultural and social narratives? Beyond the technical strategies implemented by computer scientists to mitigate biases, what approaches can counteract these inbuilt distortions? How should we understand and critically engage with datasets? Can creative or activist interventions reappropriate AI-generated imagery to challenge structural bias? Voto's essay explores the role of datasets as archives, in the Foucauldian sense, in visualising and giving presence to marginalised communities. Calling for the integration of gender and decolonial perspectives, she explores dataset curation as a crucial area for intervention, emphasising that the selection and omission of data fundamentally shape what AI can imagine and represent. To illustrate her theoretical reflections, Voto presents Felipe Rivas San Martín's artistic project *Un archive inexistente*, demonstrating how AI can be creatively harnessed to visualise queer lives erased from the historical record.

One of the most critical issues raised by generative AI in the realm of content creation concerns the increasingly blurred boundaries between tool and creator, between invention and replication, challenging conventional notions of authorship and originality. How is generative AI redefining what it means to be an author? How do meaning and authorship operate at the various stages of image production and reception? Andrea Valle's contribution analyses the different modes in which algorithmic creative practices are changing the nature of creative production as much as the status and agency of those who create with and through these technologies. Valle contrasts genetic and semiotic approaches to authorship and analyses how digital images fragment authorial subjectivity across various levels of enunciation and materiality. Thus, through a semiotic analysis of enunciative practices in text-to-image models, he examines how digital and algorithmic practices generate intricate configurations of authorship. When agency is apportioned across human, machinic, and procedural levels, new theoretical tools are required in order to fully capture its semiotic and epistemological implications.

The final section of the volume is devoted to the role of artificial intelligence in shaping *Posthuman/Non-human Imagery.*

Algorithmic enhancement tools are becoming more and more common in our day-to-day lives, especially when it comes to manipulating images on social networks. Massimo Leone examines the nature of digital filters as semiotic and existential operators, tracing their connection to analogue counterparts while also emphasising the novelty introduced by post-capture algorithmic manipulation. Through the analysis of emblematic case studies, filters are interpreted both as tools for playful experimentation and as instruments that convey deep-seated cultural prejudices, either consciously or inadvertently. This raises important questions: in what ways can filters be seen as both conveyors of authenticity and as instruments of seduction and control? How might they be employed to either reinforce or challenge social biases and deconstruct stereotypical cultural narratives? Which ideologies and assumptions embedded in beauty standards and historical representations are filters inviting us to confront? What does the use of filters reveal about the complex interplay between aesthetics, technology, culture, and both individual and collective identity?

When it comes to the aesthetic and socio-political impact of algorithmic image manipulation and the attribution of truthfulness to images, hyperrealistic photographs and videos produced using AI technologies are certainly a poignant case in point. Remo Gramigna focuses on AI-generated faces and the concept of "algorithmic similarity" that they introduce. How does resemblance manifest from the perspective of a non-human observer? What distinguishes human perception from machine vision? And what constitutes a "face" in the eyes of artificial intelligence? If our perception of a photograph's fidelity—its ability to reflect its subject—depends not only on its indexical origins (i.e., the underlying chemical process) but, more crucially, on our recognition of its resemblance to the subject, how are notions of indexicality, accuracy, and objectivity being reshaped by increasingly powerful machine learning technologies? In addressing these questions, Gramigna also highlights how AI-generated imagery compels us to re-evaluate long-standing assumptions about authenticity, authorship, and artistic intention. In doing so, he sheds light on the complex issues surrounding misinformation, intellectual property, and the evolving role of human creativity in an era increasingly dominated by machine-generated content.

Another crucial aspect of the impact of artificial intelligence on the development of non-human agents which can reliably interact with humans in shared tasks concerns Human–Robot Interaction, a field that is undergoing a profound transformation due to the emergence of learning algorithms designed to enable robots to learn through observation. These algorithms facilitate real-time interaction in dynamic and unpredictable environments, reducing reliance on preprogrammed rules or explicit instructions. Central to this shift is the paradigm of Learning from Demonstration, a machine learning approach in which robots acquire new skills by imitating human behaviour. This framework is, in turn, grounded in cognitivist theories of social cognition and imitation, particularly those informed by Theory of Mind or "mind-reading," which hold that social understanding relies on inferring human intentions by modelling internal mental states. But are these theoretical foundations truly well-suited to fostering smoother human–robot interactions? Is imitation primarily the result of abstract inference, or is it rooted in embodied, sensorimotor engagement? By raising such questions, Martina Bacaro advocates for an alternative, semiotic perspective on interaction and learning—one that is more attuned to the contextual and embodied dimensions of real-world human–robot collaboration.

In the concluding chapter of the volume, Pietro Conte and Maria Serafini explore how artificial intelligence is reshaping afterlife imagery and the associated practices of memory and mourning. At an individual level, machine learning is employed to construct neural networks capable of generating digital doppelgängers of both the living and the deceased, raising profound questions about the persistence and transformation of identity within the virtual realm. At a collective level, the authors investigate how images embedded in training datasets seem to vanish once absorbed into AI models, only to reemerge, altered and recombined, in new automatically generated images. These spectral traces, or *revenants*, continue to "haunt" the models, resurfacing with each iteration. These ghostly presences—both personal and collective—suggest that AI is reconfiguring memory and remembrance by entangling the dead in novel networks of visibility, control, and reproduction. In doing so, AI reinforces existing structures of power, exclusion, and surveillance, ultimately prompting us to undertake a critical re-evaluation of how memory, representation, and digital afterlives are constructed and regulated in the age of artificial intelligence.

Part I
The Black Box of Computational Imaging

Chapter 2
AI and Algorithms: What Digital Technology Can Teach Us About Content

Bruno Bachimont

Abstract If we are concerned about the consequences of using algorithms on content such as text, sound, and video, it is because content has become calculable. By considering only their syntactic materiality, these forms of content are translated into formats, that is, into formal tokens to be manipulated solely based on their syntactic type, independently of their meaning. In this sense, AI is the project of translating content into algorithms that generate meaningful and relevant transformations, despite their restriction to formal structures. After outlining the historical and conceptual roots of formalisation as a means to render content computable, this chapter reviews the main classes of tools used in AI. It concludes with a discussion of the status of the results produced by these tools, particularly the content created by generative AI. This content appears meaningful, and thus seems to address us and represent the world, which in turn raises fundamental questions about enunciation (who is speaking) and modelling (what world is being represented). Upon closer examination, generative AI appears to offer a calculable meta-archive of everything that has been said, so that it can be said again, differently.

Keywords Generative AI · Symbol · Sign · Interpretation · Modelling

2.1 Introduction

Artificial intelligence (AI) has become so pervasive in our practices that we often use the term when we simply mean the use of algorithms applied to digital content. However, the most important thing is probably not so much that we use AI, but that we rely on algorithms and calculation to explore and interpret content. The real tension that arises is between, on the one hand, content that is subject to interpretation insofar as it expresses a semiotic form and, on the other hand, calculations that are subject to formalisation and manipulation. There is an irreducible gap between these two approaches to content. But rather than rejecting in principle the use of algorithms in

B. Bachimont (✉)
Compiègne University of Technology, Compiègne, France
e-mail: bruno.bachimont@utc.fr

P. Conte et al. (eds.), *Algomedia. The Image at the Time of Artificial Intelligence*,
Lecture Notes in Morphogenesis, https://doi.org/10.1007/978-3-032-08726-3_2

the interpretation of content, it is rather this gap that makes it interesting and enriches its interpretative potential.

Instead of closing this gap, AI is widening it still further. It does this in several ways, because AI is not a single term referring to a homogenous class of technology. It corresponds to different approaches that are all paradigms seeking to enrich the possibilities of calculation in the execution of tasks that, for human beings, are a matter of interpretation.

In this article, which aims to provide an introduction to these issues, we return to the concepts of algorithms, AI, and content to better grasp the current issues and the prospects opened up by the innovations grouped under the name of AI.

2.2 Algorithms and Content. Defining Manipulation Tools: Calculations and Algorithms

Once the preserve of specialists, the term algorithm has become a commonplace concept in everyday use. It is worth taking a closer look at its definition and understanding its various aspects.

An algorithm is a calculator. But by calculator we mean a human calculator, i.e. what a person does when carrying out a calculation. Our computer-populated world has made us forget that calculations used to be carried out by specialised people, setting up calculation workshops to produce the usual tables (logarithms, trigonometry, etc.) as well as calculations dedicated to solving technical and scientific problems (differential equations). As Copeland [4] points out, a calculation made by a human being is characterised by the following properties:

- A finite sequence of instructions, themselves made up of a finite number of symbols.
- Producing the expected result in a finite number of steps.
- The only tools and techniques needed are paper and a pencil.
- No intuition, inventiveness or creativity is required to carry out the calculation. In other words, (human) calculators are interchangeable; all that is required is the ability to follow instructions and apply them.

In his famous 1936 article, Alan Turing formalised this notion of human computation to derive that of the calculating machine [25]. This is what motivates and justifies Turing's thesis (also known as Church's thesis, for Church carried out equivalent work in the same year on the characterisation and definition of computations): anything that a human can compute can be computed by a Turing machine, as the machine simply translates the principles that make up computations and their execution [17].

A calculation is therefore a particular type of symbolic manipulation, in which interpretation is excluded in favour of the application of rules and instructions without recourse to intuition, imagination or invention. If we use the term *interpretation* to refer to these latter characteristics—namely, the ability to consider a sign not for what it is, but for what it means or refers to, that is, what it is not—then we can draw a radical opposition between calculation and interpretation, between calculative manipulation and interpretative understanding.

Yet this opposition is undoubtedly too simplistic, and the formalisation of computation into a machine, as proposed by Turing, provides a better understanding of what we are talking about here. Turing's machine comprises a memory strip that represents the sheet of paper of the human calculator. This strip is one-dimensional, and the signs making up the calculation are written in sequence on it. To distinguish the sheet from the signs inscribed, we define boxes on the sheet that can contain no more than a single sign, and which contain a blank symbol by default. Only a finite number of boxes contain anything other than a white symbol. Instead of the pencil of the human calculator, we have a reading head that points to a single cell at a time. It can do the following: read the contents of the square, write another sign in it, move one square to the right or left or stay where it is. Finally, this reading head has an internal state, which Turing describes as a state of mind: it is a sort of memory of what has been done, a memory of operations already carried out. To remain consistent with the definition we have given of a calculation (made by a human), the number of states of mind is finite, as is the number of signs that can be written on a square.

This device ends with instructions which are executed by the read head. The term "instruction" is a misnomer because it suggests that we have a machine that reads a program to execute it, whereas the instructions are the machine: if we change the instructions, we get another machine, and we have to build a new machine. These "instructions" are defined as follows: if the read head points to a square and reads a given content (a particular symbol), and if it is in a given internal state, then it executes a write operation (writing a different symbol or the same symbol as the one already on the square) and moves one square to the right or left or not.

So we have an elementary mechanism that reproduces the behaviour of a human calculator in a mechanical way. The important point here is not so much that it is a machine as that it is a formalisation that allows us to deduce what can and cannot be done by the machine. Formalisation allows us to study the limit of computability.

Regardless of the mathematical results obtained by Turing, it is interesting to understand what the principle of such a machine implies at the semiotic level. Up to this point, we have used the notions of *symbols* and *signs* interchangeably. We will now distinguish between them.

The boxes in the Turing machine contain symbols: it is therefore assumed that one box can be uniquely and univocally identified from all the others and that the contents of this box can be categorised, recognised as one of the types of symbol that can be written (a 0 or a 1, or one of the 10 decimal numbers if we are using base 10). There is therefore an assumption of typicality (each content of a cell is the occurrence of a type of symbol) and distinguishability (all types of symbols can be distinguished unambiguously from one another). The cells of the memory strip contain what we call symbols (their occurrences, in fact), and symbols are characterised by their typicality and distinguishability. A symbol, by its type, is therefore an ideal position in an alphabet or repertoire, a position that can be distinguished from all the others, and its occurrence is a concrete position in the memory strip, also locatable among the others (the read head can position itself on it). The symbol is therefore an abstract and ideal entity: the unit of the multiplicity of its occurrences, or the equivalence class of the latter.

But if the occurrence refers to its type, this does not mean that we have a reference that could be called semiotic, where the symbol refers to something other than itself, to an interpretation or signified. If the symbol is recognised and categorised, this does not mean that it is interpreted. The symbol of the Turing machine is devoid of meaning; it is merely a categorised and recognised form. We can therefore define a sign as a symbol with a meaning, referring to a signified. Conversely, a symbol is a sign without a signified, a sign that is recognised and categorised as such but remains uninterpreted.

This is why, although it may seem surprising at first sight, the fact that we have chosen a binary alphabet for the numeric does not mean that what is on the memory strip are numbers: they are simply positions, symbols. An occurrence of 0 is recognised as a 0 symbol, and an occurrence of 1 is recognised as a 1 symbol. But they are not numbers in the arithmetical sense of the term: they are only positions. Considering them as numbers is already an interpretation, where the symbol 0 as a position in the alphabet refers to an abstract entity—the number 0—which obeys laws and can take part in arithmetic operations. In other words, the mathematical sign 0 of binary arithmetic is implemented in the Turing machine using the symbol 0 as a position in a two-valued alphabet or a formal two-state system.

This distinction between symbol and sign is important to us because it allows us to emphasise the fact that the entities we consider in digital tools are devoid of meaning and semantics, even if they already carry an ideality in that they refer to an abstract system of positions in a virtual space. It should be remembered that the very term virtual, in the digital context, means independence from the material support, i.e. indifference to the material variations that occur during the physical realisation of symbols. Similarly, to virtualise means to become independent of the material support and its contingent variations, retaining only the formal differences of position in the alphabet or repertoire of symbols.

But if the symbol and the sign are different, how can they be articulated? Digital techniques offer a solution in the form of formats. A format is an arbitrary convention that allows symbols to be mapped to a signified or represented entity. For example, the ASCII format can be used to match the binary sequence 0101001 with the letter "A". An intense red could be coded on 8 bits by 11111111 (255 in decimal). The essence of the format is that it allows you to move from a series of meaningless symbols to an entity that carries a meaning or perception (the letter "A" or the red colour of a pixel).

But if the sequence of symbols is indeed meaningless, there is nothing to prevent us from reading 0101001 as a letter or as a colour: the letter A or a very dark red (on the red component of the three fundamental components of a pixel). In other words, when faced with a binary sequence, I can use it either as a text or an image. Obviously, there is nothing to say that the binary sequence will produce an intelligible text, but it will produce a sequence of letters, or that the image will show anything recognisable, but we will see something. From a formal point of view, there is nothing to prevent one format rather than another being used to read a binary sequence; it is simply a matter of convention.

So, as we said, there is an opposition between calculation and interpretation. But this difference is between two idealities or abstractions: one from the occurrence to its type, the other from the signifier to its signified. This difference will be fundamental in grasping the complexity of the notion of content, to which we will now turn.

2.3 Making Content Manipulable: From Meaning to Codification

Content is a material object with a meaning or interpretation. More precisely, it is a physical object conveying a semiotic form that allows it to be seen as a signifier, an injunction to consider it not for what it is but for what it means. Even if the physical object never completely gives way to its signified, it is the pretext for it, and the signified is its *raison d'être*. So, recognising a sound as speech draws attention not to the words spoken but to the meaning interpreted and understood. We can, of course, be sensitive to the musicality of the verb and the choice of words, but all these considerations will be mobilised with a view to interpretation and comprehension. Content is therefore an address, a call for meaning, the reception of which is the interpretative response, demonstrating that the content has been received and understood for what it claims to be, i.e. to convey meaning.

Content has a mixed nature, between the materiality that constitutes it and the meaning it conveys. In this respect, it refers to two ways of understanding it that are essential in the contemporary digital context, even if they are not limited to it. As materials, content has a physical nature that allows for manipulation and technical transformation. This is why, when faced with content whose material support is natural, i.e. not constructed but used, such as sound waves for voice or sand for traces, we can always construct new material supports that can be manipulated and shaped, such as writing supports (clay, parchment, papyrus, etc.). In this way, the sign inscribed on a technical medium can see its way of signifying determined or reconfigured by the technical manipulation to which the medium or its material nature is subjected. Quite simply, by manipulating the letters of a written medium, we can transform the signs into other signs and modify the significance of the inscription (materiality as a reference to a signified) and consequently its reference to a signified.

The material support of content, its expressive and signifying face, thus becomes a means of acting on the meaning signified. But this possibility does not imply that it is possible to control this meaning and deduce what the transformation of meaning will be from the manipulation of the signifying materiality. There is an arbitrariness between technical manipulation and semiotic interpretation.

Following Rastier's [20] considerations—and without going back to the seminal observations of Hjelmslev [7]—we can see that the logic of physical manipulation and that of interpretation are independent, or even opposed. While the former operates from the local to the global, by way of combinatorial manipulation, combining the occurrences that make up the signifying materiality, interpretation remains holistic and moves from the global to the local, where the anchoring in a concrete situation, the determination of a being in the world or a being thrown into a context, determine a contextual meaning. This double movement in the opposite direction is reflected, on the one hand, by the fact that technical manipulations necessarily rely on the a priori provision of basic manipulable units, which will be composed and assembled, and, on the other hand, by the fact that the meaning-bearing units in the content are released by interpretation rather than being the condition for it. Whereas in manipulation, the units are given a priori to make manipulation possible, in interpretation, the units are identified a posteriori on the basis of their anchorage in the context or situation.

These opposing movements are easily seen when we consider digital text. Whereas, materially, we have to start from a prior digital alphabet, such as UniCode or ASCII, in order to be able to compose the text as a sequence of occurrences of these characters chosen from the alphabet, the signs that carry meaning in the text result from the interpretation that is made of them. We might naively believe that any sequence of non-blank characters constitutes a meaningful sign or word, and that the technical conditioning of characters naturally leads to meaningful units. However, a phrase such as "à la queue leu leu" is not a sequence of words, but a single word in contemporary French, whereas technically we have 5 words or sequences of non-white characters separated by white characters. Technical conditioning is not a necessary or sufficient condition for identifying meaningful units in content.

If technical manipulation is not the condition for interpretation, reconfiguring the signifying material in an arbitrary relationship to the meaning produced, we understand that there will be profound difficulties in creating and analysing content using technical tools.

These difficulties stem from the two arbitrarinesses we have identified:

- The arbitrariness of formats or the arbitrariness of the symbol: the a-semantic nature of symbols, which only become signs through external conventions. If the same binary content can be decoded as an image, sound or text, it is because the symbol offers no resistance to any interpretation whatsoever; any convention can do the job as long as it indicates how a sequence of occurrences on the memory tape is translated into a meaningful or perceptible trace (a character, a pixel, a dot, etc.).
- The arbitrariness of interpretation or the arbitrariness of the sign: once the arbitrariness of the format has been fixed by choosing a given decoding convention, the material configuration obtained does not impose a unique or univocal organisation of the perceptible material into a signifying organisation; only the interpretation carried out contextually makes it possible to identify the signs and their associated meanings.

For meaning to become controllable and determinable by formatted technical materiality, it needs to be formalised so that semantics is reduced to syntax. The reason is that formalisation makes it possible to constrain contextual variation and force interpretation to become univocal and repeatable. It is the purpose of formal logic to control meaning from syntax, and to identify syntaxes that allow the semantics to be expressed in such a way that the manipulable is interpretable and the interpretable manipulable in a principled isomorphism. This situation is encountered in databases and the formalisation of structured data content (notably the semantic web), but in practice, most content escapes this parallel between syntax and semantics, between formal manipulation and contextual interpretation.

This tension between technical manipulation and formal interpretation can be found in all types of content. It is particularly evident in the contemporary digital context insofar as all content can now be coded digitally. Whereas originally the binary code and the first computers were dedicated to the manipulation of numbers (scientific calculation) or letters (business computing), non-textual and non-symbolic content has gradually been taken on board. This is a major development, and one that is easy to understand. Whereas it was sufficient, so to speak, to find an appropriate coding system for alphabets that were already fixed in the case of symbolic documents (it was enough to use the characters of the alphabet already codified by the printing press, for example), such a solution was not possible in the case of non-symbolic content such as images and sounds. The latter are not made up of discrete entities already organised in a finite alphabet. They refer to physical signals, such as an acoustic wave or light waves.

It was therefore necessary to understand how to move from the infinite variety of continuous signals to a limited, finite alphabet enabling the content to be coded. The operations required to do this are well known and mastered, so much so that we have too often forgotten just how remarkable and important they are: we need to move from the infinite reality we perceive to a manipulable materiality that reconstitutes a simulacrum of our perception of reality. The three operations are:

- Sampling, which reduces the content to a finite set of measurements:
  - A sound becomes a finite set of 44,100 measurements per second for sampling at 44.1 kHz;
  - A visual scene[1] becomes a set of 2048 × 1080 pixels or light points (2 K cinema) whose luminance and chrominance, or the three fundamental components red, green and blue (RGB), are measured.
- Quantification, which relates the continuous value obtained from the sample to one of a repertoire of possible discrete values:
  - For sound captured in Hi-Fi or CD quality, there are 65,536 possible values.
  - A light component of a point or pixel (red, green or blue) is generally considered to have 4096 possible values.

[1] It is not appropriate to talk about the image here, because the issue here is what is seen, not the image that renders it, the image that shows what is seen. The digital image will therefore be the result of the capture of a visual scene by an appropriate device (a digital camera). The analogue image is a non-binary, analogue and continuous codification of a visual scene (real or imaginary, recorded or composed).

- The codification that translates this value into the binary alphabet that is ultimately manipulated by the machine:
  - For a sound measured and set to one of the 65,536 possible values considered, 16 bits are all that's needed to encode and represent this value (in fact, $2^{16} = 65{,}536$).
  - For a light component set to one of the 4096 possible values, all we need to do is code it on 12 bits ($2^{12} = 4096$).

When decoded using the appropriate formats, the binary code obtained enables an image or sound to be perceived that corresponds to the digital (syntactic) coding of the original image or sound.

2.4 The Two Paradigms of AI: Knowledge and Data

Artificial intelligence has been given many definitions and characterisations, reflecting both a research programme and assumptions about the nature of the intelligence or the machine. For our part, we will stick to a more cautious or modest characterisation: artificial intelligence is the field of computing techniques that enable a machine, the computer, to perform tasks that require cognitive skills on the part of humans, particularly those considered to be the basis of intelligence. These skills include reasoning, particularly counterfactual, logical and modal reasoning, anticipation, planning, introspection, interpretation and so on. This does not mean that the computer will solve these tasks in the same way as we humans, or that it will have to demonstrate the same skills as we do. When Deep Blue, IBM's computer, beat the then world chess champion Garry Kasparov in 1996,[2] it did so using significant computing power and methods which, from a cognitive point of view, are not based on any of the skills deemed to be the basis of intelligence. Artificial intelligence can therefore be defined as that which enables the same results to be obtained as human intelligence, possibly or even mainly by using other means.

Defined in this way, artificial intelligence can therefore refer to very different technical and scientific paradigms. Traditionally, there are two main paradigms, the first based on knowledge modelling, the second on data processing. The first paradigm consists of characterising the knowledge that we, as human beings, demonstrate or mobilise in solving the problems we encounter [23]. This characterisation is made explicit in a more or less formalised language, usually natural or specialist language. The challenge is then to be able to completely formalise the statements obtained so that they can be calculated or manipulated by a computer. The formalisation obtained is made up of a fact base, which codifies the states of the world, and a rule base, which enables changes in the world to be deduced from the facts observed. The system is considered to be efficient when it makes forecasts or calculations that are consistent with observed changes in the world or with human reasoning based on the same

[2] See for example https://en.wikipedia.org/wiki/Deep_Blue_versus_Garry_Kasparov/.

facts. This was the era of the famous expert systems, the most emblematic of which was undoubtedly Mycin [24].

This approach therefore consists of simulating and reproducing human reasoning via its logical formalisation, and therefore has the advantage of being intelligible: the results obtained are explicable because they derive from the inference rules used (the knowledge) on assertions (the facts) in an order reproducing the reasoning that a human could have used. In addition to this explicability, such an approach can be used on restricted facts, only those that reflect the situation under consideration. They do not have to represent the range of different possible situations, whether already encountered or only possible, underlying the knowledge formalised in rules. In fact, the knowledge held by humans is already a compilation or sedimentation of the various relevant situations and experiences, so there is no need to characterise it as such. So we give the system a knowledge base that summarises and compiles the world's experience, and only the facts that characterise the situation it needs to reason about. In short, we are working with a complex model but little data.

But these advantages can become disadvantages, or even limitations, when we have to deal with problems for which we are unable to spell out our know-how in rules. Know-how does not refer to any theoretical knowledge (know-what). This is the case for many expert tasks, such as interpreting images, managing complex systems and uncertain reasoning. In such cases, we reverse the approach: since we are unable to provide the system with the rules of reasoning, we will provide it with the data enabling it to induce them. Instead of having a small amount of data and a complex model, we will then have a large amount of data, all of which is relevant and necessary to represent the diversity of situations where the rules to be obtained are supposed to apply, and a very simple model characterising the type of knowledge to be induced.

From then on, the approach is in two stages: in the first stage, we have a learning algorithm (the very simple model already mentioned) which makes it possible to induce a knowledge model from the data supplied. In this first stage, we have the training data, which is the data representing the world of the task to be processed (the base of all the images that may be encountered in an interpretation task, the set of all the relevant texts for a translation task), and two models: the training model using the data, and a learned model produced by the training model. The training data is often broken down into two batches: data to infer the learned model and other data to evaluate or test it. In the second stage, the knowledge model is used on the data characterising the situation under study for which we want to solve a problem or perform a task (interpret an image, translate a text, diagnose a disease). In this second stage, we have the data from the situation (a text to be translated, an image to be interpreted) and the application of the learned model to this data.

The following table summarises the two paradigms of artificial intelligence.

	Data	Model	Model (data)
Knowledge paradigm	Situation data	Human knowledge model, formalised for computer processing	Solution
Data paradigm	World data	Learning model A general form of model specified using world data	Learned model Knowledge modelling corresponding to the processing of data from the world
	Situation data	Learned model	Solution

This table teaches us an informal but invariable law: Data $\times$ Model $=$ constant. This law shows that you can't do anything for nothing—it is the principle of conserving information. But it is necessary to articulate the forms in which the information is presented and to ensure that these forms do indeed refer to the whole or totality of what needs to be considered. So, when data was scarce (the Internet was still confidential, the Web did not yet exist, and only textual content was mostly digitised), we needed a model produced from what was available and accessible, i.e. human knowledge. When the Web made its appearance, when the Internet deployed its possibilities, and when content, including non-textual content, became available in massive quantities and assembled in databases or warehouses [3], and when the power of machines finally made mass processing possible, it became possible to envisage a reversal and to have mass data for simpler models making it possible to abstract knowledge that was not generally available in an explicit or intelligible form.

The image processing discussed in this book is based essentially on the second paradigm. We will describe its main characteristics and innovations, focusing on its most recent and undoubtedly most impressive form, generative AI.

2.5 Approaches to Data-Driven AI: The Basic Model

We are not going to go into detail here about the many algorithms used to process data. We will confine ourselves to presenting their operating principles in order to make clear what we consider to be both their strengths and their limitations. We will first describe the general form of the most common learning model, the neural network (Fig. 2.1), and then consider some of its uses.

The neural network is a computer metaphor for the biological neuron. A metaphor, not a model, because biology has long since shown how different the human brain can be from a formal neural network. Nevertheless, the metaphor is powerful and remains a constant source of inspiration. The formal neuron was conceived in a seminal paper by Walter McCulloch and Walter Pitts [10], which set out what would be systematically adopted in subsequent architectures, notably Frank Rosenblatt's Perceptron ([21]; Fig. 2.2).

The principle is simple:

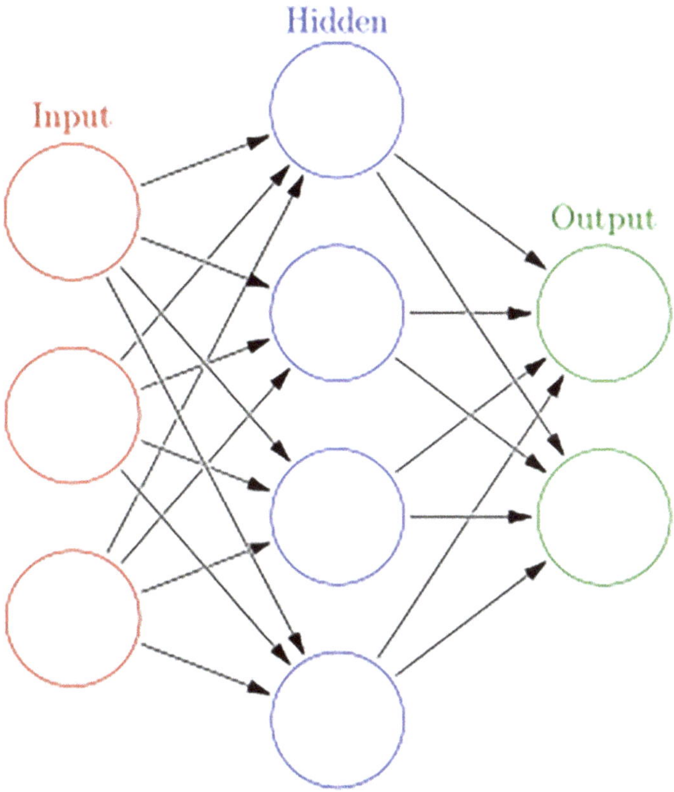

Fig. 2.1 Neural network (from Wikipedia.en)

- We have n input links, each worth 1 or 0 depending on whether it is activated or not. In practice, these links are considered to be input neurons whose activity level (0 or 1) is all that is considered, the latter being given.
- A neuron receives each of its links with a given weight.
- The neuron performs two successive operations:
  - The weighted sum of link values (their value times the associated weight)
  - The application of an activation function, generally a threshold function, to determine whether or not a neuron sends an activation value as an output.

This is the learning model we mentioned earlier: this model includes all the neurons that can be built by setting values for the weights of the input links. The challenge is then to learn the values for these weights using training data, the data from the world of the task. The learned model is the network obtained with these values. It will be applied for a set of input values on each link, the data from the situation being evaluated. The solution is a yes or no answer (activation or non-activation of the output): this is why a perceptron is characterised as a classifier

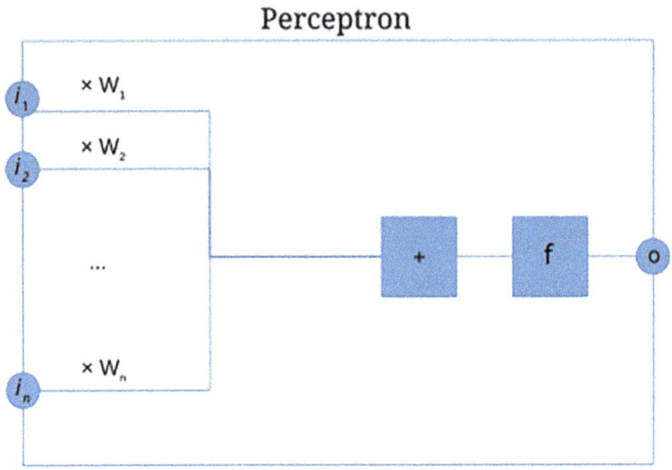

$$o = f\left(\sum_{k=1}^{n} i_k \cdot W_k\right)$$

Fig. 2.2 Perceptron (from Wikipedia.en)

which determines whether or not the data being evaluated belongs to a given class (i.e. whether or not it is a radioactive cancer).

This simple model was denounced as too simple by Minsky and Papert [12] because it could only learn a limited type of classifier, corresponding to a linear function separating two populations. But it was later shown that by inserting so-called hidden layers of neurons between the input links and the output classifier neuron, it was possible to model any separation (classification) function [9, 22].

Since then, there has been an abundance of literature on the different network architectures that can be designed, depending on the number of hidden layers involved and the type of activation functions used. The introduction of hidden layers in formal neural networks has led to the development of deep learning, the term used to describe all the techniques that make use of such layers. The main techniques developed include Convolutional Neural Networks (CNNs), which can recognise local structures in content, and Recurrent Neural Networks (RNNs), which can analyse sequences. Algorithms using transformers combine the respective qualities of CNNs (recognising local structures) and RNNs (recognising dependencies in a sequence), while making it possible to overcome some of their shortcomings, for example the fact of being able to process the components of a sequence in parallel rather than successively, so as to keep processing time independent of the length of the sequence.[3] Without going into the details of these developments, let's just mention two essential properties:

[3] For an overview of the techniques used in deep learning, see the reference work by Bengio [2].

- The learning model is a fundamental factor in determining the type of knowledge that can be learned. For example, Rosenblatt's perceptron could only learn linear functions.
- The initial data determines the universe from which to parameterise the learning model. For an AI system, its world, its universe, is reduced to its data. Put another way, an AI system has no context or environment, only data. It does not matter when or where it is used, only the data on which it is configured and then used.

2.6 Variations on the Model

Since we are interested in data-based AI, the question is to understand what we want to do with it. Data has properties and implicitly belongs to a common space. This space is prima facie defined based on the ways in which data is declared and defined. For example, an image will belong to the space of images composed of $n \times p$ pixels, each associated with $r \times g \times b$ bits characterising the colours red, green and blue respectively. A text belongs to the space of character sequences that can be constructed from an alphabet, and so on. From this space, we can define three types of processing or questions posed [13]:

- Recognition of new data: using existing data deemed to belong to pre-existing, explicit categories, how can we define the category of new data? This task is often presented as a prediction, because it involves predicting and recognising the category of a piece of data, knowing the categories of the training data. This is supervised learning (using classes that are already known).
- Data structuring: using existing data, we learn categories to classify and differentiate it. New data will then be classified according to these learned categories. This learning is unsupervised (the classes are not known a priori) but it can also make predictions.
- Data generation: using existing data, we learn how to produce new data that is as close to it as possible, and as similar as possible. This last task differs from the first two in that it involves inventing data, whereas the first two work with real, supplied data.

In what follows, we are going to focus on this last category because it allows us to question precisely the power and the limits of data-driven AI.

2.7 Generative AI

Here, too, we will confine ourselves to schematic indications that are sufficient for our purposes. The range of available tools on offer is almost infinite, but in recent years, three models have come to the fore.

The first model corresponds to Generative Adversarial Networks (GAN [5]). The principle, which is very clever, is to have two neural networks working in competition: the first (the "faker") learns to generate content that is likely to deceive the second (the "expert"), which is trained to discriminate between false and genuine content. In both cases, this is the usual supervised learning process, but the expert learns from real content whose quality (true or false) is known, whereas the counterfeiter learns from the answers provided by the expert. The faker does not produce real content, but invents content to fool the expert.

The second model corresponds to diffusion models. The principle is to add noise, i.e. random information to the data, and then learn how to remove it in order to recover content close to the original data. In other words, we move through the space of the initial data, arbitrarily distancing ourselves from the training data thanks to the noise, before returning to the subspace specific to this data. Learning, therefore, involves the ability to denoise images. Starting with a given image, denoising leads not to one of the initial training data sets but to a neighbouring, invented data set. This model can be made more complex by taking into account image categories, and therefore types of denoising (one class for denoising to obtain images of cats, another of dogs, etc.), or even more complex categories mixing several characteristics. To do this, we no longer reason in terms of the data space, but in terms of a different space, a latent space, where the data are projected in such a way as to introduce proximities between them reflecting their common membership of categories. Denoising then reasons on the basis of this space and these categories. This is how we can obtain images of a cat on a carpet rather than in the garden.

Finally, the last model is the one corresponding to the so-called "transformer" models and to the algorithms managing attention, i.e. the dependencies between the data in the situation being processed. The principle behind these models is inspired by neural networks, which are capable of tracking or generating new data sequences from input sequences. The best-known example is ChatGPT, which returns strings of characters in response to, or in addition to, sequences of characters provided as input (requests or prompts).

These tools are also based on the principle of latent spaces, i.e. another representation of the initial data in a space of smaller dimensions, which also allows the data to be coded in such a way that their mutual distance reflects their semantic proximity (in any sense you like, but most of the time this means that they belong to the same classes). The steps are as follows:

Formatting the initial data, which is called tokenisation because it involves putting it into tokens or elementary units. In the case of text, this involves encoding textual content (in the broad sense, with graphics, characters, computer code, etc.) using these tokens: it will correspond to a succession of such tokens. The advantage is in considering only a few tens of thousands of possible tokens (100,000 in ChatGPT4 [1]) rather than all the possible textual forms of a language and associated content (on the order of 500,000 words, adding those from different computer languages, graphics, etc.). This reduces the dimensionality of the problem (one dimension per word to one dimension per token). But not by enough.

For this reason, a latent space with dimensions generally set at 512 is used. This latent space is also constructed by learning so as to code the tokens according to their context of use: two tokens will be all the more similar if they are used in the same contexts. There are various algorithms for doing this, the best known being Word2Vec [11]. This approach is based on a classic semantic theory of natural language, namely Harris's distributional semantics, where the challenge is to model the semantics of a term by its distribution in corpora of use [6]. The semantic axes learned in this latent space are not always easy to interpret, if ever. But the principle is that the 512 dimensions of this space are as many semantic dimensions, making it possible to describe the meaning of a token, which is therefore a vector of this space.
An encoding device is then used to learn how to encode in the latent space.
A decoding device learns to produce a sequence based on the encoded sequence. The two devices, encoding and decoding, are trained by means of sequences supplied as input and the relevant responses expected (a translation, a continuation of the input sequence, etc.).
The originality of the encoding and decoding is to be able to make the tokens of the input and output sequences depend on each other by means of a mechanism known as attention [26] allowing long dependencies to be processed on the one hand and in parallel on the other (all the words of the input sequence are processed at the same time, which makes it possible to have a virtually constant response time whatever the length of the input sequence).
A translation of the vectors obtained (produced by decoding) into the target language space (which can be as varied as you like, as long as you know how to move from the latent space into it).

To understand the status of what is produced by such tools, we need to bear in mind the following points:

- Coding input data: tokenisation;
- Embedding tokens in a latent space;
- The vectors making up the response are calculated in this space on the basis of the vector proximities observed;
- Translation of the vectors obtained into the target language space.

Nowhere does the process depend on anything other than the initial data and the encoding in the latent space, the latter having a structure that remains opaque to interpretation because its dimensions are too numerous to be labelled and combined on the one hand, and because the link with the initial data is not preserved on the other. The result produced can only be assessed by its relevance and not by the way it was produced. In conclusion, this raises the question of the status of this production.

2.8 Interpretation: The Status of Artificial Intelligence Productions

Data AI, particularly in its generative form, analyses and produces data. This production is read as data carrying a meaning that we confer on it and designating a potential reference that we associate with it. However, in the processes we have outlined above, nowhere is meaning invoked; we are only talking about vector distances in a latent space, nor is reference mobilised—the data are sufficient in themselves. There is, therefore, a gap between how these results are produced and how we interpret them. Understanding this discrepancy enables us to understand the power of these approaches, but also their inherent limitations. If, as we will argue, these limits are beyond our control, the question that will arise is whether, in the end, these limits do not impose themselves on us too, bringing us closer to machines when these limits seemed to distinguish us from them.

To refine our analysis of these productions, we propose to analyse the content produced along three dimensions: what is stated, what is said and what is designated.

Stating or enunciating something is the process by which content is enunciated by a person. The utterance is then a commitment to the world, a being thrown into a situation in which we take part by this utterance. As such, this utterance is a call, intended to be interpreted by those who receive it. Saying is therefore a doing, a commitment to the world and a promise of meaning.

What is said is what is stated—it is the statement. It is not a process, but something whose materiality can be grasped: a word that I can record, a piece of writing that I can copy down, an image that I can transmit.

Finally, the designated corresponds to the entities of the world, real or fictional, to which the spoken word refers.

There is a complex relationship between what is stated and what is said. A priori, every content emanates from an utterance: the utterance refers to an intention calling for an interpretation. It cannot be reduced to a simple thing that can be decontextualised from the doing that produced/enunciated it and the doing that interprets/receives it. However, what is said is also a thing, and therefore producible by doings. But what relationship can be established between the doing that is a stating and the doings of which the world of the designated is constituted? Would any statement resulting from any doing be analogous to a statement arising from a stating? Would every doing that produces a statement be analogous to stating?

Similarly, what is said and what is designated do not seem to be on the same level. What relationship can be established between things that are words (what is stated or said) and things that are not? Statements are things that possess relations of succession (syntagmatic) and juxtaposition (paradigmatic): we understand how one word leads to another, one saying to another. But a statement is not intended to go beyond the order of discourse and refer to something other than itself. Inspired by a Percian semiosis, we can consider that these successions are the only reality of what is said, in an unlimited or infinite semiosis [15], which is not exhausted in an object of the world but in an action (doing). The reference to the designated is related to the inference towards an object that becomes a sign (everything is a sign). In other words, there is only a relationship between signs; a sign can only refer to another sign.

In this context, AI-generated productions are easy to position, but we also understand why they leave us perplexed. In fact, AI only deals with what is said: as we have emphasised several times, AI only manipulates the data it has been provided with. There are no contents or promises of meaning (a saying or stating), only manipulated codes, in other words, a doing that is not a saying.

However, AI implements processes: in what way is producing what is said by calculation not a matter of saying? This is the problem of enunciation specific to AI. Similarly, in what way does the said produced not refer to a designated of which it would be the expression? This is the problem of modelling posed by AI: in what way are AI-generated productions a model of the world since they are produced from data which, in turn, refers to it?

2.9 Modelling

First of all, the problem of modelling. If we go back to the stages of the processes involved in data-driven AI, there are indeed models, in particular a learning model and a learned model. They are models in the sense that they allow data to be queried without having to consult it, instead of the data itself. In the same way that the digital model of an aircraft can be used to simulate its flight without having to build it, the learned model can be used to process new data (create it, recognise it, classify it) without having to reconsider all the initial data but only what the learned model has retained.

The learned model is therefore a model, but a model of the initial data, not of the world it represents. In particular, the latent space at the foundation of certain learning processes constitutes a compiled archive of the initial data, the corpus of texts or images under consideration. Antonio Somaini[4] refers to it as a meta-archive, which is highly relevant: an archive because it refers to documents from the past, but a meta-archive insofar as the initial documents are rendered inaccessible, drowned in the latent space.

But if the AI of the data only models the data, and if the model obtained corresponds to the knowledge that makes it possible to manipulate this data in order to respond to its place, this model is not a model of the world or of the entities whose data it is. Yet we cannot help reading them in this way: any semiotic production, if it is perceived or received as such, is understood as an expression about a world. We are subject to a transcendental illusion [8] where, although we know it to be an illusion, we interpret the result produced not as an extrapolation of data, but as data about the world. Following the example of Plato's Phaedrus [18, 19], the texts

[4] Séminaire international de sémiotique de Paris, https://www.fmsh.fr/agenda/intelligence-artificielle-generative-et-nouveaux-enjeux-semiotiques-5.

and images produced seem to be statements; they seem to think about what they are saying or showing, whereas all they do is repeat the same thing as soon as they are questioned. However, the contemporary problem is compounded by the fact that, bypassing Plato's critique, tools such as generative AI based on LLM give the illusion of dialogue.

2.10 Enunciation

When it comes to enunciation, the confusion is also potentially significant. How can we reproach the machine for not expressing itself when we are talking to it and asking it questions? The answer we can give is simple. We have already pointed out that an AI's world is reduced to its data, and that this data has no background or context. In other words, the data representing the world has no world in which it is inscribed and interpreted. When the machine responds or produces an utterance, it does not depend on who it is interacting with, nor on the situation, nor on place or time. Machine interaction is not situated and has no "milieu" [16]. There is no enunciation unless we commit ourselves and assume, for better or for worse, the statement produced: we can be held responsible for it. A machine that only manipulates information coded in a latent space is neither committed to nor responsible for what it produces.

However, it could be argued that this argument is based on the myth of the signified and intentionality. Paolucci [14] points out that cognitive capacity cannot be recognised by the ability to say "I" or to be an ego totalising lived experiences, but rather by the ability to refer to utterances other than one's own, when one speaks in the manner of someone else. Now, generative AI tools have shown that they can produce statements belonging to an author's own stylistic and argumentative repertoire, expressing themselves according to particular canons. Shouldn't we see this as proof that the machine is just as good at enunciating as we are, and that the enunciative commitment we mentioned is merely an existential modality, an experience that does not in itself constitute a cognitive skill or define thought? This remains debatable, because here we would be replacing a totalisation by the ego with a symbolic totalisation by the machine. To disqualify the former is also to disqualify the latter. What is more, enunciation is not so much a matter of the ego, which would be the transcendental principle of experience, but of an engagement with a situation or a milieu that encompasses and contains us, in relation to which we determine and position ourselves. The environment, or context, is what means that our data is constantly changing, that it is formatted and selected according to these constantly evolving criteria, and that ultimately the rules of reasoning are determined according to the horizon of our engagement. But the machine has fixed data, an intangible learning model and no horizon of engagement.

2.11 Conclusion

For the first time in our history, we can combine the interpretation of content with algorithmic manipulation based on the memory and assimilation of a dizzying array of past and present content. However, human interpretation and mechanical assimilation are not on the same level. Human interpretation is contextual, reconfiguring data and manipulation rules. Assimilation is based on formalised and formatted initial data, which remains unchanged throughout the process. Processing is blind manipulation, where data is reduced to meaningless symbols. This semantic reduction guarantees the repeatability of the calculation and the certainty of the result: as the data consists solely of its signifier, it cannot be redefined by the interpretative process, nor can it be manipulated [20].

But this formalised memory can be compiled into a learned model, a data model that constitutes a meta-archive of this memory. To interrogate content via these models is to summon the memory of the world in order to analyse, classify and produce data. But this informed production does not mean that a person is making a statement or modelling the world. AI data only speaks for itself, and the tools based on it can only respond to what it says.

References

1. Alammar, J., Grootendorst, M.: Hands-on Large Language Models: Language Understanding and Generation. O'Reilly, Sebastopol (2024)
2. Bengio, Y., Goodfellow, I., Courville, A.: Deep Learning. MIT Press, Cambridge, MA (2016)
3. Ceruzzi, P.E.: A History of Modern Computing, 2nd edn. MIT Press, Cambridge, MA (2003)
4. Copeland, B.J.: The Church-Turing Thesis. In: Zalta, E.N., Nodelman, U. (eds.) The Stanford Encyclopedia of Philosophy (Spring 2024 ed.). Metaphysics Research Lab, Stanford University, Stanford (2024)
5. Goodfellow, I., Pouget-Abadie, J., Mirza, M., Xu, B., Warde-Farley, D., Ozair, S. et al.: Generative Adversarial Networks. In: Proceedings of International Conference on Neural Information Processing Systems (NIPS 2014) (2014)
6. Harris, Z.S.: Distributional structure. WORD **10**(2–3), 146–162 (1954). https://doi.org/10.1080/00437956.1954.11659520
7. Hjemslev, L.: Prolégomènes à une théorie du langage. Minuit, Paris (1971)
8. Kant, I.: Critique of Pure Reason (1781). Cambridge University Press, Cambridge, MA (1998)
9. Le Cun, Y.: Quand la machine apprend. La révolution des neurones artificiels et de l'apprentissage profond. Odile Jacob, Paris (2019)
10. McCulloch, W., Pitts, W.: A logical calculus of the ideas immanent in nervous activity. Bull. Math. Biophys. **5**, 115–133 (1943)
11. Mikolov, T., Chen, K., Corrado, G.S., Dean, J.: Efficient estimation of word representations in vector space. In: Proceedings of Workshop at ICLR (2013). arXiv:1301.3781
12. Minsky, M., Papert, S.: Perceptrons: An Introduction to Computational Geometry. MIT Press, Cambridge, MA (1969)
13. Mitchell, T.M.: Machine Learning. McGraw-Hill, Redmond (1997)
14. Paolucci, C.: The myth of meaning: generative AI as language-endowed machines and the machinic essence of the human being. Semiotica **262**, 5–23 (2025). https://doi.org/10.1515/sem-2024-0204

15. Peirce, C.S.: Collected Papers. Harvard University Press, Cambridge, MA (1960)
16. Petit, V., Guillaume, B.: We have never been wild: towards an ecology of the technical Milieu. In: Handbook of Sensory Physiology, vol. 3, pp. 81–100. Springer, New York (2018)
17. Petzold, C.: The Annotated Turing: A Guided Tour Through Alan Turing's Historic Paper on Computability and the Turing Machine. Wiley, Indianapolis (2008)
18. Plato: Phaedrus. Cambridge University Press, Cambridge (1972)
19. Plato: Phèdre, suivi de la Pharmacie de Platon par Jacques Derrida. Garnier–Flammarion, Paris (2006)
20. Rastier, F.: Sémantique et Recherches Cognitives. Presses Universitaires de France, Paris (1991)
21. Rosenblatt, F.: The perceptron: a probabilistic model for information storage and organization in the brain. Psychol. Rev. **65**(6), 386–408 (1958). https://doi.org/10.1037/h0042519
22. Rumelhart, D.E., Hinton, G.E., Williams, R.J.: Learning representations by back-propagating errors. Nature **323**, 533–536 (1986). https://doi.org/10.1038/323533a0
23. Russell, S., Norvig, P.: Artificial Intelligence: A Modern Approach, 3rd edn. Pearson, Edinburgh (2016)
24. Shortliffe, E.H.: MYCIN: Computer-Based Consultations in Medical Therapeutics. American Elsevier, New York (1976)
25. Turing, A.: On computable numbers, with an application to the *Entscheidungsproblem*. Proc. Lond. Math. Soc. **42**(2), 230–265 (1936)
26. Vaswani, A., Shazeer, N., Parmar, N., Uszkoreit, J., Llion, J., Gomez, A.N. et al.: Attention is all you need. In: Proceedings of 31st International Conference on Neural Information Processing Systems (NIPS'17), Long Beach (2017)

Bruno Bachimont is full professor at Compiègne University of Technology (UTC). He specialised in the fields of logic, documentary computing, and digital philosophy. He was director of research at UTC from 2006 to 2017, and director of research and development at the Sorbonne University's Faculty of Science and Engineering from 2017 to 2021. From 1997 to 2012, he worked at the Institut National de l'Audiovisuel (INA, France) as Director of Research and, later, as Scientific Adviser, contributing to the digitisation of content and archival materials. He is interested in questions concerning the consequences of digital technology in general and AI in particular, on knowledge and culture, especially in relation to history, memory, and archives.

Chapter 3
How Machines See and Generate Images

Sergio Picascia and Alfio Ferrara

Abstract The integration of Artificial Intelligence (AI) into artistic domains has grown significantly in recent years, particularly through the advancements in Computer Vision (CV). CV is a field of AI focused on enabling machines to interpret and understand visual information trough automatic methods. In this work, we present a comprehensive overview of CV techniques applied to image analysis and generation. We begin by tracing the evolution from classical approaches of feature extraction and colour processing, to contemporary deep learning methods, such as convolutional neural networks (CNNs) and Transformers, achieving high-level semantic understanding of visual content. The study then shifts to image generation, examining different paradigms and exploring the various architectures involved in the task. Additionally, we explore multimodal models that integrate textual and visual data, highlighting their role in the conditional image generation. Finally, the paper touches on the ethical considerations surrounding AI-generated imagery, including concerns about authenticity, bias, and misuse. This work aims to provide a cohesive understanding of how AI is shaping the visual dimension of artistic and computational practices.

Keywords Computer vision · Image processing · Image generation · Multimodality · Data ethics

The human visual system effortlessly distinguishes objects in daily life by converting, through the eyes, the light reflected from these objects into electrical signals. These signals are then transmitted to the brain, where they are processed to localise and classify the objects within our field of view. This capability is developed through years of experience and ongoing exposure to visual stimuli. Similarly, computer vision systems operate on a comparable paradigm, substituting human eyes with

S. Picascia (✉) · A. Ferrara
University of Milan, Milan, Italy
e-mail: sergio.picascia@unimi.it

A. Ferrara
e-mail: alfio.ferrara@unimi.it

P. Conte et al. (eds.), *Algomedia. The Image at the Time of Artificial Intelligence*, Lecture Notes in Morphogenesis, https://doi.org/10.1007/978-3-032-08726-3_3

sensor devices and employing interpreting devices in place of the human brain. Sensor devices, e.g. cameras, play a pivotal role in data acquisition, handling the capture and calibration of images or video footage. On the other hand, the interpreting device, typically a computer, employs sophisticated algorithms to extract relevant information from the data acquired.

Computer vision is the field of artificial intelligence that empowers computers to interpret and understand the visual world through digital images or videos. In contrast to human cognition, knowledge acquisition consists of a significantly more intricate challenge for computers. The challenge lies in the inherent complexity of vision, often described as an inverse problem, where machines strive to capture and reconstruct from images various properties of the visual world, such as shapes, colours and edges. For this reason, researchers rely on sophisticated machine learning algorithms in an attempt to overcome these challenges and perform these tasks. While even a child would find it trivial to recognise a dog after seeing one, for machines it demands thousands, if not millions, of observations to sufficiently train algorithms which, ultimately, are confined to executing very narrowly defined tasks. However, the rise in availability of substantial data volumes and computational resources required for algorithm training has drawn considerable attention to computer vision applications over the last decade. From autonomous driving to biometric scanning, computer vision algorithms are seamlessly integrated into everyday activities, often without conscious recognition of their presence or utilisation.

In the upcoming sections, we will dive into how machines perceive images in practice and explore the latest advancements in computer vision. We will then take a closer look at image generation, examining the models employed and discussing the social impact of their applications.

3.1 How Computers Handle Images

In understanding how computer vision systems decode the visual world, we can break down their process into three crucial stages: image acquisition, image processing, and image recognition. Image acquisition marks the initial step, where a sensor captures the incoming light and converts it into digital signals, creating a digital representation of the scene. Following this, image processing employs various techniques to refine the overall image quality, enhance certain visual characteristics, or isolate specific regions of interest. Lastly, image recognition utilises these refined features to tackle complex tasks, such as locating and classifying objects or understanding the overall context and semantic meaning of a scene. Through these steps, computer vision systems are able to extract insights from the visual world. In this section, we will cover these three steps in detail, exploring from fundamental concepts to state-of-the-art approaches to computer vision tasks.

3.1.1 Image Acquisition

A digital image serves as a visual representation of a scene, captured, stored and presented in digital format. Digital images are composed of picture elements, known as pixels. Each pixel contains discrete and finite numerical quantities, representing its colour or intensity; as we will see, a higher value typically indicates a greater strength of a colour. The total number of pixels is commonly used to specify the resolution of the image, indicative of its level of detail: the resolution is either denoted as the number of pixel rows (height) by the number of pixel columns (width), e.g. 1920 × 1080, or as the total number of pixels within the image, e.g. 2,073,600 or 2.07 megapixels. Therefore, one can claim that the height and width of a digital image define the size of a matrix, where each value is represented by the pixel intensity.

The colour depth of an image, also referred to as its bit depth or pixel depth, represents the number of bits assigned to store the colour of each pixel. This measurement is commonly expressed in terms of bits per pixel (bpp), with frequently encountered colour depths being 1-bit monochrome, 8-bit grayscale, and 24-bit colour. In 1-bit or binary images, pixels can manifest solely two colours, typically black (0) and white (1); due to the compact file size, binary images are often utilised in fax transmission or document scanning. Meanwhile, in grayscale images, pixels represent varying intensities of light, with values spanning from 0 (representing black, total absence of light) to 255 (representing white, maximum presence of light). Finally, in 24-bit colour images, each pixel consists of three 8-bit colour channels, each representing a grayscale image of the same size as the original image. Typically, colour images are depicted using the RGB colour model, in which the three channels represent the three primary colours: red, green, and blue (Fig. 3.1). By combining these channels, a wide spectrum of colours, approximately 16.7 million, can be reproduced. The number of channels represents the third dimension (together with height and width) of a digital image; for this reason, we refer to colour images as 3-dimensional tensors,[1] and grayscale images as 2-dimensional tensors, due to the presence of a single channel.

3.1.2 Image Processing

It is now evident that, for machines, images are merely arrays of numbers corresponding to pixel values, lacking any further information about their content. To extract meaningful knowledge from these numerical representations, it is essential to employ techniques that highlight specific features of the image. Image processing encompasses a diverse range of these techniques, manipulating digital images through computer algorithms. These methods serve as crucial preprocessing steps in numerous applications, as we will explore in later sections. Our focus will include key image transformations aimed at enhancing various aspects of an image,

[1] A tensor is a mathematical object defined as a multi-dimensional array of numerical values, which can be used to represent data in various dimensions.

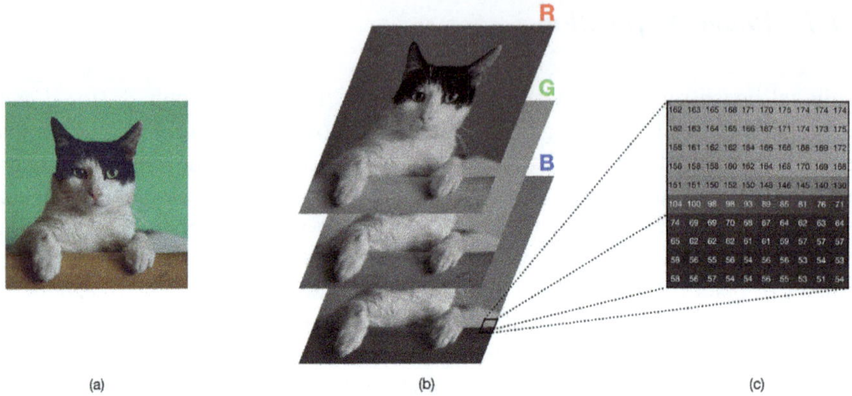

Fig. 3.1 Focus on a digital image. The colour image **a** consists of three channels, **b**: red (R), green (G) and blue (B). Each channel is a grayscale image, where each pixel is represented by light intensity (**c**). A high pixel value represents a high presence of that channel colour

along with feature extraction techniques designed to retrieve specific characteristics that serve as valuable inputs for algorithms.

3.1.2.1 Image Transformations

A digital image transformation involves modifying the appearance or structure of a digital image using mathematical operations or algorithms. These transformations alter the pixel values or spatial arrangement of the image, resulting in changes to its visual appearance or information content.

Apart from feature enhancement, combinations of these transformations are commonly utilised also in data augmentation, a technique for generating new data samples from existing datasets. This approach has been extensively employed in computer vision, especially during the early deep learning applications, to mitigate overfitting[2] in algorithms trained on limited datasets. Indeed, while humans can effortlessly recognise rotated or flipped versions of an image as equivalent to the original one, machines perceive any transformation as a significant alteration in pixel values, effectively creating entirely new data samples. These transformations are instrumental in enhancing the robustness and generalisability of machine learning models by diversifying the training data and improving the models' ability to generalise from the training set to real-world applications.

The spectrum of digital image transformations is extensive, ranging from adjustments to brightness, contrast, and colour balance to more complex alterations. In this section, we will focus more in depth on affine transformations, filtering and thresholding.

---

[2] Overfitting happens when a model performs extremely well on training data, but fails to generalise to new, unseen data.

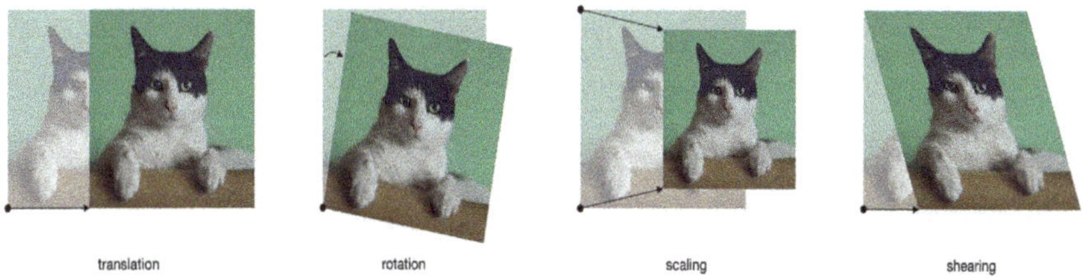

Fig. 3.2 Examples of affine transformations

Affine Transformations. These are a class of geometric transformations that preserve lines and parallelism in an image (Fig. 3.2).

- Translation entails shifting all points within an image by a designated distance along both the x and y axes, repositioning the entire image. • Rotation involves rotating an image around a specified point, known as the rotation centre. The angle of rotation dictates the degree and direction of the applied rotation.
- Scaling involves adjusting the size of an image by multiplying the coordinates of each pixel by a scale factor along both the x and y axes. A scale factor greater than 1 enlarges (scales up) the image, while a scale factor between 0 and 1 shrinks (scales down) it. Additionally, a negative scale factor results in a reflection (flip) of the original image.
- Shearing entails distorting an image along one axis while keeping the other axis unchanged, causing the image to tilt in a specified direction.

Filtering. Image filtering involves modifying the pixel values of an image by applying a convolution operation between the image and a kernel, also known as a filter. Depending on the values inside the kernel, this process enables the enhancement of image features, noise reduction, and edge detection, among other applications.

The *convolution operation* is a mathematical process that combines the input image with a kernel. The kernel is a small, two-dimensional array of coefficients that slides across the entire image, one pixel at a time. As the kernel overlays the input image, each pixel value is multiplied element-wise with the corresponding kernel coefficient. The resulting products are then summed up to produce which becomes a new pixel value in the output image. This process is repeated for every pixel in the input image, with the kernel sliding across the entire image (Fig. 3.3).

The effect of image filtering depends on the specific kernel used: a simple averaging kernel can be used to blur an image, reducing noise and softening edges, while a sharpening kernel can enhance the contrast between adjacent pixels, making edges appear more pronounced (Fig. 3.4).

Thresholding. Image thresholding involves converting a grayscale or colour image into a binary image, with the goal of separating objects of interest from the background. Pixels with values above the threshold are set to a maximum value (usually white), while those below the threshold are set to a minimum value (usually black).

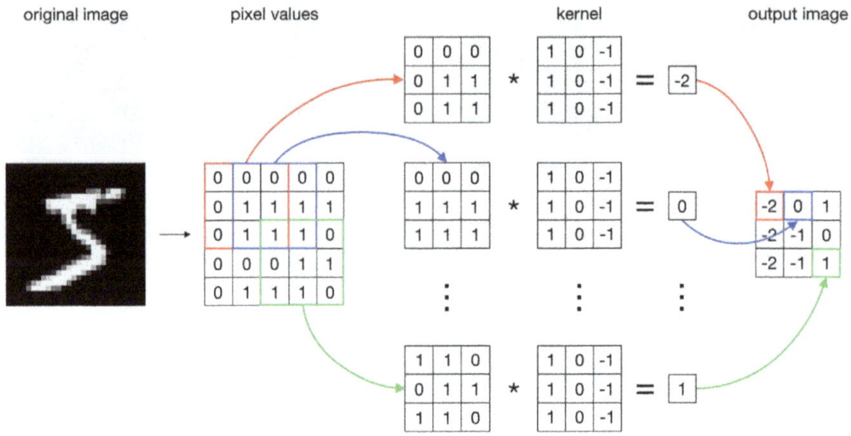

Fig. 3.3 The convolution operation. A 3×3 kernel is being applied throughout an image representing the digit 5 (on the left). It moves one pixel at a time from left to right, and from top to bottom, being multiplied by the underlying pixel values of the image. The multiplication result represents the intensity of one pixel in the output image

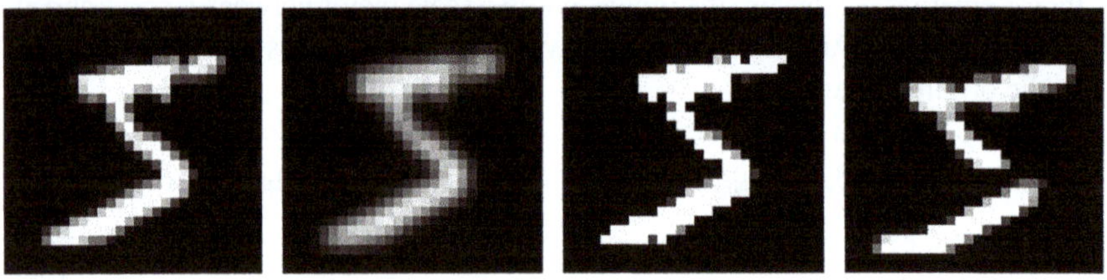

Fig. 3.4 Examples of filtering operations. From left to right: identity (does not alter pixel intensities), blurring (smooths edges), sharpening (enhance edges), vertical Sobel operator (enhancing vertical edges)

There are two main approaches to image thresholding: global thresholding and adaptive thresholding (Fig. 3.5). Global thresholding applies a single threshold value to the entire image, whereas adaptive thresholding adjusts the threshold value based on the local characteristics of the image, dealing with varying contrast or texture.

3.1.2.2 Feature Extraction

If image transformations allow for the alteration and enhancement of digital image features, the process of feature extraction consists in automatically selecting and extracting relevant information from images, such as colours, objects and shapes, to represent the image in a more meaningful and compact way. Image recognition algorithms can operate more efficiently and effectively when directly applied to the extracted refined features rather than to raw pixel intensities. Those features enable

Fig. 3.5 Application of two thresholding techniques on the image on the left: global thresholding (in the middle), and adaptive thresholding (on the right). The ability to clearly outline the cat in the picture shows the clear advantage of employing adaptive thresholding in this situation

algorithms to focus on the most relevant aspects of the image, instead of being overwhelmed by the vast amount of pixel data.

Colour Histogram. A colour histogram is a graphical representation of the distribution of pixel intensities within an image: it plots each intensity value along the horizontal axis, and the number of pixels with that intensity value along the vertical axis (Fig. 3.6a). Colour histograms are descriptors robust to noise and invariant to transformations, such as rotation or scale; however they are more sensitive to illumination changes.

Edge Detection. It is a technique used to identify and highlight the boundaries or edges of objects within an image that typically correspond to significant changes in intensity or colour (Fig. 3.6b). The most popular edge detector [1] involves the following steps: smoothing the image to remove noise, compute the gradient of the image to highlight the areas of significant change; apply thresholding on the gradient

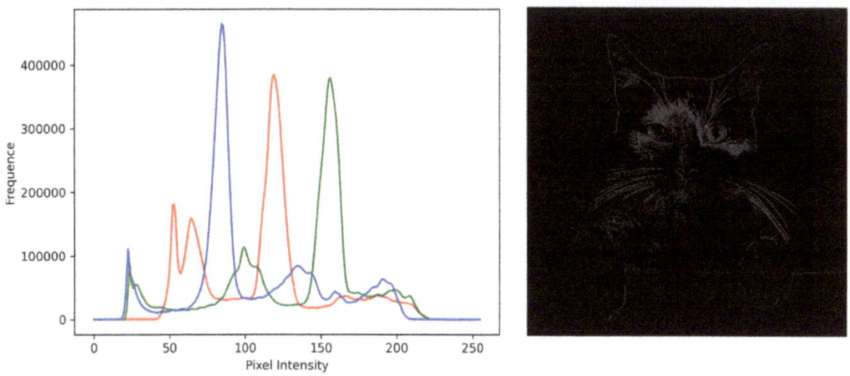

(a) (b)

Fig. 3.6 Two feature extraction techniques applied to the picture of a cat: colour histogram on the left, edge detection on the right

values to determine edges; link together edges to form a continuous boundary; refine the edges to improve overall quality.

Semantic Representation. In recent years, the advent of deep neural networks has revolutionised the feature extraction process, enabling the development of models that can effectively convert images into meaningful vector representations, known as embeddings. These embeddings transcend mere summaries of pixel intensities, capturing instead the rich semantic meaning and context of the image. By learning to represent images in a more abstract and meaningful way, deep neural networks have opened up new possibilities for image analysis, understanding, and manipulation. We will have a closer look to these models in the following section.

3.1.3 Image Recognition

In the final stage of the computer vision pipeline, the computer leverages the refined features to perform sophisticated reasoning and tackle complex tasks. In this section, we will explore the prevailing model architectures and the most prominent computer vision tasks that they try to solve.

3.1.3.1 Deep Learning Models

Computer vision researchers have been trying for years to solve complex tasks, achieving significant results using deep learning models only starting from the last decade. This success was partially due to the rise in resource availability, both in terms of computational power and annotated datasets. One of the first attempts at employing deep learning models for computer vision tasks was done by Lecun et al. [11], who defined a convolutional neural network architecture called *LeNet-5* to predict hand-written digits. However, due to limitations in computational power, the training of that kind of architectures was inefficient at the time. The turning point for the field came in 2012 with AlexNet [10], a deep convolutional neural network outperforming any other traditional approach in the ImageNet Large Scale Visual Recognition Challenge [17]. ImageNet [2] is one of the first large-scale visual datasets, consisting of millions of annotated images, categorised into thousands of classes, today, it is still a benchmark dataset for evaluating the performance of image classification models. The availability of large collections of annotated data, such as ImageNet, together with the idea of training neural networks on graphics processing units (GPUs rather than on the less efficient central processing units (CPUs, made AlexNet the very first example of successfully employing deep learning for a computer vision task.

Artificial Neural Networks. In the previous section, it has been shown how mathematical transformations applied to images allow the extraction or enhancement of certain visual features. This step is necessary, since the machine is only capable of interpreting numbers which, by themselves, do not bring any additional information

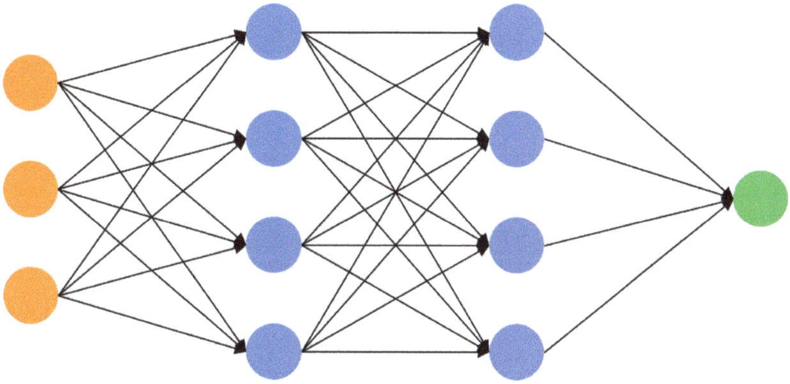

Fig. 3.7 Architecture of a deep neural network. Nodes in the input layer contain values of the input data, i.e. pixel values. Hidden layers receive and transform these values, propagating the new information. The output layer delivers the final prediction, which varies in form depending on the task addressed

about the content of the image itself. These transformations are at the core of artificial neural networks (ANNs), whose architecture emulates that of biological neural networks: neurons are arranged in layers, with each neuron receiving input information from the preceding layer, transforming it and transferring the new information to every neuron in the next layer. Thus, the information is passed through the various layers until it reaches the output layer, which delivers the final prediction. Neural networks with multiple hidden layers, meaning all the layers apart from the input and output ones, are referred to as *deep neural networks* (Fig. 3.7).

The transformations employed in a neural network involve the multiplication of input values by a matrix of weights. The strength of ANNs resides in the ability to learn these matrices of weights by themselves, in such a way that they maximise, to a certain extent, their ability to perform a certain task. For instance, if the ANN is asked to detect cats in images, it will try to estimate the weights in such a way that it makes the least possible mistakes. These mistakes are caught comparing the values predicted by the ANN in the output layer and the *ground truth*, the real values, i.e. the real location of the cat in the image.

In transforming the original input multiple times, hidden layers of ANNs generate an internal representation of the input data, called an *embedding*. An embedding is an *n*-dimensional vector, where *n* represents the number of values it consists of. Typically, the higher the number of dimensions, the better the model is able to represent the input data. When effectively trained, the embeddings generated by a neural network are capable of capturing the semantic meaning of the data and finally delivering the intrinsic information that was missing from bare pixel values. The newly learnt *latent space* allows not only the machine to effectively perform the task it is being trained on, but can also be exploited for comparing images by computing

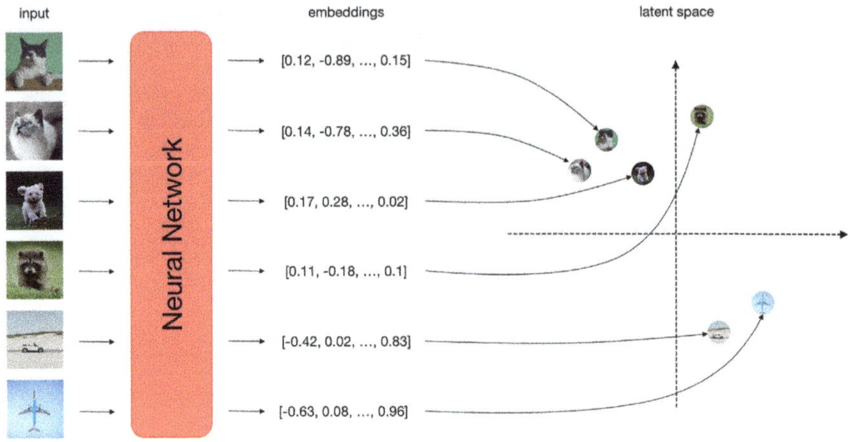

Fig. 3.8 Input images are represented by a neural network in the form of n-dimensional embeddings placed in a common latent space. Here, the semantic meaning of the input data is captured, with images representing similar objects being close to each other

the distance between their vectors. In doing so, we will discover that embeddings of images containing similar objects are placed close to one another in the latent space (Fig. 3.8).

However, simple ANNs have proven not to be optimal image processing architectures for several reasons. Firstly, as previously mentioned, images are typically represented as 2 or 3-dimensional tensors containing pixel values. However, to be employed as input for an ANN, it is necessary to flatten these objects into 1-dimensional arrays. In practice, this translates to taking the values row by row and placing them in a single, continuous sequence; dropping dimensions results in a significant loss of spatial information that the model will not be able to capture. Additionally, even modestly sized images lead to large input sizes when flattened, resulting in a high number of parameters (weights) that need to be estimated by the model. Furthermore, ANNs struggle with handling even slight variations in images, such as flips or rotations, which can cause the model to produce inaccurate predictions. Consequently, ad hoc architectures have been developed over the years to effectively capture and encode the spatial information contained in images.

Convolutional Neural Networks. A convolutional neural network (CNN) is a deep learning model capable of encoding spatial and temporal dependencies within the data. Thanks to these capabilities, CNNs have proven to be fundamental for image and video processing tasks, with the first successful example being the previously mentioned *LeNet-5*, whose architecture is shown in Fig. 3.9.

CNNs usually consist of three types of layers: convolutional, pooling, and fully connected layers (Fig. 3.10).

A *convolutional layer* is the core of CNNs: it exploits learnable kernels, or filters, applied over the input image through the convolution operation, the same operation

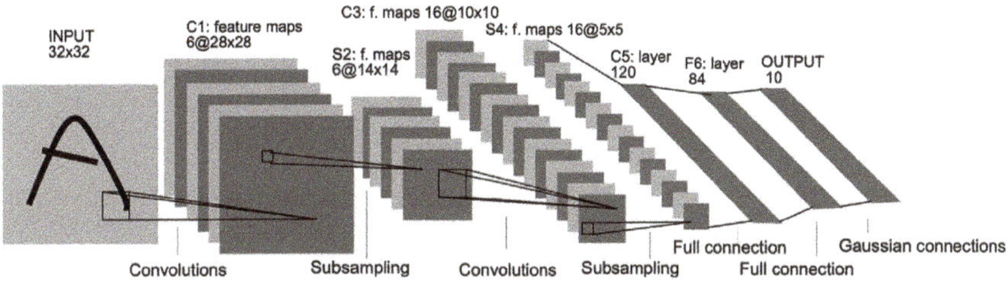

Fig. 3.9 The architecture of LeNet-5. It presents 7 layers: a convolutional layer followed by a subsampling layer, also known as pooling layer; another convolutional layer followed by a subsampling layer; finally three fully connected layers. Image from Lecun et al. [11]

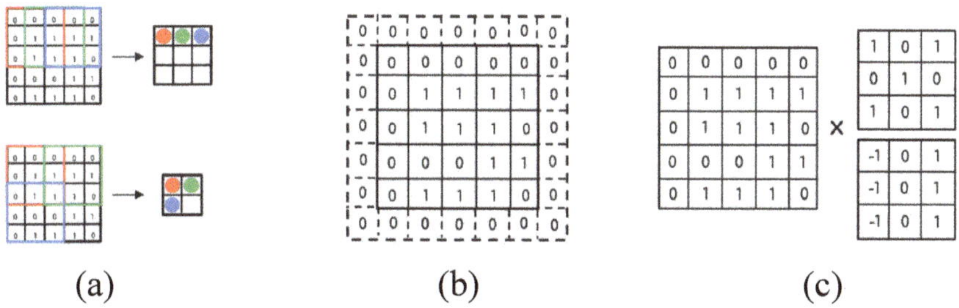

Fig. 3.10 The parameters of a convolutional layer. A stride (left), size of step taken by the kernel at each convolution, equal to 1 (top) and 2 (bottom). A zero-padding (centre), amount of zeros around the image, equal to 1. A depth (right), number of kernels, equal to 2

employed in filtering and introduced in Sect. 3.1.2.1. The objective of convolutional layers is to significantly reduce the complexity of the model, optimising the convolution process. This optimisation is achieved through different parameters, such as stride, zero-padding, and depth, which profoundly impact the convolution behaviour. *Stride* represents how many pixels the kernel moves over the image at each convolution step: a stride of one translates to one-pixel movement, while increasing the strides results in a reduction in size of the output image, also called feature map (Fig. 3.10a). *Zero-padding* consists of filling with zeros the border of the input image, in order to increase its size and give more relevance to edge pixels (Fig. 3.10b). Finally, *depth* simply denotes the number of kernels employed in a convolution layer (Fig. 3.10c).

After a convolutional layer, there typically is a pooling layer, which aims at gradually reducing the dimensionality of the feature map, thereby further reducing the number of parameters and the computational complexity of the model. This reduction is achieved through the application of a *max* or *average* function on a region of the image (Fig. 3.11).

Finally, at the end of the network are located fully connected layers, traditional neural network layers in which each neuron is directly connected to every other neuron in the two adjacent layers. These layers typically culminate with a SoftMax activation function, which computes the probabilities of the output classes.

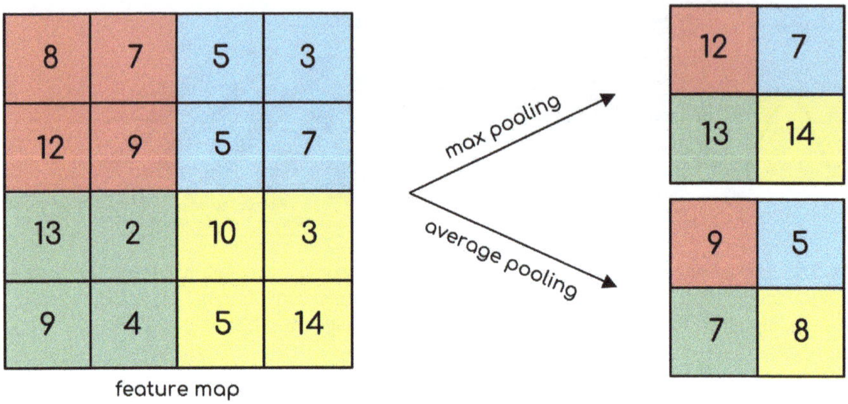

Fig. 3.11 The pooling layer. An example of max pooling (top) and average pooling (bottom): the two functions are applied on the coloured region of the feature map, with the corresponding results shown in the reduced output

Vision Transformers. Recent studies have defined a new deep learning architecture based on the attention mechanism: transformers [20]. The innovation resides in the usage of self-attention, an approach that allows the model to weight input elements differently, in order to focus on relevant information during data processing. Transformers have become the state-of-the-art approach, especially for natural language processing tasks [3], with large transformer-based models commonly pre-trained on large corpora and then fine-tuned for specific applications. Naive application of the self-attention mechanism to images would require each pixel to attend to every other pixel: this is unfeasible with respect to the time required for training the models. Thus, many approximations have been proposed in order to apply transformers in the context of image processing. In Vision Transformers (ViTs) [4] the input image is reshaped in a sequence of flattened 2-dimensional patches, whose linear projections, concatenated with a positional embedding, are provided to a transformer encoder (Fig. 3.12). ViTs have been proven to outperform CNNs on image classification and their encoding capabilities are now exploited in other contexts, such as image generation.

3.1.3.2 Computer Vision Tasks

A multitude of different tasks define the field of computer vision, among which the most prevalent are image classification, object detection, and semantic segmentation. The typical outputs of models performing these tasks are illustrated in Fig. 3.13.

Image Classification. Image classification is perhaps the most intuitive computer vision task, where the objective is to categorise an image into a specific label or category based on its visual features. The annotation of the training data determines

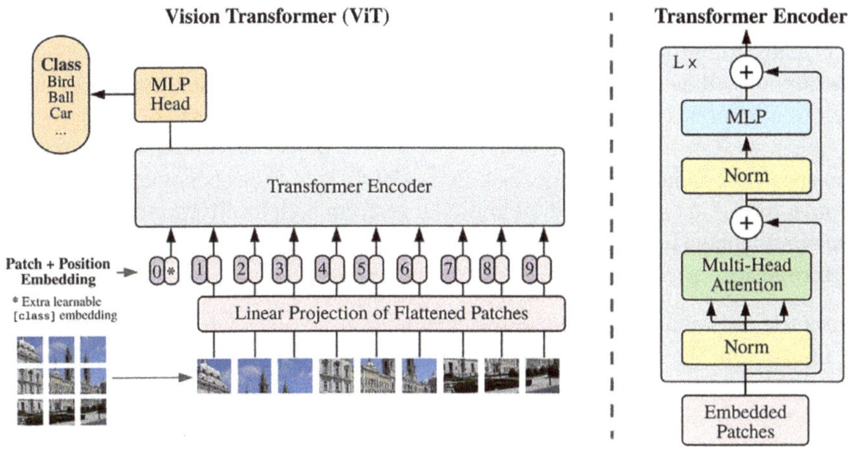

Fig. 3.12 The architecture of a Vision Transformer. Image from Dosovitskiy et al. [4]

Fig. 3.13 The typical output of the main computer vision tasks

the type of classification task, which can be categorized into three types: binary classification, where the model assigns one of two labels to the image, i.e. "cat" and "not cat"; multi-class classification, where the label set comprises multiple classes, i.e. "cat," "airplane," "person"; multi-label classification, where a single image is annotated with multiple labels, i.e. "dog," "person," "outdoor." Image classification is a fundamental task that extends beyond mere image tagging and retrieval, as it also serves as a crucial component for more complex tasks, such as object detection.

Object Detection. Object detection involves identifying and locating objects of interest within an image determining their precise spatial locations, typically accomplished by drawing bounding boxes around them.

It combines two essential tasks: object localisation and object classification. Object localisation regards the prediction of bounding boxes that enclose the objects of interest; while some methods rely on predefined anchor boxes of various aspect ratios and scales, others directly predict the coordinates of the bounding boxes using a regression model. Object classification assigns to each identified bounding box a label, chosen among a set of predefined categories, representing the object it contains. Traditionally, these tasks are executed sequentially, initially identifying candidate regions of interest (RoI) and then classifying them. However, this two-step approach has been proven not to be suitable for real-time object detection, where prediction speed is crucial. You Only Look Once (YOLO) [14] has become an extremely popular model in the field, with its ability to perform both localisation and classification simultaneously, achieving unmatched prediction speed while maintaining high performance.

Semantic Segmentation. Semantic segmentation involves labelling each pixel in an image with the corresponding class label, partitioning the image into semantically meaningful regions. Unlike object detection, semantic segmentation provides a detailed understanding of the scene at the pixel level. The output of an image segmentation model is typically a mask of the same dimensions as the input image, defining the different semantically meaningful regions. In this output, each pixel is assigned a class label indicating the category it belongs to.

A similar task is *instance segmentation*, where the output image not only includes class labels for each pixel, but it also distinguishes between individual instances of the same class. This means that pixels belonging to different instances of the same class are assigned different labels.

3.2 Computers as “Artists”

In recent years, significant advancements in artificial intelligence have transformed computers from mere tools for computation and automation into dynamic collaborators in the creative process. From the generation of visual art and music composition to crafting compelling narratives and interactive experiences, computers have emerged as formidable artists’ and content creators’ assistants. This section discusses the emerging field of generative AI, focusing on the generation of images and the models involved in this process. Additionally, we examine the integration of diverse data modalities and explore the ethical considerations inherent in the intersection of AI with creativity.

3.2.1 Image Generation

Image generation is the task of producing new unseen images, either resembling those from a given dataset or adhering to certain specifications. Within this framework, it is possible to distinguish two main paradigms: unconditioned and conditioned image generation (Fig. 3.14). Unconditioned image generation refers to the generation of images in a purely unsupervised manner, with the model being asked to generate an

Fig. 3.14 Examples of unconditioned image generation and different kinds of conditioned image generation

image without any specific constraints or guidance. With this approach, models learn to capture and replicate the underlying distribution of training images, transforming random noise into novel and plausible outputs solely relying on their inner representation and understanding of the training set. In contrast, conditioned image generation involves the introduction of specific conditions or input constraints provided to the model. These descriptions, them being textual descriptions, class labels, or other forms of structured inputs, are employed to condition the generation process. Therefore, conditioned generation demands not only the understanding of the underlying image representation, as in unconditioned generation, but also the ability to leverage supplementary data to tailor the generated outputs to the desired specifications.

The latter approach has gathered particular attention due to its versatility and possibilities provided to users. The methodology allows for various applications, such as initiating the generation process from a source image with the idea of modifying it or applying its style to a target image (img2img). One could also decide to modify a specific portion of a given image, providing the model with an additional mask depicting the region to modify (inpainting). Another common approach is to generate a new image based on a textual description or caption (txt2img).

3.2.2 Generative Models

In this section, we will present three models mainly employed for the task of unconditional image generation, namely: Generative Adversarial Networks, Variational Autoencoders, and Diffusion Models.

3.2.2.1 Generative Adversarial Networks

Introduced by Goodfellow et al. [6], Generative Adversarial Networks (GANs) have emerged as an innovative framework in the field of deep learning, particularly for image generation and synthesis. GANs consist of two competing neural networks: the *generator*, creating new images, and the *discriminator*, evaluating whether an image is real or created by the generator. These networks continuously improve through competition, refining their outputs over multiple iterations (Fig. 3.15).

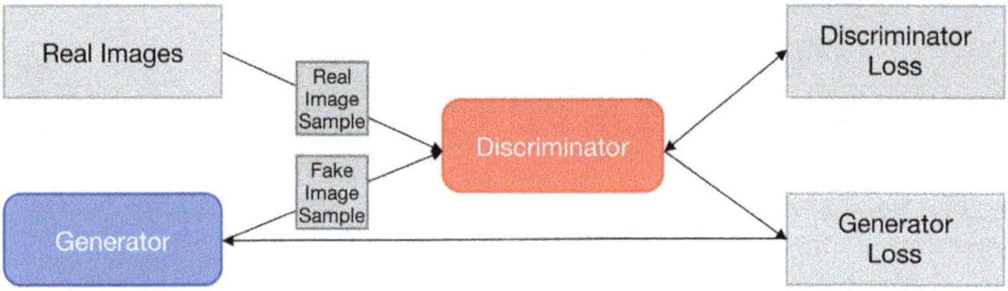

Fig. 3.15 The architecture of a generative adversarial network

While the objective of the generator is to produce newly fake images resembling as closely as possible real images, the discriminator's objective is to identify among a set of provided images which ones are real (taken from an existing dataset) and which ones are fake (created by the generator).

The two models are trained together: a sample of real images taken from an existing dataset, and a sample of fake images produced by the generator, are provided to the discriminator. Then, the discriminator has to assign the labels "real" and "fake" to the provided images: each time the discriminator fails in this assignment, its parameters are updated in order to improve its classification performance; each time one fake image is correctly classified by the discriminator as "fake," the parameters of the generator are updated in order to improve its generation performance.

Convergence is a crucial aspect in GANs training. As the generator improves, the discriminator will become less effective in distinguishing real from fake images. If the discriminator starts making random guesses, it can no longer provide meaningful feedback to the generator, which may hinder further improvement. Therefore, it is crucial to closely monitor the training process and stop once the discriminator begins to lose its ability to differentiate between real and generated images.

3.2.2.2 Variational Autoencoders

Variational Autoencoders (VAEs) [9] are built upon two fundamental concepts: dimensionality reduction and autoencoders.

Dimensionality reduction refers to the process of diminishing the number of features describing the data. This can be achieved through the selection of existing features or through the extraction of new features derived from old ones. This process can be particularly useful when dealing with high-dimensional data that, in such a way, can be described in a more concise form, ideally without significant loss of information.

An autoencoder is a model that can be used to perform dimensionality reduction. It consists of two neural networks, acting respectively as an encoder and as a decoder. The encoder compresses the data from the initial highdimensional space into a latent space, a low-dimensional space capturing the essential features of the input data; the decoder does the exact opposite, trying to reconstruct the initial observation from the

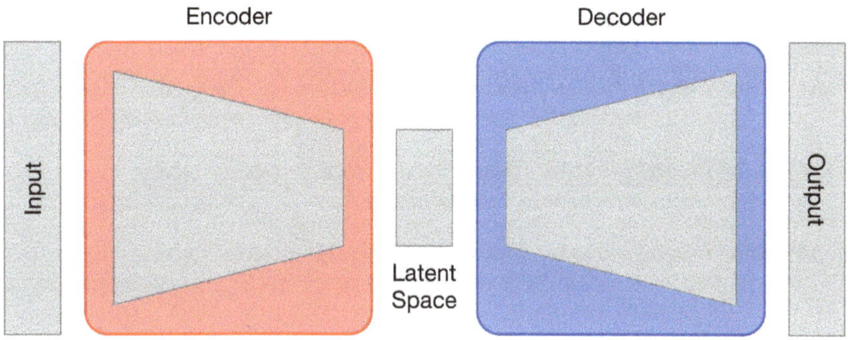

Fig. 3.16 The architecture of an autoencoder

latent space data (Fig. 3.16). By minimising the reconstruction error, autoencoders aim at reducing the dimensionality of the input data while retaining the majority of the information within the reduced representations of the latent space. However, unlike GANs, the latent space of an autoencoder lacks regularisation during training, precluding its direct use for sampling seed vectors.

A variational autoencoder keeps the same architecture of an autoencoder, with the objective of not only minimising the reconstruction error, but also of regularising the latent space to make it suitable for sample generation. The encoder does not transform the input data into a single data point in the latent space, as with autoencoders, but it transforms it into an entire probabilistic distribution. A new data point is sampled from this distribution and converted into the final output by the decoder. The encoder learns the mean and covariance matrix of this distribution, trying to keep it as close as possible to a standard normal. The learned distribution has two main properties: continuity, where close points give similar results when decoded, and completeness, where a point sampled gives meaningful content when decoded. Consequently, the learned distribution can be utilised for generating data by drawing seed vectors.

3.2.2.3 Diffusion Models

Diffusion models [7] have gained significant attention in recent years, resulting in being fundamental components of modern image generators. Their approach involves destroying the training data by incrementally adding noise to it and then learning to recover the input image via reversing this process. This enables diffusion models to effectively learn to generate coherent images starting from randomly sampled noise.

At the heart of diffusion models lies the diffusion process. Initially, random noise is added repeatedly to the input image in subsequent steps. After a sufficient number of iterations, the original image will eventually be transformed into pure noise. The goal of the model is to learn how to convert this noise into a coherent image, by reverting this process (Fig. 3.17). The model usually employed in this process is a U-Net [16], a fully convolutional neural network typically involved in semantic

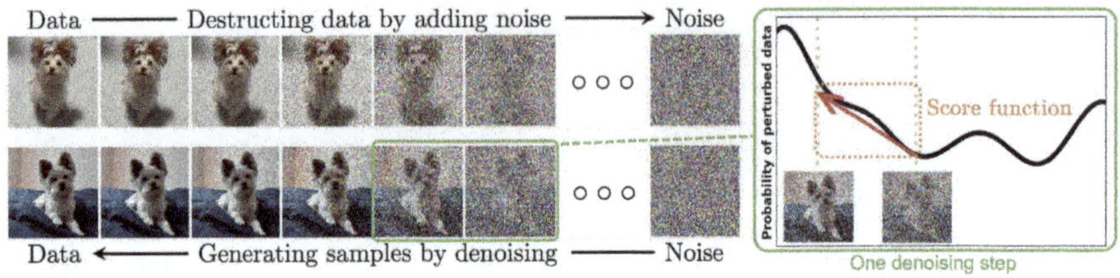

Fig. 3.17 The diffusion process. Image from Yang et al. [22]

segmentation. The task of the U-Net in the diffusion process is to transform an input image with noise into the same image with less noise: this is called one diffusion step. After several steps, the model returns an image resembling a real one.

3.2.3 Multimodality

Multimodality refers to the presence of multiple distinct modes or forms within a given dataset. This phenomenon has become significantly frequent lately: the content consumed and produced by individuals on a daily basis comprises a mix of images, texts, and audio, blurring traditional distinctions between activities such as reading, watching, or writing. Over the years, researchers have dedicated significant efforts to investigating methods for integrating and processing data in different modalities.

In the field of computer vision, one natural manifestation of multimodal data analysis can be found in the image captioning task, the ability of generating faithful image descriptions. In this regard, Contrastive Language-Image Pretraining (CLIP) [13] has emerged as a cutting-edge model obtaining outstanding results and gathering significant attention from the scientific community. Its utilization has extended beyond image captioning, serving as a foundational component in various text-to-image applications. The contribution of this work lies in the departure from conventional multimodal learning techniques: instead of instructing a model to replicate the exact description of an image, it learns a new latent space in which corresponding image and text representations coexist. In order to do so, it employs state-of-the-art transformer-based vision and language models for computing the representations of images and texts respectively. It then learns the semantic relatedness among the data by leveraging a contrastive learning framework in which the similarity between an image representation and its corresponding text representation is maximised, while the similarity with all the other texts is minimised (Fig. 3.18). The model has been proved to generalise well across a wide range of tasks, without the need for further fine-tuning.

The contributions of CLIP and other related models, introduced in previous sections, have been crucial for the advancements in capabilities of modern text-to-image generators, such as StableDiffusion (SD) [15]. In SD, the textual prompt is transformed with the CLIP text encoder in a vector representation, which will be

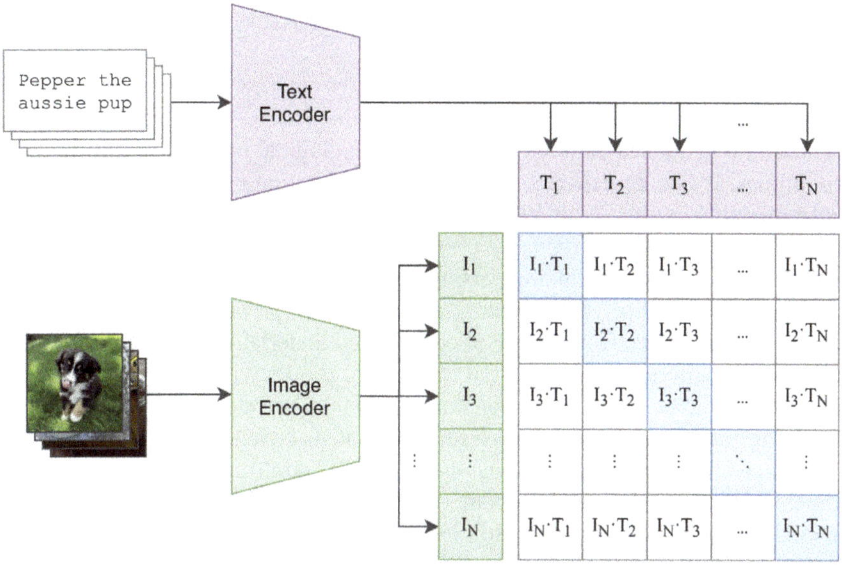

Fig. 3.18 The CLIP learning framework. Image from Radford et al. [13]

leveraged in the diffusion process for conditioning the generation of the image on the textual input (Fig. 3.19). In contrast with typical diffusion approaches, the diffusion process in SD is applied on an object having smaller dimensions with respect to the final image. This reduction in size, which allows for faster diffusion steps, is one of the key characteristics of SD. The output resulting from the diffusion steps is then processed by a autoencoder decoder, that converts it into the final image having the requested size.

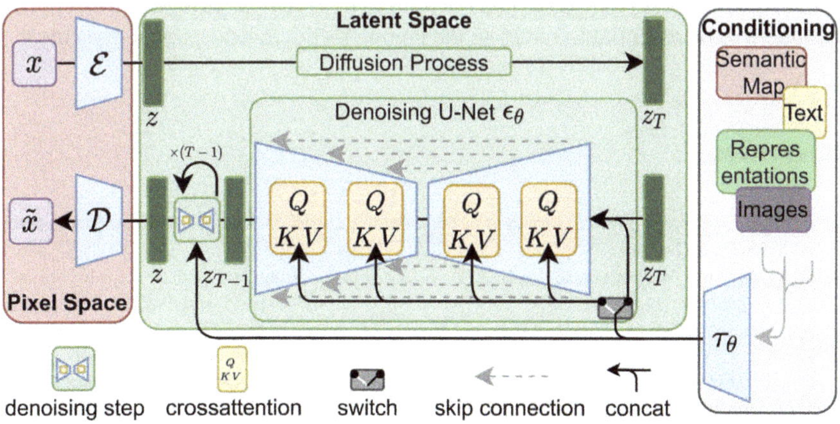

Fig. 3.19 The StableDiffusion framework. Image from Rombach et al. [15]

3.2.4 Ethical Implications

The recent advancements in generative models have reached surprising performance levels, making it increasingly challenging to distinguish between human-generated and machine-generated content. While this technological progress opens up many possibilities, it also necessitates a thorough consideration of the associated ethical implications.

Generative models are typically trained on vast amounts of data scraped from the internet, potentially including personal or sensitive information [18]. This practice poses significant privacy concerns, as individuals' private data may be utilised without consent. Moreover, the use of copyrighted material without proper authorisation or attribution raises substantial intellectual property issues [19]. It is crucial for individuals to be aware of the potential use of their private data or creative works in training models capable of generating derivative content [8]. The reliance on web-sourced training data means that these models learn from the content available online, which frequently contains biases [12]. Consequently, there is a risk of having the model perpetuate or even amplify existing biases, leading to harmful outcomes in contexts where fair and equitable representation is crucial. Significant efforts are being devoted to understanding the internal workings of these models, with the aim of identifying and mitigating bias sources [5]. Furthermore, the outputs of generative models can be exploited for malicious purposes, such as creating misleading or harmful content, including fake news, manipulated evidence, and deceptive political propaganda [21]. The dissemination of such content not only endangers individuals consuming or being depicted in the content but also undermines public trust in media and institutions.

In summary, while generative models in computer vision offer tremendous potential, it is imperative to address their ethical implications comprehensively.

Ensuring responsible development and deployment involves safeguarding privacy, respecting intellectual property rights, mitigating biases, and preventing the misuse of generated content. As this technology continues to evolve, ongoing interdisciplinary dialogue will be essential to navigate the complex ethical landscape it presents.

References

1. Canny, J.: A computational approach to edge detection. IEEE Trans. Patt. Anal. Mach. Intell. PAMI **8**(6), 679–698 (1986)
2. Deng, J., Dong, W., Socher, R., Li, L.-J., Li, K., Fei-Fei, L.: Imagenet: A large-scale hierarchical image database. In: 2009 IEEE Conference on Computer Vision and Pattern Recognition, pp. 248–255 (2009)
3. Devlin, J., Chang, M.-W., Lee, K., Toutanova, K.: BERT: pre-training of deep bidirectional transformers for language understanding. In: Burstein, J., Doran, C., Solorio, T. (eds.) Proceedings of the 2019 Conference of the North American Chapter of the Association for Computational Linguistics: Human Language Technologies, vol. 1 (Long and Short Papers), pp. 4171–4186. Association for Computational Linguistics, Minneapolis, Minnesota (2019)

4. Dosovitskiy, A., Beyer, L., Kolesnikov, A., Weissenborn, D., Zhai, X., Unterthiner, T., Dehghani, M., Minderer, M., Heigold, G., Gelly, S., Uszkoreit, J., Houlsby, N.: An image is worth 16×16 words: transformers for image recognition at scale. International Conference on Learning Representations (2021)
5. Ferrara, E.: Fairness and bias in artificial intelligence: A brief survey of sources, impacts, and mitigation strategies. Science **6**(1), 3 (2024). https://doi.org/10.3390/sci6010003
6. Goodfellow, I., Pouget-Abadie, J., Mirza, M., Xu, B., WardeFarley, D., Ozair, S., Courville, A., Bengio, Y.: Generative adversarial nets. In: Ghahramani, Z., Welling, M., Cortes, C., Lawrence, N., Weinberger, K.Q. (eds.) Advances in Neural Information Processing Systems, vol. 27, pp. 2672–2680. Curran Associates, Red Hook, NY (2014)
7. Ho, J., Jain, A., Abbeel, P.: Denoising diffusion probabilistic models. In: Proceedings of the 34th International Conference on Neural Information Processing Systems, NIPS '20, pp. 6840–6851. Curran Associates, Red Hook, NY (2020)
8. Jiang, H.H., Brown, L., Cheng, J., Khan, M., Gupta, A., Workman, D., Hanna, A., Flowers, J., Gebru, T.: AI art and its impact on artists. In: Proceedings of the 2023 AAAI/ACM Conference on AI, Ethics, and Society, pp. 363–374. Association for Computing Machinery, New York, NY (2023)
9. Kingma, D.P., Welling, M.: Auto-encoding variational bayes. In: 2nd International Conference on Learning Representations. Banff, AB, Canada (2014)
10. Krizhevsky, A., Sutskever, I., Hinton, G.E.: Imagenet classification with deep convolutional neural networks. In: Pereira, F., Burges, C.J., Bottou, L., Weinberger, K.Q. (eds.) Advances in Neural Information Processing Systems, vol. 25. Curran Associates, Red Hook, NY (2012)
11. Lecun, Y., Bottou, L., Bengio, Y., Haffner, P.: Gradient-based learning applied to document recognition. Proc. IEEE **86**(11), 2278–2324 (1998)
12. Luccioni, S., Akiki, C., Mitchell, M., Jernite, Y.: Stable bias: Evaluating societal representations in diffusion models. In: Oh, A., Naumann, T., Globerson, A., Saenko, K., Hardt, M., Levine, S. (eds.) Advances in Neural Information Processing Systems, vol. 36, pp. 56338–56351. Curran Associates, Red Hook, NY (2023)
13. Radford, A., Wook Kim, J., Hallacy, C., Ramesh, A., Goh, G., Agarwal, S., Sastry, G., Askell, A., Mishkin, P., Clark, J., Krueger, G., Sutskever, I.: Learning transferable visual models from natural language supervision. In: Meila, M., Zhang, T. (eds.) Proceedings of the 38th International Conference on Machine Learning, vol. 139 of Proceedings of Machine Learning Research, pp. 8748–8763 (2021)
14. Redmon, J., Divvala, S., Girshick, R., Farhadi, A.: You only look once: unified, real-time object detection. In: Proceedings of the IEEE Conference on Computer Vision and Pattern Recognition (CVPR), Las Vegas, NV, USA, pp. 779–788. https://doi.org/10.1109/CVPR.2016.91
15. Rombach, R., Blattmann, A., Lorenz, D., Esser, P., Ommer, B.: High-resolution image synthesis with latent diffusion models. In: Proceedings of the IEEE/CVF Conference on Computer Vision and Pattern Recognition (CVPR), pp. 10684–10695 (2022)
16. Ronneberger, O., Fischer, P., Brox, T.: U-Net: Convolutional Networks for Biomedical Image Segmentation. Medical Image Computing and Computer-Assisted Intervention—MICCAI 2015, pp. 234–241 (2015)
17. Russakovsky, O., Deng, J., Su, H., Krause, J., Satheesh, S., Ma, S., Huang, Z., Karpathy, A., Khosla, A., Bernstein, M., Berg, A.C., Fei-Fei, L.: ImageNet Large scale visual recognition challenge. Int. J. Comput. Vision **115**(3), 211–252 (2015)
18. Schuhmann, C., Beaumont, R., Vencu, R., Gordon, C., Wightman, R., Cherti, M., Coombes, T., Katta, A., Mullis, C., Wortsman, M., Schramowski, P., Kundurthy, S., Crowson, K., Schmidt, L., Kaczmarczyk, R., Jitsev. J.: Laion-5b: An open large-scale dataset for training next generation image-text models. In: Koyejo, S., Mohamed, S., Agarwal, A., Belgrave, D., Cho, K., Oh, A. (eds.) Advances in Neural Information Processing Systems, vol. 35, pp. 25278–25294. Curran Associates, Red Hook, NY (2022)
19. Smits, J., Borghuis, T.: Generative AI and intellectual property rights. In: Custers, B., Fosch-Villaronga, E. (eds.) Law and Artificial Intelligence: Regulating AI and Applying AI in Legal Practice, pp. 323–344. T.M.C. Asser Press, The Hague (2022)

20. Vaswani, A., Shazeer, N., Parmar, N., Uszkoreit, J., Jones, L., Gomez, A. N., Kaiser, L., Polosukhin, I.: Attention is all you need. In: Guyon, I., Von Luxburg, U., Bengio, S., Wallach, H., Fergus, R., Vishwanathan, S., Garnett, R. (eds.). Advances in Neural Information Processing Systems, vol. 30, pp. 5998–6008. Curran Associates, Red Hook, NY (2017)
21. Westerlund, M.: The emergence of deepfake technology: a review. Technol. Innov. Manag. Rev. **9**(11), 39–52 (2019)
22. Yang, L., Zhang, Z., Song, Y., Hong, S., Xu, R., Zhao, Y., Zhang, W., Cui, B., Yang, M.-H.: Diffusion models: a comprehensive survey of methods and applications. ACM Computing Surveys **56**(4), pp. 1–39 (2023) (Article 105)

Sergio Picascia received the B.Sc. degree in Economics from the University of Naples (2019) and the M.Sc. degree in Data Science and Economics from the University of Milan, Italy (2021). He is a Ph.D. candidate in Computer Science at the University of Milan, where he also participated in the NextGenerationUPP project, which focused on the application of artificial intelligence and advanced information management techniques for the digital transformation of Italian legal processes and digital justice. He has published in international journals and conferences, focusing his research interests in natural language processing and its applications to the humanities and social sciences.

Alfio Ferrara is full professor of Computer Science at the University of Milan, where he also serves as the Rector's Delegate for AI Literacy. He teaches Natural Language Processing, Reinforcement Learning, and Computational Methods in the Humanities, and is one of the founding members of the new research centre on Digital Humanities and Applied AI, which aims to bridge the gap between the human sciences and information technology.

Chapter 4
From Representation to Prediction: Theorising the AI Image

Lev Manovich

Abstract This chapter explores the emergence and implications of AI-generated media, with a focus on images. AI generative media is situated within historical context, drawing connections to earlier artistic movements and techniques like photomontage, postmodern bricolage, and database art. The analysis emphasises how AI continues the tradition of creating new art from accumulations of existing cultural artifacts but does so through novel methods of pattern extraction and prediction. The chapter examines several key aspects of AI generative media, including how it enables new forms of "media translation" between different types like text and images. I analyse the tension between stereotypical/idealised outputs and unique human creations, noting AI's tendency to generate less unique and specific details in many cases. The relationship between subject and style in AI-generated images is also explored. I demonstrate how engagement with AI generative tools offers new perspectives on longstanding questions in art and media theory. AI generative media continues the tradition of artists expanding creative possibilities through emerging technologies while also challenging our understanding of style, content, and the nature of artistic creation.

Keywords AI images · Generative media · Database art · Digital art · Generative AI

An earlier version of this essay was previously published online on the website of the Streaming Museum (https://www.streamingmuseum.org/post/manovich-artificial-aesthetics). I thank Nina Colosi, founder and creative director of the Museum, for granting permission to rework and reference the material.

L. Manovich (✉)
City University of New York, New York, NY, USA
e-mail: lmanovich@gc.cuny.edu

P. Conte et al. (eds.), *Algomedia. The Image at the Time of Artificial Intelligence*, Lecture Notes in Morphogenesis, https://doi.org/10.1007/978-3-032-08726-3_4

4.1 Introduction

In this chapter, I will describe a number of characteristics of visual generative media in its current form that I believe are particularly significant or novel. Some of my arguments also apply to generative media in general, but mostly I focus on AI images. My approach to thinking about AI media reflects not only my previous theoretical work on digital media (I published my first articles on computer graphics and animation already in 1991–1992) but also practical experience of using popular AI image tools such as Midjourney almost daily for two years, from August 2022 to July 2024.

I analyse AI-generated visual media through multiple lenses, drawing parallels between current AI practices and historical artistic movements. I contextualise AI art within the broader history of media creation, examining how it builds upon and diverges from previous methods of image generation. My methods involve comparing AI media with traditional artistic practices, conducting extensive experiments with AI image generation tools, and examining and testing the capabilities and limitations of these systems. The theories of AI images offered in the chapter synthesise ideas and perspectives from several fields: history of modern art, history of digital media, media theory, and software studies.

4.2 The Terms

Let us begin by defining the terms. In this text, "artist" or "creator" refers to any skilled person who creates cultural objects in any media or their combinations. The terms "generative media," "synthetic media," "AI media," "generative AI," "GenAI" are all interchangeable. They refer to the process of creating new media artefacts with software tools that use certain types of artificial neural networks (e.g., AI models) trained on vast collections of media objects already in existence. The artefacts these models can generate include images, animation, video, songs, music, text, 3D models and scenes, code, synthetic data, and other types of media. While the generation of new media objects has received lots of public attention since 2022, today, the more common use of these tools in the Creative Industry is for media editing. For example, a writer can ask ChatGPT or another AI language bot to help edit an article or generate its abstract, while a photographer can use "generative fill" tools in Photoshop to replace any area of an image with other content that fits visually with the content outside of this area.

This chapter's focus is on a particular type of media created with GenAI tools—images. This type of media is referred to as "generative images," "synthetic images," "AI images," and "AI visuals"—all these terms mean the same thing. The word "generative" can be used in different ways to refer to making cultural artefacts using any algorithmic processes (as opposed to only generative AI models) or, more generally, any rule-based process that does not use computers. This is how the phrases *generative art* and *generative design* are typically used today in cultural discourses and popular media. In this chapter, I am using *generative* in a more restrictive way to designate AI models and GenAI apps for media generation and editing that use these methods.

4.3 "AI" as a Cultural Perception

There is not one specific technology or a single research project called "AI." However, we can follow how our cultural perception of this concept evolved over time and what it was referring to in each period. In the last fifty years, when an allegedly uniquely human ability or skill is being automated using computer technology, we refer to it as "AI." Yet, as soon as this automation is seamlessly and fully successful, we tend to stop referring to it as an "AI case." In other words, "AI" refers to technologies and methodologies that automate human cognitive abilities and are starting to function, but are not quite there yet. "AI" was already present in the earliest computer media tools. The first interactive drawing and design system, Ivan Sutherland's *Sketchpad* (1961–1962), had a feature that would automatically finish any rectangles or circles you started drawing. In other words, it knew what you were trying to make. In the very broad understanding just given, this was undoubtedly "AI" already.

My first experience with a desktop paint program running on an Apple II was in 1984, and it was truly amazing to move your mouse and see simulated paint brushstrokes appear on the screen. However, today we no longer consider this to be "AI." Another example would be the Photoshop function that automatically selects an outline of an object. This function was added many years ago—this, too, is "AI" in the broad sense, yet nobody would refer to it as such today. The history of digital media systems and tools is full of such "AI moments"—amazing at first, then taken for granted and forgotten as "AI" after a while (in AI history books, this phenomenon is referred to as the "AI effect"). At the moment, "creative AI" refers only to recently developed methods where computers transform some inputs into new media outputs (e.g., text-to-image models) and specific techniques (e.g., certain types of AI models). However, we must remember that these methods are neither the first nor the last in the long history and future of simulating human art abilities or assisting humans in media creation.

4.4 "Make It New": AI and Modernism

After training on trillions of text pages or billions of art and photographic pictures taken from the web, AI models can generate fresh texts and visuals on the level of highly competent professional writers, artists, photographers, or illustrators. These capacities of the AI systems' networks are distributed over trillions of connections between billions of artificial neurons rather than being determined by standard algorithms. In other words, we developed a technology that, in terms of complexity, is

Fig. 4.1 *Unsupervised*, Refik Anadol Studio [1]. Selected frames from the animation

extremely similar to the human brain. We do not fully grasp how our AI technology works, just as we do not fully comprehend human intellect and creativity.

The current generation of generative AI models and tools, such as Midjourney and Stable Diffusion, has been trained on very large and diverse datasets consisting of billions or even trillions of individual texts, or image and text pairs. It is, however, equally interesting to limit the training data set to a specific area within the larger space of human cultural history, or to a specific set of artists from a specific historical period. *Unsupervised* by Refik Anadol Studio [1][1] is an AI art project that exemplifies these possibilities (Fig. 4.1). The project uses AI models trained on the image dataset of tens of thousands of artworks from the MoMA collection. This collection, in my opinion, is one of the best representations of one of the most creative and experimental periods in human visual history—a hundred years of modern art (1870–1970)—as well as many important examples of artistic explorations in the subsequent decades. It captures modernist artists' feverish and relentless experiments to create new visual and communication languages and "make it new."

On the surface, the logic of modernism appears to be diametrically opposed to the process of training generative AI systems. Modern artists desired to depart from classical art and its defining characteristics, such as visual symmetry, hierarchical compositions, and narrative content. In other words, their art was founded on a fundamental rejection of everything that had come before it (at least in theory, as expressed in their manifestos). AI models are trained in the opposite manner, by learning from historical culture and art created up to now. An AI model is analogous to a very conservative artist studying in the "meta" museum without walls that houses historical art.

But we all know that art theory and art practice are not the same thing. Modern artists did not completely reject the past and everything that came before them.

[1] Refik Anadol Studio [1].

Instead, modern art developed by reinterpreting and copying images and forms from old art traditions, such as Japanese prints (van Gogh), African sculpture (Picasso), and Russian icons (Malevich). Thus, the artists only rejected the dominant high art paradigms of the time, realistic and salon art, but not the rest of human art history. In other words, it was deeply historicist: rather than inventing everything from scratch, it innovated by adapting certain older aesthetics to contemporary art contexts. In the case of geometric abstract art created in the 1910s, these artists drew upon images that were already widely employed in experimental psychology to study human visual sensation and perception. Thus their supposedly revolutionary visual language was also less radical than it appeared.

When it comes to artistic AI, we should not be blinded by how these systems are trained. Yes, AI models are trained on previously created human art and cultural artefacts. However, their newly generated outputs are not mechanical replicas or simulations of what has already been created. In my opinion, these are frequently genuinely new cultural artefacts with previously unseen content, aesthetics, or styles.

Of course, simply being novel does not automatically make something culturally or socially interesting or significant. Indeed, many definitions of "creativity" agree on this point: it is the creation of something that is both original and worthwhile or useful. However, estimating what percentage of all novel artefacts produced by generative AI are also useful and/or meaningful for a larger culture is not a feasible project at this time. For one thing, I am not aware of any systematic effort to use such systems to "fill in," so to speak, a massive matrix of all content and aesthetic possibilities by providing millions of specifically designed prompts. Instead, it is likely that, as in every other area of popular culture, only a small number of possibilities are realised over and over by millions of users, leaving a long tail of other possibilities unrealised. So, if only a tiny fraction of the vast universe of potential AI artefacts is being realised in practice, we cannot make broad statements about the originality or utility of the rest of the universe.

4.5 Generative Media and Database Art

Some AI artists, such as Anna Ridler, Sarah Meyohas and Refik Anadol, utilised AI models trained on specific datasets. Many other artists, designers, architects, and technologists use models released by other companies or research institutions that were already trained on very large datasets (e.g., Stable Diffusion), and then fine-tune them on their own data.

For example, artist Lev Pereulkov[2] (Fig. 4.2) fine-tuned the Stable Diffusion model 2.1 using 40 paintings by well-known "non-conformist" artists who worked in USSR starting in the 1960s (Erik Bulatov, Ilya Kabakov, and others). Pereulkov's image series *Artificial Experiments 1–10* (2023),[3] created with this customised AI model, is

---

[2] Pereulkov [2].

[3] Pereulkov [3].

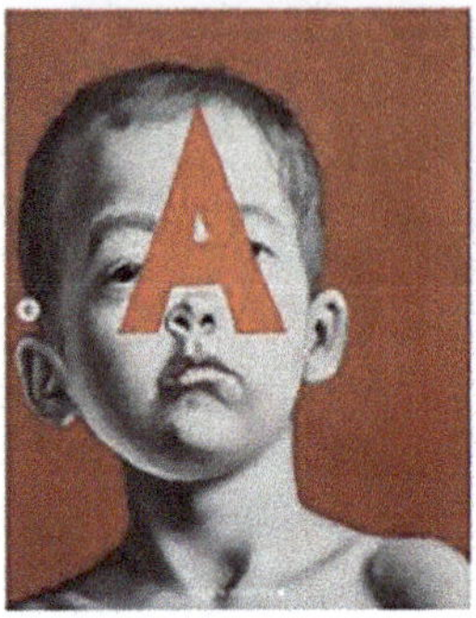

Fig. 4.2 Lev Pereulkov, *Artificial Experiments 1–10*, 2022. Three images from the series of 10 shared on Instagram

an original piece of art that captures the artistic characteristics of these artists as well as their unique surreal and ludicrous semantics without closely repeating any of their existing works. Instead, their "DNA" captured by the model enable new meanings and visual concepts.

Most of the millions of everyday people and creative professionals who employ generative media tools use them as is, and don't fine-tune them further. This may change in the future as training AI models using our own data may become easier to use. But regardless of these specifics, all newly created cultural artefacts produced by AI models have a common logic.

Unlike traditional drawings, sculptures, and paintings, generative media artefacts are not created from scratch. They are also not the result of capturing some sort of sensory phenomenon, such as photos, videos, or sound recordings. They are instead built from large archives of already existing media artefacts. This generative mechanism links generative media to earlier art genres and processes. We can compare it to film editing, which first appeared around 1898, or even earlier, composite photography, which was popular in the nineteenth century. We can also consider specific artworks that are especially relevant, such as experimental collage film *A Movie* (Bruce Conner, 1958) or many Nam June Park installations that feature bits of television programs playing across hundreds of TV monitors. Seeing projects like *Unsupervised* or *Artificial Experiments 1–10* in the context of this media creation method and its historical variations will help us understand this and many other AI artworks as art objects engaged in dialogues with art from the past, rather than as purely technological novelties or works of entertainment.

I see many relevant moments and periods when I scan the history of art, visual culture, and media for other prominent uses of this procedure. They are relevant to the current generative media not only because artists working at these times used the procedure, but also because the reason for this use was consistent in all cases. A new accumulation and accessibility of masses of cultural artefacts led artists to create new forms of art driven by these accumulations. Let me describe a few of these examples.

Net and digital artists created a number of works in the late 1990s and early 2000s in response to the new and rapidly expanding universe of the World Wide Web. Health Bunting's *_readme* (1998), for example, is a web page containing the text of an article about the artist, with each word linked to an existing web domain corresponding to that word. Mark Napier's *Shredder 1.0* (also 1998) presents a dynamic montage of elements that comprise numerous websites—images, texts, HTML code, and links.

Going further back in time, we find a broad cultural paradigm that was also a reaction to the accumulation of historical art and culture artefacts in easily accessible media collections. This paradigm is known as "post-modernism." Post-modern artists and designers frequently used bricolage and created works consisting of quotations and references to art from the past, rejecting modernism's focus on novelty and breaking with the past.

While there are many possible explanations for the emergence of the post-modern paradigm in the 1960s and 1980s, one is relevant to our discussion. The accumulation of earlier art and media artefacts in structured and accessible collections such as slides libraries, film archives, art history textbooks with many photos of the artworks, and other formats—where different historical periods, movements, and creators were positioned together—inspired artists to begin creating bricolages from such references as well as extensively quoting them (Fig. 4.3).

What about the "modernism" of the 1910s and 1920s? While the overall emphasis was on originality and novelty, one of the procedures it developed in search of novelty was direct quotations from the vast universe of contemporary visual media that was rapidly expanding at the time. Large headings, for example, and the inclusion of photos and maps made newspapers more visually impactful; new visually oriented magazines, such as *Vogue* and *Times*, were also launched in 1913 and 1923, respectively; and of course, a new medium of cinema continued to develop.

In response to this visual intensification of mass culture, in the early 1910s, Georges Braque and Pablo Picasso began incorporating actual newspaper, poster, wallpaper, and fabric fragments into their paintings. A few years later, John Heartfield, George Grosz, Hannah Hoch, Aleksandr Rodchenko, and a handful of other artists began to develop photo-collage techniques. Photo-collage became another method of creating new media artefacts from existing mass media images.

Contemporary artworks that employ AI models trained on cultural databases, such as *Unsupervised* or *Artificial Experiments 1–10*, continue a long tradition of creating new art from accumulations of images and other media. In this way, these works of art keep opening up new possibilities for art and its techniques, particularly those of what I referred to earlier as "database art" [4]. The introduction of new methods for reading cultural databases and creating new narratives from them is part of this expansion.

Thus, *Unsupervised* neither creates collages from existing images, as did modernist artists of the 1920s, nor quotes them extensively, as did postmodern artists of the 1980s. Instead, the members of Anadol Studio train a neural network to extract patterns from tens of thousands of MoMA's artworks. The trained net then generates new images that share the same patterns but don't look like any specific paintings.

Fig. 4.3 Photomontage by John Heartfield, 1919

Throughout the animation, we travel through the space of these patterns (e.g., "latent space"), exploring various regions of the universe of contemporary art.[4]

Pereulkov's *Artificial Experiments 1–10* use a different technique to generate new images from an existing image database. He chose only forty paintings by artists who share key characteristics. They developed their oppositional art in late communist society (USSR, 1960s–1980s). They also lived in the same visual culture. In my memories, this society was dominated by two colours: grey (representing the monotony of urban life) and the red of propaganda. In addition, Pereulkov chose paintings that share something else: "I chose, as a rule, paintings that conceptually

[4] For a more details about GAN net training methods used by Refik Anadol Studio, see Carina [5].

relate in some way to the canvas—or to the space on it. I obtained the photo of a painting, *New Accordion*, from Kabakov, which features paper applications on top of the canvas".[5] Pereulkov also crafted custom text descriptions of each painting used for fine-tuning the Stable Diffusion model. To teach the model the specific visual languages of the chosen artists, he added terms such as "thick strokes," "red lighting," "blue background," and "flat circles" to these descriptions.

Clearly, each of these steps represents a conceptual and aesthetic decision. In other words, the key to the success of *Artificial Experiments 1–10* is the creation of such a database. This work demonstrates how fine-tuning an existing neural network that was trained on billions of image and text pairs (such as Stable Diffusion) can make this network follow artists' ideas; the biases and noise of such a massive network can be overcome and minimised, and do not need to dominate our own imagination.

4.6 From Representation to Prediction

Historically, humans created images of existing or imagined scenes by a number of methods, from manual drawing to 3D CG (see below for explanation of the methods). With AI generative media, a fundamentally new method emerges. Computers use large datasets of existing representations in various media to predict new images (still and animated).

One can certainly propose different historical paths leading to visual generative media today, or divide one historical timeline into different stages. Here is one such possible trajectory:

1. Creating representations manually (e.g. drawing with a variety of instruments, carving, etc.). More mechanical stages and parts were sometimes carried out by human assistants, typically training in their teacher's studio—so there is already some delegation of functions.
2. Creating manually but using assistive devices (e.g. perspective machines, camera lucida). From *hands* to *hands + device.* Now some functions are delegated to mechanical and optical devices.
3. Photography, x-ray, video, volumetric capture, remote sensing, photogrammetry. From *using hands* to *recording information using machines.* From *human assistants* to *machine assistants.*
4. 3D CG. You define a 3D model in a computer and use algorithms that simulate effects of light sources, shadows, fog, transparency, translucency, natural textures, depth of field, motion blur, etc. From *recording* to *simulation.*
5. Generative AI. Using media datasets to predict still and moving images. From *simulation* to *prediction.*

[5] My personal communication with Pereulkov, 04/16/2023.

"Prediction" is the actual term often used by AI researchers in their publications describing visual generative media methods. So, while this term can be used figuratively and evocatively, this is also what actually happens scientifically when you use image generative tools. When working with a text-to-image AI model, the artificial neural network attempts to predict the images that best correspond to your text input. I am certainly not suggesting that using all other already accepted terms, such as "generative media", is inappropriate. But if we want to better understand the difference between AI visual media synthesis methods and other representational methods developed in human history, employing the concept of "prediction" and thus referring to these AI systems as "predictive media" captures this difference well.

4.7 Media Translations

There are several methods for creating "AI media." One method transforms human media input while retaining the same media type. Text entered by the user, for example, can be summarised, rewritten, expanded, and so on. The output, like the input, is a text. Alternatively, in the image-to-image generation method, one or more input images are used to generate new images.

However, there is another path that is equally intriguing from the historical and theoretical perspectives. "AI media" can be created by automatically "translating" content between media types. This is what happens, for example, when you are using Midjourney, Stable Diffusion or other AI image generator services and enter a text prompt, and the AI generates one or more images in response. Text is "translated" into an image.

Because this is not a literal one-to-one translation, I put the word "translation" in quotes. Instead, input from one medium instructs the AI model to predict the appropriate output from another. Such input can also be said to be 'mapped' to some outputs in other media. Text is mapped into new styles of text, images, animation, video, 3D models, and music. The video is converted into 3D models or animation. Images are "translated" into text, and so on. Text-to-image method translation is currently more advanced than others, but various forms will catch up eventually.

Translation (or mapping) between one medium and another is not a new concept. Such translations were done manually throughout human history, often with artistic intent. Novels have been adapted into plays and films, comic books have been adapted into television series, a fictional or non-fictional text was illustrated with images, etc. Each of these translations was a deliberate cultural act requiring professional skills and knowledge of the appropriate media. Some of these translations can now be performed automatically on a massive scale thanks to AI models, becoming a new means of communication and culture creation. Of course, artistic adaptation of a novel into a film by a human team and automatic generation of visuals from novel text by a net is not the same thing, but for many simpler cases, automatic media translation can work well. What was once a skilled artistic act is now a technological capability available to everyone. We can be sad about everything that might be lost as a result of the automation—and democratisation—of this critical cultural operation: skills, something one might call "deep artistic originality" or "deep creativity," and so on. However, any such loss may be only temporary if the abilities of "culture AI" are, for example, even further improved to generate more original content and understand context better.

Because the majority of people in our society can read and write in at least one language, text-to-another media methods are currently the most popular. They include text-to-image, text-to-animation, text-to-3D, and text-to-music models. These AI tools can be used by anyone who can write, or by using readily available translation software to create a prompt in any of the languages these tools understand best at a given point. However, other media mappings can be equally interesting for professional creators. Throughout human cultural history, various translations between media types have attracted attention. They include translations between video and music done by VJs in clubs; long literary narratives turned into movies and television series; texts illustrated with images in various media such as engravings; numbers turned into images (digital art); texts describing paintings (*ekphrasis* tradition, which began in Ancient Greece), mappings between sounds and colours (especially popular in modernist art); etc.

The continued development of AI models for mappings between all types of media, without privileging text, has the potential to be extremely fruitful, and I hope that more tools will be able to accomplish this. Such tools will be very useful both to professional artists and other creators alike. However, being an artist myself, I am not claiming that future "culture AI" will be able to match, for example, innovative interpretations of Hamlet by avant-garde theatre directors such as Peter Brook or astonishing abstract films by Oskar Fishinger that explored musical and visual correspondences. It is sufficient that new media mapping AI tools stimulate our imagination, provide us with new ideas, and enable us to explore numerous variations of specific designs.

4.8 The Stereotypical and the Unique

Both the modern human creation process and the predictive AI generative media process seem to function similarly. An AI model is trained using unstructured collections of cultural content, such as billions of images and their descriptions or trillions of web and book pages. The neural net learns associations between these artefacts' constituent parts (such as which words frequently appear next to one another) as well as their common patterns and structures. The trained net then uses these structures, patterns, and "culture atoms" to create new artefacts when we ask it to. Depending on what we ask for, these AI-created artefacts might closely resemble what already exists, or they might not.

Similarly, our life is an ongoing process of both supervised and unsupervised cultural training. We take art and art history courses, view websites, videos, magazines, and exhibition catalogues, visit museums, and travel in order to absorb new

Fig. 4.4 Examples generated in Midjourney version 4 using the short text prompt "morning sky"

cultural information. And when we "prompt" ourselves to make some new cultural artefacts, our own biological neural and networks (infinitely more complex than any AI nets to date) generate such artefacts based on what we have learned so far: general patterns we have observed, templates for making particular things, and often concrete parts of existing artefacts. In other words, our creations may contain both exact replicas of previously observed artefacts and new things that we represent using templates we have learned, such as the golden ratio or the use of complementary colours.

AI neural nets used for image generation frequently have a default "house" style. This is the actual term used by Midjourney developers. If one does not specify a style explicitly, the AI will generate it using this "default" aesthetic (Fig. 4.4).

To steer away from this default, you need to add terms to your prompts specifying, a description of the medium, the kind of lighting, the colours and shading, or a phrase like "in the style of" followed by the name of a well-known artist, illustrator, photographer, fashion designer, or architect. Here are two examples (Figs. 4.5 and 4.6) of such prompts I made, and the images that Midjourney generated from these prompts. The terms used to define particular style characteristics are highlighted in *italics*.

Because it can simulate many thousands of already-existing aesthetics and styles and interpolate between them to create new hybrids, AI is more capable than any single human creator in this regard. However, at present, skilled and highly experienced human creators also have a significant advantage. Both humans and artificial intelligence are capable of imagining and representing non-existent and existing objects and scenes alike. Yet, unlike AI image generators, human-made images can include very particular content, unique minuscule details, and unique aesthetics (i.e. very particular use of image's visual variables, so to speak) that is currently beyond the capabilities of AI. In other words, today, a large group of highly skilled and experienced illustrators, photographers, and designers can represent everything an

Fig. 4.5 Lev Manovich. An image generated with Midjourney v3. The prompt: "giant future 1965 modern airport in Siberia made from water and ice, painted on large wood panel by Hieronymus Bosch, bright pastel colours with white highlights, 23f lens, very detailed—ar 4:3—s 1250—test"

AI model can do—although it will take much, much longer. But they can draw all kinds of images that currently even the best AI model cannot generate equally well. For example, in 2022–2023, AI models often struggled to generate realistic hands. By 2024, this has been solved, but the models still have difficulty generating very complex compositions with dozens of fully realistic human figures and faces shown in detail—something that skilled human artists were able to do at least since the Renaissance.) Equally importantly, very skilled human professionals can create unique objects, faces, compositions, and so on—as opposed to often more commonplace or idealised versions generated by AI.

What is the cause of this aesthetic and content gap between human and artificial creators? "Cultural atoms," structures, and patterns in the training data that occur most frequently are very successfully learned during the process of training an artificial neural network. In the "mind" of a neural net, they gain more importance. On the other hand, "atoms" and structures that are rare in the training data or may only appear once are hardly learned or not even parsed at all. They do not enter the artificial culture universe learned by AI. Consequently, when we ask AI to synthesise them, it is unable to do so. Due to this, text-to-image AI tools such as Midjourney, Stable Diffusion or RunwayML are not currently able to generate perfectly well drawings in my style, expand my drawings by adding newly generated parts, or replace specific portions of my drawings with new content drawn perfectly in my style (e.g., they cannot perform

Fig. 4.6 Lev Manovich. An image generated with Midjourney v4. The prompt: "Photo of two Russian high-school students, clear skin, *very soft studio light, 50 mm lens, monochrome, silver tones, high quality, ultra realistic—v 4—q 2*." This image also illustrates the point I am making later in the chapter: "AI frequently generates new media artefacts that are more stereotypical or idealised than what we intended"

useful *outpainting* or *inpainting* on the digital photos of my drawings.) Instead, these AI tools generate more generic objects than what I frequently draw, or they produce something that is merely ambiguous yet uninteresting.

I am certainly not claiming that the style and the world shown in my drawings are completely unique. They are also a result of specific cultural encounters I had, things I observed, and things I noticed. But because they are uncommon (and thus unpredictable), AI finds it difficult to simulate them, at least without additional training using my drawings (Figs. 4.7 and 4.8).

Here we encounter what I see as the key obstacle creators face when using generative AI generative tools—and this holds for both AI images and all other types of media: GenAI frequently generates media that are more stereotypical or idealised than what we intended. This can affect any image dimensions—elements of content, lighting, crosshatching, atmosphere, spatial structure, and details of 3D shapes, among others. Occasionally, it is immediately apparent, in which case you

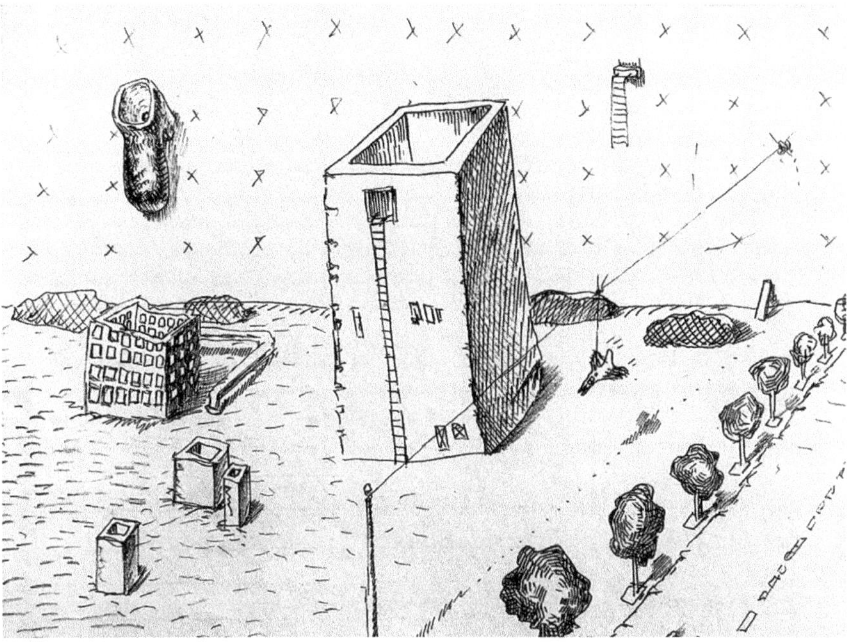

Fig. 4.7 Lev Manovich, untitled drawing, pen on paper, 1981

can either attempt to correct it or disregard the results. Very often, however, such "substitutions" are so subtle that we cannot detect them without extensive observation or, in some cases, the use of a computer to quantitatively analyse numerous images. In other words, new AI generative media models, much like the discipline of statistics since its inception in the eighteenth century and the field of data science since the end of the 2010s, deal well with frequently occurring items and patterns in the data but do not know what to do with the infrequent and uncommon. We can hope that AI researchers will be able to solve this problem in the future, but it seems so fundamental that we should not anticipate a solution immediately.

4.9 Subject and Style

In the arts, the relationship between content and form has been extensively discussed and theorised. This brief section does not attempt to engage in all of these debates or to initiate discussions with all relevant theories. Instead, I would like to consider how these concepts play out in AI's "generative culture." However, instead of using *content* and *form*, I'll use a different pair of terms that are more common in AI research publications and online conversations between users: *subject* and *style*.

Fig. 4.8 One of my attempts to generate a new version of this image with Stable Diffusion AI, Fall 2022

At first glance, AI media tools appear capable of clearly distinguishing between the subject and style of any given representation. In text-to-image models, for instance, you can generate countless images of the same subject. Adding the names of specific artists, media, materials, and art historical periods is all that is required for the same subject to be represented differently to match these references. Photoshop filters began to separate subject and style as soon in the 1990s, but AI generative media tools are more capable. For instance, if you specify "oil painting" in your prompt, simulated brushstrokes will vary in size and direction across a generated image based on the objects depicted. AI media tools appear to "understand" the semantics of the representation as opposed to earlier filters that simply applied the same transformation to each image region regardless of its content. For instance, when I used "a painting by Malevich" and "a painting by Bosch" in the same prompt, Midjourney generated an image of space that contained Malevich-like abstract shapes as well as many small human and animal figures like in popular Bosch paintings that were properly scaled for perspective (Fig. 4.9).

AI tools routinely add content to an image that I did not specify in my text prompt in addition to representing what I requested. This frequently occurs when the prompt

Fig. 4.9 Lev Manovich. An image generated in Midjourney using prompt "painting by Malevich and Bosch," Fall 2022

includes "in the style of" or "by" followed by the name of a renowned visual artist or photographer. In one experiment, I used the same prompt with the Midjourney AI image tool 148 times, each time adding the name of a different photographer. The subject in the prompt remained mostly the same—an empty landscape with some buildings, a road, and electric poles with wires stretching into the horizon. Sometimes, adding a photographer's name had no effect on the elements of a generated image that fit our intuitive concept of style, such as contrast, perspective, and atmosphere. But every now and again, Midjourney also modified the image content. For example, when well-known works by a particular photographer feature human figures in specific poses, the tool would occasionally add such figures to my photographs. (Like Malevich and Bosch, they were transformed to fit the spatial composition of the landscape rather than mechanically duplicated.) Midjourney has also sometimes changed the content of my image to correspond to a historical period when a well-known photographer created his most well-known photographs.

According to my observations, when we ask Midjourney or a similar tool to create an image in the style of a specific artist, and the subject we describe in the prompt is related to the artist's typical subjects, the results can be very successful. However, when the subject of our prompt and the imagery of this artist are very different, "rendering" the subject in this style frequently fails. Using the prompt "decaying peonies by Caspar David Friedrich" in Midjourney generates images that simulate important features of the artist's style, such as combinations of cool colours and dramatic atmosphere. But in other ways, generated images depart significantly from the artist's style (Figs. 4.10, 4.11 and 4.12). The types of lines, rendering of details, and symmetrical compositions in these AI images would never appear in Friedrich's actual paintings. AI can also often insert some generic-looking objects, such as the rock formations in the upper right corner of first image.

Fig. 4.10 Using prompt "by Caspar David Friedrich—v 5" in Midjourney generates images that capture the artist's style sufficiently well. Image source: https://www.midlibrary.io/styles/caspar-david-friedrich

Fig. 4.11 Images generated in Midjourney v5 using a prompt: "decaying peonies by Caspar David Friedrich—v 5." Images source: https://www.midlibrary.io/styles/caspar-david-friedrich

To summarise, to successfully simulate a given visual style using current AI tools, you may need to change the content you intended to represent, or to accept the fact that AI will insert some details you do not want. In other words, not every "subject" can be rendered successfully in any "style." And what exactly AI can learn also varies from case to case: it can often successfully learn some features of an artist's style but not others.

In fact, after many months of using these tools and also analysing endless images generated by others, I came to realise that the idea that GenAI can perfectly extract a "style" from a collection of aesthetically similar images is not correct. Instead, it appears that in the process of training, visual AI models learn associations between visual characteristics of the works of a given artist (i.e., what we often call a visual language) and the content of their artworks (unless given artworks are fully abstract). This puts into question the whole idea that style and content can be neatly separated.

These observations, I believe, complicate the binary opposition between the concepts of "content" and "style." For some artists, AI can extract at least some aspects of their style from examples of their work and then apply them to different types of content. Sometimes, more aspects of a given artist's style can be learned

Fig. 4.12 Images generated in Midjourney v5 using a prompt: "decaying peonies by Caspar David Friedrich—v 5." Images source: https://www.midlibrary.io/styles/caspar-david-friedrich

and used for a new generation, and sometimes, fewer. But for other artists, it seems, their style and content cannot be separated. For me, these kinds of observations and reflections are one of the most important reasons for using new media technologies like AI generative media and learning how they work. Of course, as a practicing artist and art theorist, I had been thinking about the relationships between subject and style (or *content* and *form*) for a long time—but being able to conduct systematic experiments like the one I described can lead to new ideas, and allow us to look back at cultural history and our familiar concepts and theories of art and culture in new ways.

References

1. Refik Anadol Studio: Unsupervised (2022). https://refikanadol.com/works/unsupervised/. Accessed 21 June 2025

2. Pereulkov, L.: Artist's Telegram Channel. Telegram. t.me/pereulye (2025). Accessed 21 June 2025
3. Pereulkov, L.: Artificial Experiments 1–10. Telegram. t.me/pereulye (2023). Accessed 21 June 2025
4. Manovich, L.: Database as a Symbolic Form (1998). https://manovich.net/index.php/projects/database-as-a-symbolic-form. Accessed 21 June 2025.
5. Carina, Y.: Creating art with Generative Adversarial Network: Refik Anadol's *Walt Disney Concert Hall Dreams*. In: Medium (2022). https://medium.com/@ymingcarina/creating-art-with-generative-adversarial-network-refik-anadols-wdch-dreams-159a6eac762d. Accessed 14 May 2025

Lev Manovich is a Presidential Professor at The Graduate Center, City University of New York. After studying painting, architecture, and filmmaking, Manovich began using computers to create digital art in 1984. He has played a key role in creating a number of new research fields: new media and digital culture (1991–), software studies (2001–), cultural analytics (2005–), and AI aesthetics (2017–). His writings include 200 articles that are widely cited across multiple disciplines. Among his 17 books are *Artificial Aesthetics*, *Cultural Analytics*, *Instagram and Contemporary Image*, *Software Takes Command*, and *The Language of New Media.*

Part II
Algorithmic Mediascapes

Chapter 5
Artificial Intelligence, VR, and the Virtual

Andrea Pinotti

Abstract Drawing on theology, Andrea Pinotti's contribution explores the concept of virtuality and its manifold implications in the age of artificial intelligence and virtual reality. Beginning with the etymological roots of "virtual" in *virtus*, the author traces a lineage through Scholastic philosophy, Aristotelian metaphysics, and modern thinkers like Deleuze and Lévy. The notion of virtual containment (*continentia virtualis*) is central, positing that the virtual is not unreal but a realm of latent potentialities awaiting actualisation. This metaphysical framework is then applied to AI, especially in relation to latent space in image generative models, where virtualities (e.g. styles) are harboured and actualised in particular pictures through prompts. Through experiments with text-to-image model DALL·E and referring to Neural Style Transfer, the essay highlights how generative AI seems to presuppose a clean separation between form and content: this leads to a problematic extrapolation of the style as a universal from a given corpus of images, which ends up reviving the ancient Scholastic debate on universals. In the last section, the essay considers the adjective "virtual" as it appears in the strictly technical sense of the term "Virtual Reality" and highlights analogies and differences in AI and VR as technologies of actualising virtualities.

Keywords Generative artificial intelligence · Latent space · Neural style transfer · Virtual reality · Virtuality

5.1 Virtualisation

In recent years, we have witnessed a general process of virtualisation of the human experience in its most diverse aspects. Imposing severe restrictions on close relationships, the *SARS-CoV-2* pandemic radically reconfigured the human relationship to the near-far dichotomy, leading to a dramatic increase in long-distance "virtual" relationships. In this context, digital tele-technologies have represented a fundamental tool in ensuring the maintenance of not only intersubjective experiences, but

A. Pinotti (✉)
University of Milan, Milan, Italy
e-mail: andrea.pinotti@unimi.it

P. Conte et al. (eds.), *Algomedia. The Image at the Time of Artificial Intelligence*, Lecture Notes in Morphogenesis, https://doi.org/10.1007/978-3-032-08726-3_5

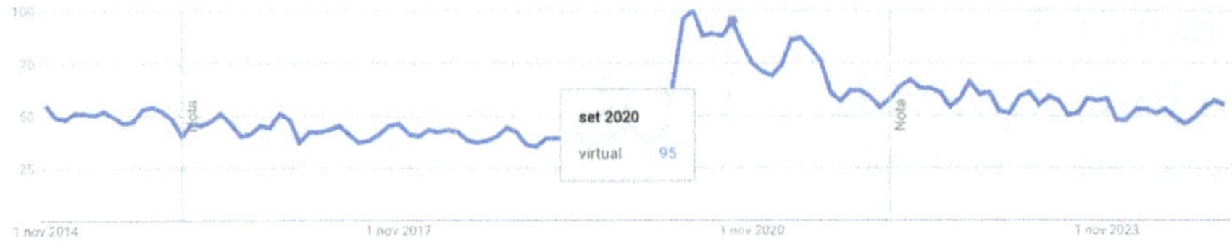

Fig. 5.1 Google trends graph on the term "virtual" on a global scale; accessed 23 November 2024

also those between subjects and objects. Let us think (just to limit ourselves to two paradigmatic examples), for the former, of the massive implementation of remote learning strategies; for the latter, of the pervasive diffusion of "virtual tours" in the open-access online collections of museums all over the world. As this Google Trends graph on the number of Google searches of the term "virtual" over the last ten years shows (November 2014 to November 2024), interest in this word peaked during the most acute phase of the pandemic and then stabilised once the emergency had subsided at a level that is still higher than in the pre-pandemic period (Fig. 5.1).

As is always the case, a general growth of interest in a term corresponds to an effect of inverse proportion: the *extension* of the term increases, while its *intension* is diluted. Thus, the term "virtual" and the concept behind it have been extended to encompass a broad semantic area that includes the remote, the online and even the digital per se. This expansion has ended up blurring the boundaries of the precise definition of the term "virtual." Any Zoom call is characterised as a virtual call. Trying to counter this trend, which is now deeply entrenched in everyday language usage, would be like tilting at windmills. Nevertheless, it is always useful to try to dot the I's and cross the T's, to at least develop a critical awareness of that tendency and the reasons behind it.

In order to do this, it is necessary to place alongside the generic and undifferentiated use of the term "virtual" two more specific meanings, which I intend to investigate (consistent with the general theme of this book) in their relationship with algorithmic machines and AI: "virtual" as "the virtual language system on which generative artificial intelligences rely" ([12]: 2), and "virtual" as it appears in the locution "virtual reality" (VR), a technology which today is increasingly hybridised with artificial intelligence systems. The examination of these two more specific meanings will allow me to explore the connections with algorithmic machines in general, and with image-generative AI in particular.[1]

[1] My gratitude goes to Alfio Ferrara for his valuable comments on AI-related technical implications in what follows.

5.2 Virtual Versus Actual, Latent Versus Manifest

The corpus of data employed by generative AI as a virtual archive has recently been investigated from the perspective of the semiotics of enunciation: "Generative AIs, in this respect, are machines intrinsically linked to enunciative praxis, because their work is precisely a work of summoning, actualising and realising the potential and virtual possibilities of the dataset" ([12]: 3). Here, however, in what follows, I would like to adopt a different perspective, one that complements the semiotic approach, to try to highlight the theological and philosophical background behind this characterisation of the dataset as "virtual" and "potential."

If we trace the adjective "virtual" back to its etymological root, we find two Latin terms: the adjective *virtualis* and the noun *virtus*. The former appeared at the beginning of the twelfth century in theological terminology to describe the multiple capacities of the undivided and simple soul, while the latter had already been employed in the second century in the first Latin translations of the Septuagint (the Greek Old Testament) to render the notion of *dynamis*, the divine power [23]. The usage was later adopted in Latin translations of the New Testament to indicate the miraculous powers of Christ. In imperial times, the term was extended to also encompass the power of the emperor.

A fascinating and complex history—thoroughly reconstructed by Parmentier [31] in his monumental work—links the early theological applications of the term "virtual" to its modern and contemporary uses in disciplines such as optics, mechanics, quantum physics and computer science, and to twentieth-century philosophical reflections around virtuality, for example in thinkers such as Gilles Deleuze and Pierre Lévy.

Of the many meanings that the term "virtual" has taken on in the course of this long history, both terminological and conceptual, for the purposes of the theme I propose to investigate here (i.e. the points of contact between virtuality and AI), the term *continentia virtualis* ("virtual containment" or "virtual inclusion") is particularly relevant as it specifically refers to *mereology* (the study of the relations between parts and wholes). Parmentier ([31]: 129–167, 266–285) shows how this locution was produced at the convergence of the Neoplatonic tradition of Pseudo-Dionysius the Areopagite (fifth to sixth century, later taken up in the twelfth century by Hugh of Saint Victor) and the Aristotelian tradition, which provided a full systematisation of it in the thirteenth century thanks to Thomas Aquinas.

In his *Questions on Truth* (*Quaestiones disputatae de veritate*), written circa 1256–1259 in Paris, Aquinas states:

> There are two ways in which a number of things can come to exist in one thing [*multa esse in uno*]. In one way, they exist there in separation and multiplicity, as for instance, with many forms, each is reflected separately in a mirror, and as many men are in one house. In the other way, they are there according to one simple form, as many effects exist virtually in a cause [*virtute sunt in causa*], as conclusions in a principle [*conclusiones in principio*], and as bodily members in seed [*membra in semine*]. [...] One who sees some one thing does not have to see all the things which exist in it as united in one form, except when he comprehends the total power [*totam virtutem*] of that one thing. Thus, one who sees some principle does

not have to see all the conclusions which exist virtually [*virtute*] in it, unless he comprehends the principle ([2]: question 20, art. 4, reply).

These three areas of application of virtuality—ontological (many effects exist virtually in a cause), logical (conclusions are implied virtually in a principle) and biological (bodily members are virtually contained in the seed)—are cited by Aquinas to answer an exquisitely Christological question. Christ is a finite creature, and as such can only know finite things like any human being. But because of his union with the Word of the Creator, capable of infinite knowledge, should we not admit that he knows infinite things as well? Among others, Paul, for example, argues that "It hath well pleased the Father, that all fullness of divinity [*omnem divinitatis plenitudinem*] should dwell" in Christ (Colossians I: 19).

The solution adopted by Aquinas in his reply consists in establishing exactly what is meant when he says that one sees and knows in the Word. Unlike any human being, Christ possesses the ability to know the past, the present and the future and the secrets of hearts insofar as he participates in the infinite divine Word. However, his also finite nature as a man does not allow him to know the divine infinity in its entirety, the prerogative of which remains the ability to do many more things than Christ's soul can know. But here what interests us is not so much the solution itself, but rather the mereological premises that make it possible, and the consequences for us humans as producers of artificial things such as images.

It is particularly the analogy with biological processes that gives us an important clue. Aristotle in Book VII of *Metaphysics* had already proposed a parallelism between the way in which the seed generates and the way in which the artist/artisan creates, that is, the one who possesses the "techne" understood as the ability to produce according to rules: "The productive effect of the seed is not different from that from skill [*techne*], given that the seed possesses the form potentially [*dynamei*]" (*Metaph.* VII 9, 1034a; [4]: 198).

While Parmentier ([31]: 789–790) asserts that the Scholastic notion of "virtus" has nothing to do with the Aristotelian notion of "potentiality [*dynamis*]," it seems to me that this passage from Aristotle, when placed side by side with Aquinas' argument, proves exactly the opposite. This point is further clarified if we consider the notion of potentiality in its dialectical relationship with that of actuality (*energeia*) as addressed in Book IX. Here Aristotle expands on the famous "*dynamis/energeia*" dyad, illustrating it with iconic examples significantly drawn from the fields of sculpture and geometrical representations: "We say that something exists in potentiality [*dynamei*] if it is like a statue of Hermes existing in a block of wood or like a half-line existing in a whole line, given that the half-line could be detached" (*Metaph.* IX 6, 1048a; [4]: 267).

The example taken from sculpture was already cursorily anticipated in *Metaph.* V 7, 1017b, with the difference that here Hermes was potentially not in wood but in stone. In both cases, the argument is crystal clear. The piece of wood or the stony mass is currently what it is, with its material and its form. They can, however, become something else (the figure of the god Hermes), the moment the sculptor gives their material a new form.

The history of the effects of the terminological-conceptual association *dynamis-virtus* is centuries old, extending to the philosophical formulations of the second half of the twentieth century. In this regard, I will limit myself to two paradigmatic authors: the aforementioned Deleuze and Lévy.

In his book dedicated to Bergson (his main source for the conception of virtuality), Deleuze distinguishes the possible (as opposed to the real) from the virtual (as opposed to the actual): "The 'virtual' can be distinguished from the 'possible' from at least two points of view. From a certain point of view, in fact, the possible is the opposite of the real, it is opposed to the real; but, in quite a different opposition, the virtual is opposed to the actual" ([9]: 96). In this sense, far from being unreal, the virtual possesses full reality in its own right: "The virtual is opposed not to the real but to the actual. The virtual is fully real in so far as it is virtual" ([10]: 208). (We note, en passant, that the years in which Deleuze draws attention to the term virtual in the twentieth-century philosophical debate are the same years in which the inventor of VR headsets, Ivan Sutherland, conducted his first experiments and published his seminal articles: [37, 38]).

Lévy places this quote from Deleuze as an exergue to his book on the virtual ([24]: 15). By reaffirming the reality of the virtual, Lévy insists, however, on an aspect that is particularly relevant to the topic I wish to address here, namely the *événementialité* of the actualisation process: "Actualization is an *event*, in the strongest sense of the term. An act is accomplished, but not predetermined, and it in turn modifies the dynamic configuration in which it assumes meaning. The articulation of the virtual and the actual animates the very dialectic of the event, the process, of *being as creation*" ([24]: 171). The seed virtually contains the tree, the tree is latent in the seed. But the seed does not "know" exactly what form the plant that will develop from it will take: "Based on its internal limitations, the seed will have to invent the tree, coproduce it together with the circumstances it encounters" ([24]: 24). Actualisation (of the plant virtually contained in the currently existing dormant seed) is thus a creative process that leads from the *latent* to the *manifest*, in a negotiation between the structural genetic constraints of the seed itself and the external (today we would say "epigenetic") circumstances in which the process takes place.

But what happens when we move from natural virtuality to artificial virtuality? It is significant that to illustrate this transition, Lévy makes explicit reference to Aristotle's doctrine of the four causes (material, formal, efficient, final) to argue that the manifestation of the figure emerging from the latency of matter takes place thanks to a creative process whereby the sculptor does not limit himself mechanically to making explicit what was already implicit but invents solutions by responding to unforeseen events. While Aristotle refers to bronze statues (*Phys.* II 3, 195 a 6–8; *Metaph.* V 2, 1013 b 6–9), Lévy prefers to speak of a marble *kouros* becoming actualised in the creative process of sculpting: "Actualization creates a solution for the problem presented by the virtual. In doing so, it does not simply replenish its resources or provide a form for the mechanism of realisation. Actualisation *creates form*. It creates a radically new kind of information. Efficient cause is related to actualisation because the labourer, sculptor, or demiurge, if it is a living and thinking being, can never be reduced to a simple executant, for it interprets, improvises, and resolves problems" ([24]: 174).

5.3 AI Latent Space as Virtual Space

The way in which Lévy characterises the virtual as a reservoir of latent potentialities waiting to actually manifest themselves through a process that is not merely mechanical but intrinsically creative in its negotiation with unpredictable factors shows remarkable similarities with the way in which, in the context of AI, the movement from the so-called "latent space" to the required output is usually described.

Firstly, latent space refers to the abstract, multidimensional mathematical space where AI models represent data, in which each dimension corresponds to a feature or pattern learned by the model (for example, in image generative AI models, the latent space might encode abstract properties like the style or shape of an object). As such, the latent space is not a physical space but a potential—and therefore in our terms *virtual* –space that organises relationships between data points based on their learned features: "In the case of visual generative AI, we can say that the training phase consists in the construction of the traits that constitute its system of virtualities" ([12]: 42). Its virtuality consists in its capacity to represent potential outputs, enabling through computational processes the generation of new objects (e.g., coherent images resulting from noise introduced during training in diffusion models). Each point in the latent space can correspond to a potential but not yet actualised object: "Taken together, the images corresponding to all the points within a latent space constitute all possible visualizations of the space. We can consider all these images as a sort of complete cartography or atlas of the latent space" ([35]: 93).

Secondly, the affinity between the *creative* dimension identified by Lévy in his characterisation of the virtual and the properly *generative* dimension of AI models should be emphasised. Unlike *discriminative* models, which perform the function of classifying or labelling data (e.g. to recognise whether a picture contains a dog or a cat), *generative* models like Generative Adversarial Networks (GANs) and Transformer-based architectures are able to create or produce new content, data, or outputs—such as text, images, audio, or video—which are not merely copied from their training data but are instead synthesised or generated based on patterns learned from large datasets (which are of course imbued with biases). The set of latent space and machine parameters is the horizon of what the machine can produce. Within this horizon, the parameters (representing the knowledge acquired during the training process) are as relevant to this production process as the internal representations triggered by the inputs.

This creative-generative element evidently entails, on the level of both aesthetic and legal implications, a reconfiguration of the concepts of creativity and authorship. This is something that emerges with intuitive clarity in a recent book by photography theorist Ritchin [33]. The author playfully experimented with text–to-image models such as DreamStudio and DALL·E, prompting the machine to produce not generic pictures, but specifically photographs.

The types of prompts used can be summarised in the following categories:

- prompts relating to the generation of photographs that could be taken at any time (e.g. "Children with balloons": [33]: 31),
- prompts relating to photographs that could actually have been taken at a certain moment in history (e.g. "An iconic photograph from the year 1945," or "A photograph of happy people in the Warsaw Ghetto during World War II": [33]: 34 and 81),
- prompts relating to photographs that actually exist (e.g. "The first photograph ever made": [33]: 25), but which the program does not return as such (as Google Images would have done, offering us Nicéphore Niépce's *View from the Window at Le Gras*);
- prompts relating to photographs in a certain style (e.g. "Like a Stephen Shore photograph," or "A photograph of the storming of the U.S. Capitol on January 6, in the style of a twentieth-century war photographer": [33]: 59 and 82),
- prompts relating to historically impossible photographs, because preceding the invention of the medium (e.g. "A photograph of a street scene in Berlin in 1815," or "Photograph of Abraham Lincoln taking a selfie": [33]: 34 and frontispiece),
- prompts relating to ontologically impossible photographs (e.g. "A Pictorialist photograph of two Martians": [33]: 70).

In all these cases, the caption that Ritchin places beneath the image serves as both a warning and a declaration of *co-authorship*, which acknowledges the output as the outcome of cooperative work between human textual prompting and generative machine processing. For example, as regards the picture "Like a Stephen Shore photograph," its caption reads: "This is a synthetic image, not a photograph, generated by DALL·E in response to the text prompt (above) by Stephen Shore, February 2023. © Stephen Shore. Courtesy 303 Gallery, New York." As regards Ritchin's own prompt "A Pictorialist photograph of two Martians," the caption says: "This is a synthetic image, not a photograph, generated by DALL·E in response to the text prompt (above) by the author, March 2023."

Ritchin's notion of "synthetic image" (what is usually referred to as synthography) warns us that, as much as these images appear to be photographs—this was precisely the task given to the machine by the author's prompts—, they are in fact *pseudo-photographs*: they produce a photographic effect, they are "photorealistic" [36] but have not been generated by "light-writing" (photo-graphy) and are not the result of the actual presence of the object they represent in front of the lens. They lack, in other words, the *indexical* status which, starting with Peirce's reflections ([32]: 159, § 2.281), we usually believe we should attribute to the nature of the photographic image. For this reason, although phenomenologically they appear as if they were photographs, ontologically they are closer to painted images.

Among the many points of reflection offered by Ritchin's book, I intend to extrapolate a number of issues, all concerning the Pictorialist "photograph" of the two Martians (Fig. 5.2).

Fig. 5.2 In Ritchin ([33]: 70)

At first glance it reminded me of the impossible "photograph" of a dinosaur taken by Luigi Ghirri (Fig. 5.3). Dinosaurs having become extinct long before the invention of photography, this picture confronts us with a seemingly irresolvable aut aut: either dinosaurs did not disappear at all during the Cretaceous-Paleogene extinction event some 66 million years ago, and it was possible to photograph one in 1978; or photography was not invented by Nicéphore Niépce between 1826 and 1827, and Ghirri exhumed a shot taken over 66 million years ago.

On closer consideration, however, I realised that the analogy was misplaced. We have fossil evidence of large herbivorous dinosaurs with long necks (such as what appears to be a brontosaurus in Ghirri's image), while we have no evidence of what Martians look like, or even if they exist. So Ghirri's photo instead falls into the category of historically impossible photos, supposedly taken before the invention of

Fig. 5.3 Luigi Ghirri, Verona, 1978 (C-Print 1979, CSAC, Università degli Studi di Parma; ©Eredi Luigi Ghirri)

the medium, like the above-mentioned "A photograph of a street scene in Berlin in 1815," or "Photograph of Abraham Lincoln taking a selfie."

In contrast, the Ritchin and DALL·E "photograph" of two Martians is generated by a very rich human imagery relating to the hypothetical inhabitants of the Red Planet, found in countless literary and visual works of science fiction that the machine has learnt and stored as a *virtual latent reservoir* from which it creatively actualises the output we see reproduced in the book. This virtual reservoir, far from being a static storehouse, thrives on an incessant dynamic flux, like an immeasurable living organism that feeds on ever-new data, metabolised into ever-different computational outcomes, and consequently always capable of being actualised in different ways.

I gave DALL·E the same prompts provided by Ritchin, and not only did I obviously get a different output (also due to the continuous evolution of the model), but I also got two different outcomes by giving the same machine the same input a few seconds apart (Figs. 5.4 and 5.5).

The unpredictability of the different results I obtain from the same prompting corresponds to a certain extent to the unpredictability of the results I would obtain if, as a sculptor, I wanted to carve the same statue of Hermes from marble blocks of different qualities and properties (Carrara, Parian, Alabaster, Botticino...), or perhaps from different woods (oak, walnut, ebony, cedar...). I would also have to use, depending on the material, different tools for sculpting, carving, and polishing. The final outcome is not predetermined as already conceived and accomplished in the sculptor's mind, nor does it depend solely on his/her formal will, but on a complex

Fig. 5.4 "A Pictorialist photograph of two Martians". Prompt given to DALL·E by A. Pinotti on November 23, 2024, at 7.10 PM

interaction that the artist establishes with the constraints imposed by the material, the tools, the techniques employed. Only through this multi-factor negotiation can the "Hermes" form emerge and actualise itself, surfacing from the virtual latency in which it lingered in matter. Think of ceramic art, and the enormous variability of the chromatic result obtained by glaze-firing procedures: our Hermes, painted by the ceramist before entering the oven, will emerge with unpredictable colour and texture effects, as a result of chemical reactions, fluctuations in kiln temperature, oxidation variables, cooling processes, etc. If we wanted to continue along the lines of this analogy, we could suggest that today AI artists and more generally image-makers are to virtual latent space and prompt engineering as sculptors and ceramists are to matter and the tools to shape it. The difference is determined by the immeasurable mass of data and inhuman computational speed provided by AI models, together with the fact that the non-deterministic and aleatory nature of the stochastic process dramatically enhances unpredictable fluctuations as to the output (which remains always only very partially controllable).

The process of prompting interpellates the machine's latent space as a virtual space harbouring a realm of potential outputs that materialise each time in a different way because they typically rely on stochastic (randomised) processes during image generation, on the exploration of different plausible ways to visualise a description,

Fig. 5.5 "A Pictorialist photograph of two Martians". Prompt given to DALL·E by A. Pinotti on November 23, 2024, at 7.11 PM

and on the training on diverse datasets. Thus, virtuality goes hand in hand with serendipity. As Alexander Kluge puts it, "the virtual camera called AI works like an excavator. When extracting a single grid from millions or billions of existing images, it must be given a lot of room for chance" ([21]: 34).

Moreover, this virtual latent space feeds on enormous datasets of images that have been indexed, labelled, and captioned (i.e. *worded*). This fact radically reconfigures the relationship between word and image—a classic issue in the history of aesthetics since the ancient tradition of *ekphrasis*—and impacts on the very limits of what can be visualised. As has been aptly observed, in the context of text-to-image models, "natural (i.e., human) language becomes the main medium for image generation: what is *visible* is strictly correlated with what is *sayable*, since the images are generated by textual prompts. What can or cannot be said, what can or cannot be written in a prompt, determines—together with the other factors analysed above—what can or cannot be visualised and seen" ([35]: 106–107), thus also implying a reconfiguration of censorship and iconoclasm.

5.4 Neural Style Transfer, or the Virtualisation of Style

The second aspect concerning AI virtuality I would like to emphasise by continuing to explore the game played by Ritchin concerns the adjective “Pictorialist” in his caption “A Pictorialist photograph of two Martians.” This prompt instructs the machine to generate a “photographic” image according to a precise style, that of pictorialism, pioneered by Henry Peach Robinson in his *Pictorial Effect in Photography* [34]. This style aspired to affirm the artistic dimension of photography and the use of the camera as if it were a brush or chisel. Rather than merely documenting reality, pictorialist photographers aimed at evoking emotions, atmospheres, or symbolism, and employed manipulating techniques such as soft focus, special lenses and filters, or the application of pigments to achieve a painting-like effect.

I experimented with four different commands for DALL·E (Figs. 5.6, 5.7, 5.8 and 5.9): two historically possible (one photograph in the Cubist style and one in the Surrealist style: Figs. 5.6 and 5.7) and two historically impossible (one photograph in the style of 15th-century Florence painting and one in the style of Pompeian painting: Figs. 5.8 and 5.9). Here are the outputs:

Why have I defined the first two examples as historically *possible* and the second two as historically *impossible*? I mean that photographers existed who were inspired by Cubist or Surrealist poetics (visual styles born after the invention of photography), whereas it was not possible to photograph in the Pompeian style between the end of the second century BC and 79 AD, nor during the fifteenth century in Florence. But of course, albeit in a completely different sense, the second two images (which are also somewhat “pictorialist,” given the reference to Renaissance and Pompeian painting) are also perfectly possible, since they were in fact realised.

Let’s bear in mind the point already established, namely that all these pictures are not photographs but synthographic pseudo-photographs (i.e. they are all ontologically impossible photographs in the sense explained above). What I would like to highlight here is the fact that the virtualisation process also affects the status of *style* understood as unity in multiplicity (a common denominator that gathers under its label a plurality of iconic manifestations). “Cubist,” “Surrealist,” “Quattrocento Florentine painting,” “Pompeian painting”: these four styles (we could of course add dozens more) might be understood as many *universals* subsisting in the virtuality of the latent space (included in its totality in the modality of the above-mentioned *continentia virtualis*), ready to be recalled by prompting and actualised in this or that *particular* picture. Given the background in Scholasticism outlined above with Thomas Aquinas, I intentionally use here the conceptual dyad “universal/particular” to allude to the debate around the mode of existence of universals that ran through medieval philosophy [6]. In the systematisation offered in the thirteenth century by Albert the Great ([1]: 116, 44), three modes of existence of universals are given: *in re*, *ante rem*, and *post rem*. *Universalia in re* (“universals in the thing”) exist within individual, particular things as their intrinsic forms or essences (e.g. the universal “humanity” exists in each particular human as the formal principle that makes an individual a human). *Universalia ante rem* (“universals before the thing”) exist prior to

Fig. 5.6 "A Cubist photograph of two Martians". Prompt given to DALL·E by A. Pinotti on November 29, 2024, at 6.49 PM

particular things as ideal, intelligible realities in the mind of God (e.g. the universal "humanity" exists as a divine archetype in God's intellect before any individual human is created). *Universalia post rem* ("universals after the thing") exist after particular things as abstract concepts in the human mind, formed through intellectual abstraction from sense experience (e.g. after encountering individual humans, the mind abstracts the universal "humanity" as a general idea applicable to all humans).

If we transpose this taxonomy into the framework of the dataset training procedures employed by image-generative AI, we could say that Cubist, Surrealist, fifteenth century Florentine and Pompeian art styles (like all other styles) subsist in the latent space as so many *post rem* universals, extrapolated (abstracted in the sense of extracted) from thousands of images that have been labelled according to those stylistic categories, i.e. recognised as instantiating universals *in re*. But, once extrapolated a posteriori and subsisting in the virtuality of the latent space, style operates as a kind of a priori, making possible not only the discrimination between already existing images (identifying one picture as Surrealist not Cubist, another as Pompeian not Renaissance), but also the ex-novo generation of original pictures. In other words, it assumes the status of universal *ante rem*, not in the mind of God,

Fig. 5.7 "A Surrealist photograph of two Martians". Prompt given to DALL·E by A. Pinotti on November 29, 2024, at 6.57 PM

but in the "mind" of the machine (thus resulting in a kind of "artificial Platonism": [3]). This Platonism is further complicated by the fact that each machine has its own "style," a sort of meta-style (different, for example, in Midjourney and DALL·E: [8]) that influences all the stylistic outputs of individual images.

Now, how is it possible to extrapolate from a given corpus of images the style as a universal in order to constitute it as a virtuality capable of being actualised in future images that do not yet exist? To understand this, we must turn to that class of algorithmic procedures which go by the name of Neural Style Transfer (NST). NST was officially unveiled in 2015 thanks to the work of Gatys et al. [13] and has since undergone impressive development and numerous modifications to the original algorithm (as can be seen from the list mapped in the reviews offered by Jing et al. [20] and Ioannou and Maddock [19]).

Of the many issues that Neural Style Transfer procedures raise, for the purpose of my essay, I intend to emphasise here only one, which is particularly problematic: the uncritical assumption of the "form/content" distinction. Regardless of the different methods and algorithms used to perform style transfer, the common premise is that it is possible to separate the "what" of an image (in our example the two Martians) from

Fig. 5.8 "A photograph of two Martians in the style of Quattrocento Florentine painting". Prompt given to DALL·E by A. Pinotti on November 29, 2024, at 6.59 PM

its "how" (the way they are represented in form, colour, texture, volume, composition and overall perceptible qualities). This "how" would be the "form" or "style," while that "what" would be the "content" (identifiable as the subject, *sujet*, of the image). Gatys et al. apparently seem very well aware that, "Of course, image content and style cannot be completely disentangled" ([13]: 6). Nevertheless, it is precisely this assumption that, at the end of the day, steers the various NST procedures.

The idea that the "what" and the "how" can be separated is rather naive and has been long criticised and radically rejected in the fields of aesthetics and media theory alike. One need only think of Nelson Goodman for the former, Marshall McLuhan for the latter. "Obviously," Goodman remarks ([16]: 799), "subject is what is said, style is how. A little less obviously, that formula is full of faults." These faults mainly concern the fact that no expressive mode (whether iconic, verbal or other) can express the same content without at the same time bringing a different content into existence. Just as there is no possibility of perfect synonymy in verbal language, similarly, no corresponding possibility can be admitted in non-verbal forms of expression. For his part, McLuhan ([27]: 7–21) condensed the same idea into the celebrated maxim

Fig. 5.9 "A photograph of two Martians in the style of Pompeian painting". Prompt given to DALL·E by A. Pinotti on November 29, 2024, at 7.25 PM

"The medium is the message", making it a commonplace in media theory: there is no message content that can be transposed into different media without thereby becoming another message itself.

Yet it is precisely on the basis of the idea that, on the contrary, the stylistic-formal components of a corpus of images can be extracted from the corpus itself and held in latency as virtualities available for the generation of other images that NST methods operate. Their increasingly pervasive diffusion within mass culture thus ends up consolidating an inherently fallacious *idée reçue*: the separability of form and content. This separability enables, not least, the monetisation of style as virtuality in the context of a radical transformation of the visual economy: "As a pattern that can be extracted from large aggregations of digitally mobilized visual content, and thus detached from the individual image, its author, its medium, and its conditions of production, 'style' becomes a source of value" ([28]: 109).

While I disagree with the oversimplified idea that "'Style transfer' is pure mimicry—a belaboured resemblance which is also a masquerade" ([41]: 50), I nevertheless believe that only close collaboration between specialists in aesthetics and media theory on the one side, and Deep Learning and generative AI engineers on the other side, could lead to a design of NST algorithms that avoids falling into the traps of an unconsciously assumed and inherently erroneous theory eventually resulting in what has been called the "reification" of a totally decontextualised concept of style [39].

5.5 AI and VR

Let us now consider the adjective "virtual" as it appears in the strictly technical sense of the term "Virtual Reality." VR is an immersive technology that offers users an alternative world to the peripersonal space in which they physically find themselves. This space ceases to be directly perceived and is replaced by an artificial environment into which the subject is teleported, thus finding herself surrounded on all sides by an iconic landscape that unfolds seamlessly before her eyes. There are two main types of devices that enable this experience: the headset and the CAVE (Cave Automatic Virtual Environment). In the first case, the user wears a head-mounted display, which operates by blocking out the normally perceptible surrounding space and even the possibility of perceiving one's own body [22]. In the second case, the user enters a room whose walls, ceiling and floor consist of screens that wrap around her [26].

In principle, the VR experience via headset tends to be solipsistic: the experiencer finds herself, as *solus ipse*, in an environment that she perceives first-hand. In the case of CAVE, by contrast, the experience can be shared with other people, whose co-presence in the space-display is perceivable by the user. However, the headset mode may also allow for an *intersubjective* mode if the system includes a multi-player option. In this case, the other comes to meet me in the form of the *avatar*: a delegate or digital proxy that represents the user in the artificial environment. The avatar in which I embody myself can in turn be total (full-body) or partial: in this second mode, I can be represented, for example, as a half-length figure, or, as is more common, by digital hands which (guided by virtual gloves, joysticks or controllers) offer me the possibility of agency. If this is the case, these virtual environments are *interactive*, allowing me to interact with other avatars or digital entities.

An important terminological clarification borrowed from the vocabulary of robotics distinguishes between VR environments with 3 and 6 Degrees of Freedom (3DOF or 6DOF respectively): in the former I am free to rotate my gaze 360 degrees in the virtual environment, but I cannot approach or move away from the digital entities that populate it; in the latter I am also allowed to move spatially.

Just as is happening for Augmented Reality technologies [14], the implementation of AI programs is steadily increasing across Virtual Reality technologies [29, 30]. Major fields of application include: procedural content generation; predictive modelling; behavioural realism on the part of AI-driven avatars; Natural Language Processing (NLP) so as to enable realistic conversation with virtual assistants and NPC (non-player characters); personalised experiences based on the analysis of the user's preferences; simulation of real-world scenarios; emotion-recognition. Introductory manuals are already being published to make the business world aware of the risks and opportunities of these two technologies [40].

Artists, too, have begun not only to explore hybridisations between VR and AI, but also to include AI as the very subject of their artificial worlds. This is for instance the case of Kevin Mack, who in his *Namuanki* (2022) presents us with what he calls "an ancient aquatic oasis from the distant future" in which benevolent beings named The Ancient Elders are "an advanced form of self-engineered life that will emerge from the integration of humanity and artificial intelligence."[2] Drawing inspiration from the Tibetan Buddhist tradition of the *tulpa* (a kind of materialisation of thought), anthropologist Marc Da Costa and film director Matthew Niederhauser have realised *Tulpamancer* (2023): an immersive VR experience which, by means of a machine learning algorithm conjures up a world of past memories and future expectations tailored to the individual user.[3]

However, rather than examining the implementation of AI elements in the architecture of virtual immersive environments, I would like to address here what seems to me a far more challenging theoretical issue: namely, the analogy between the virtuality of AI examined above and VR, insofar as it is indeed *virtual*. Pursuing this analogy, one fundamental trait these two technologies have in common is, once again, their constitutive relation to the potential/actual dyad. Insofar as ontologically definable as programs, VR entities constitute potentialities that can be actualised in a different way each time, at the moment of the execution, producing them as events. They are veritable "object-events" ([24]: 172, [11]: 61). In this sense, if we wanted to resort to the famous distinction put forward by Nelson Goodman ([15], chap. III), rather than belonging to the sphere of *autographic* works (which cannot be replicated except as a forgery, as in the case of a painting), VR objects would rather seem to belong to that of *allographic* works (which, as in the case of a musical score, are identified through a notation—in this case the writing of a program—which needs the moment of execution for its manifestation).

It is interesting to note in this regard an analogy between image perception in immersive virtual environments and ordinary perception: their being "unframed," that is, the fact that such perceptual modalities can unfold freely in all directions without the constraints of a border (as is the case in the image perception of a painting, a photograph, a film: [5]). In this way, users of a work in VR, emancipating themselves from what has been called the "tyranny of the frame," can themselves choose the framing at will, deciding where to look and on which portion of the world-image to focus, just as happens in the real world. Here too, in fact, the phenomenological actualisation of the perceptual intention, despite being directed at the same object, yields different aspects of it each time, in that dialectic between core and adumbrations that was explored in detail by Husserl's work on perception and kinaesthetics [17, 18]. In the same way, I can repeat a hundred or a thousand times the experience

---

[2] https://www.kevinmackart.com/namuanki.html.

[3] https://marcdacosta.com/tulpamancer/index.html. I thank Margherita Fontana for drawing my attention to these two works.

of a certain work in VR, but what I will realise each time will be a perceptually different experience, depending on the direction in which I turn my gaze and on which portions of the immersive visible world I want to bring to the centre of my field of vision.

Even if the possibility of making the frame disappear altogether remains more a promise that belongs to the rhetoric and ideology of the VR device than to the factual reality (the VR headset or CAVE are in themselves, inevitably, framing devices, as is the user's own body: [7]), the point remains that the rotation of the gaze thanks to the movement of the eyes and the head entails the continuous and seamless actualisation of virtualities inscribed in the program. In line with what has been said about the latency of the AI space, in the case of VR, too, one could therefore speak of a sort of latent space waiting to be actualised through the visuomotor activity of the user.

This analogy would seem to break down when it comes to generativity. VR programs are usually pre-built coded environments which are user-responsive, but not generative (unless integrated with AI): unlike generative AI, they cannot generate novel, not predefined outputs by learning patterns from large datasets. However, if we consider the specific case of PCG (Procedural Content Generation) in VR [25], we encounter a borderline area: procedural algorithms can produce unforeseen content, such as terrains (e.g. landscapes that stretch infinitely or vary with each session) or objects (e.g. NPCs, non-player characters) in video games that emerge at runtime, without being manually crafted: the designer does not specify the exact outcome, only the rules and constraints. While it is not generative AI, such procedure does involve a form of computational creativity, and its unpredictability opens analogies to more sophisticated AI generative systems.

At the conclusion of this journey that from Aristotle's *dynamis* and Aquinas' *virtus* has led us to the latent space of generative AI, to the algorithms of NST, and finally to immersive worlds in Virtual Reality, I would like to devote a brief methodological and epistemological reflection on the vexed question *continuity/discontinuity*. Tracing the theological and philosophical roots of the notion of virtuality back to the Aristotelian and Scholastic tradition does not mean disregarding the original and unprecedented features of the technologies which are radically revolutionising all aspects of our lives, nor does it mean claiming that, after all, there is nothing new under the sun. But it would be just as wrong to want to counter this strong *continuist* position with a strongly *discontinuist* perspective, and to disregard that the mere fact that we speak of virtuality and actuality today, as in the past, is not simply lexical inertia, but is a symptom of a profound connection, which deserves further investigation.

References

1. Albert the Great: Alberti Magni ordinis fratrum praedicatorum De causis et processu universitatis a prima causa. In: Fauser, W.S.J. (ed.) Aedibus Aschendorff, Monasterii Westfalorum
2. Aquinas, T.: Truth (Quaestiones Disputatae de Veritate). Henry Regnery Company, Chicago (1952–1954)
3. Arielli, E., Manovich, L.: Artificial Aesthetics. A Critical Guide to AI, Media and Design (2021–2024). https://manovich.net/index.php/projects/artificial-aesthetics
4. Aristotle: Metaphysics. Penguin, London (1998)
5. Conte, P.: Unframing Aesthetics. Mimesis International, Milan (2020)
6. de Libera, A.: La querelle des universaux: de Platon a la fin du Moyen Age. Seuil, Paris (1996)
7. Dalmasso, A.C.: The body as virtual frame. Performativity of the image in immersive environments. Cinéma & Cie **19**(32), 101–119 (2019)
8. D'Armenio, E., Dondero, M.G., Deliège, A., Sarti, A.: For a semiotic approach to generative image AI: on compositional criteria. Sem. Rev. **9** (2025) https://doi.org/10.71743/ee5nrx33
9. Deleuze, G.: Bergsonism. Zone Books, New York (1988)
10. Deleuze, G.: Difference and Repetition. Continuum, London, New York (2001)
11. Diodato, R.: Image, Art and Virtuality. Towards an Aesthetics of Relation. Springer, Cham (2021)
12. D'Armenio, E., Deliège, A., Dondero, M.G.: Semiotics of machinic co-enunciation. About generative models (Midjourney and DALL·E), Signata **15** (2024). http://journals.openedition.org/signata/5290
13. Gatys, L.A., Ecker, A.S., Bethge, M.: A neural algorithm of artistic style (2015). arXiv:1508.06576v2 [cs.CV], https://arxiv.org/pdf/1508.06576
14. Geroimenko, V. (ed.): Augmented Reality and Artificial Intelligence. The Fusion of Advanced Technologies. Springer, Cham (2023)
15. Goodman, N.: Languages of Art. An Approach to a Theory of Symbols. The Bobbs-Merrill Company, Indianapolis–New York–Kansas City (1968)
16. Goodman, N.: The status of style. Crit. Inq. **1**(4), 799–811 (1975)
17. Husserl, E.: Ideas Pertaining to a Pure Phenomenology and to a Phenomenological Philosophy, Book 1: General Introduction to a Pure Phenomenology. Kluwer, Dordrecht–Boston–London (1983)
18. Husserl, E.: Ideas Pertaining to a Pure Phenomenology and to a Phenomenological Philosophy, Book 2: Studies in the Phenomenology of Constitution. Kluwer, Dordrecht–Boston–London (1989)
19. Ioannou, E., Maddock, S.: Evaluation in Neural Style Transfer: A review. Comput. Graph. Forum **43** (2024). https://doi.org/10.1111/cgf.15165
20. Jing, Y., Yang, Y., Feng, Z., Ye, J., Yu, Y., Song, M.: Neural style transfer: a review. IEEE Trans. Visual Comput. Graph. **26**(11), 3365–3385 (2020). https://doi.org/10.1109/TVCG.2019.2921336
21. Kluge, A.: The Dragonfly's Eye. My Virtual Camera. Spector Books, Leipzig (2024)
22. Lanier, J.: Dawn of the New Everything. Encounters with Reality and Virtual Reality. Henry Holt & Co, New York (2017)
23. Lecaudé, P.: La notion de puissance. Les équivalents latins du grec δύναμις (dynamis), thèse pour l'obtention du grade de Docteur de l'Université Paris–Sorbonne (2010)
24. Lévy, P.: Becoming Virtual. Reality in the Digital Age. Plenum Trade, New York (1998)
25. Liu, Y.: An overview of procedurally generating virtual environments. Int. J. Comput. Sci. Inf. Technol. **2**(1), 141–147 (2024). https://doi.org/10.62051/ijcsit.v2n1.16
26. Manjrekar, S. et al.: CAVE. An emerging immersive technology—a review. UKSim-AMSS **3**, 131–136 (2014)
27. McLuhan, M.: Understanding Media. The Extensions of Man. The MIT Press, Cambridge, MA (1994)
28. Meyer, R.: The new value of the archive: AI image generation and the visual economy of "style." Image. Interdisc. J. Image Sci. **37**(1), 100–111 (2023)

29. Nakamatsu, K., Patnaik, S., Kountchev, R., Li, R., Aharari, A. (eds): Advanced Intelligent Virtual Reality Technologies. Proceedings of 6th International Conference on Artificial Intelligence and Virtual Reality (AIVR 2022). Springer, Singapore (2023)
30. Oumaima, D., Mohamed, L., Hamid, H., Mohamed, H.: Application of artificial intelligence in virtual reality. In Lanka, S., Sarasa-Cabezuelo, A., Tugui, A. (eds.) Trends in Sustainable Computing and Machine Intelligence Proceedings of ICTSM 2023, pp. 67–85. Springer, Singapore (2024)
31. Parmentier, M.: Archives du virtuel. Vrin, Paris (2023)
32. Peirce, C.S.: What is a sign? In: Collected Papers, vol. 2, pp. 156–173. The Belknap Press of Harvard University Press, Cambridge, MA (1960)
33. Ritchin, F.: The Synthetic Eye. Photography Transformed in the Age of AI. Thames & Hudson, London (2024)
34. Robinson, H.P.: Pictorial Effect in Photography: Being Hints on Composition and Chiaroscuro for Photographers. Piper & Carter, London (1869)
35. Somaini, A.: Algorithmic images: artificial intelligence and visual culture. Grey Room **93**, 74–115 (2023)
36. Sunde, E.K.: From outer space to latent space. Philos. Photogr. **15**(1), 123–142 (2024)
37. Sutherland, I.E.: The ultimate display. In: Proceedings of the IFIP Congress 65, vol. 1, pp. 506–508. Washington–London, Spartan Books (1965)
38. Sutherland, I.E.: A head-mounted three-dimensional display. In: AFIPS Conference Proceeding, vol. 33, pp. 757–764 (1968)
39. Wasielewski, A.: The reification of style in AI image generation. Hertz. Stud. Art Hist. **3**, 13–20 (2024)
40. Zaralli, M.: Virtual Reality and Artificial Intelligence. Risks and Opportunities for Your Business. Routledge, New York (2024)
41. Zylinska, J.: AI Art. Machine Visions and Warped Dreams. Open Humanities Press, London (2020)

Andrea Pinotti is full professor in Aesthetics in the Department of Philosophy "Piero Martinetti" at University of Milan, where he coordinates the research centre "EXT. Extended Realities" and has directed the ERC Advanced project *An-Iconology. History, Theory, and Practices of Environmental Images* (2019–2025). His studies focus on image theories and visual culture studies, virtual and augmented reality, empathy theories, memorialisation and monumentality. He has been a fellow of various international institutions, among which The Italian Academy at Columbia University; The Warburg Institute in London; the ZFL in Berlin; the IEA, the EHESS and the FMSH in Paris. In 2018, he was awarded the Wissenschaftspreis der Aby-Warburg-Stiftung in Hamburg. In 2023 he was appointed Panofsky Professor at the Zentralinstitut für Kunstgeschichte in Munich. Among his recent publications, the volume *At the Threshold of the Image. From Narcissus to Virtual Reality* (2021, English ed. 2025).

Chapter 6
Spellbound by Technology: Augmented Reality, Artificial Intelligence, and Automagic

Sofia Pirandello

Abstract This chapter addresses the tendency to interpret the possibilities offered by Augmented Reality (AR) as magical. It shows how the functioning of AR, often associated with supernatural everyday superpowers, is actually made possible by the use of artificial intelligence (AI) and, in particular, by the constant monitoring of the user's movements and behaviour. Rather than magic, AR uses "automagic" created by image, gesture and movement recognition algorithms. Far from being supernatural, AR, implemented with AI, needs humans at all stages of its development: both for the construction of the datasets from which it draws, and as a physical substrate for the hardware and software of the devices. In this respect, the repeated use of these tools requires constant training not only of AI but also of users, with consequences for the manipulation of inanimate objects, the relationship between people and technologies and between people themselves.

Keyword Augmented reality · Artificial intelligence · Body tracking · Automagic · Philosophy of technology

This essay was written within the scope of the "Departments of Excellence 2023–2027" project, awarded to the Department of Philosophy Piero Martinetti (EXT—Coordinated Research Centre on Extended Reality) at the University of Milan by the Italian Ministry of University and Research.

S. Pirandello (✉)
University of Milan, Milan, Italy
e-mail: sofia.pirandello@unimi.it

P. Conte et al. (eds.), *Algomedia. The Image at the Time of Artificial Intelligence*, Lecture Notes in Morphogenesis, https://doi.org/10.1007/978-3-032-08726-3_6

6.1 Augmented Reality: The Ubiquitous and Responding Digital

"Hundreds of computers in a room could seem intimidating at first, just as hundreds of volts coursing through wires in the walls once did. But like the wires in the walls, these hundreds of computers will come to be invisible to common awareness" [45: 98]. This is how Mark Weiser described 21st-century computers in 1991. Embedded into objects, distributed throughout physical space and connected in networks, he imagined them involved in almost every single task of everyday life, yet remaining unseen.

Indeed, the computerisation of domestic and social environments is now so complete that it is almost "as refreshing as taking a walk in the woods" [45: 104]. One of the main technologies responsible for this process of naturalisation is Augmented Reality (AR) (cf. [49]), which adds computer graphics objects to physical space [29: 1322], with the aim of providing it with electronic properties that otherwise would not be there [48: 26].

In fact, the first device to be called "augmented reality" was a wearable headset designed by Thomas Caudell and David Mizell in 1992. The prototype would allow factory workers assembling aircraft to see the information they needed superimposed on their field of vision, enabling them to consult it without interruption. AR glasses would speed up workers' jobs by shortening the time it takes to find instructions, reducing the risk of errors and ultimately increasing productivity (cf. [8]).

Since then, many other smart glasses have been patented and launched on the market (some of which have been complete failures, such as the entire Google Glass project, which was officially abandoned in 2023 after more than ten years of enthusiastic but unsuccessful efforts) (cf. [30: 140–141], [14: 19–21]).[1] Today, however, AR refers to a larger family of tools that allow complex three-dimensional and responsive digital objects (mostly images, but also videos, text strings, audio tracks and sometimes avatars of other people) to populate both work and domestic settings (cf. [1: 356], [26: 26–27]). Although AR is widely used in the military, medical, and industrial fields, mainly in the form of wearable tools, it is currently also deployed in many other areas as diverse as video games, commercial applications, artworks, and social networks, for which it is mostly experienced through portable devices, such as smartphones and tablets. With AR, users can either select and manipulate visual information via voice commands and gestures, or use the camera in their portable devices to see computer-generated content combined with the tangible environment: the device itself does not demand attention, but rather serves to expand the number of possibilities available for real-world tasks. For this reason, AR is increasingly referred to as "spatial computing"[2] in order to distinguish it from other digital experiences that divert users from the physical world, such as the traditional desktop

[1] https://support.google.com/glass-enterprise/customer/answer/13417888.

[2] Among the major tech companies, it is Apple which stresses this concept the most: https://www.apple.com/apple-vision-pro/.

computer space or a fully artificial environment like the one produced by Virtual Reality.

Just like in ghost stories, we do not need direct contact with AR objects to interact with them. Regardless of the setting, a desire for remote control underlies the very idea of AR: digital augmentation seeks to gather as many layers of reality as possible in order to enhance the individual's operational and perceptual capabilities. In short, it makes it possible to recall information from different environments while actually navigating in a single space. When users "touch" in AR, they are simply mimicking the gestures they would make to directly manipulate an object if it were a physical thing. Depending on whether they are holding a device or wearing glasses, their fingers act like skaters on a surface or suspended in mid-air, and the device responds. Although they receive no sensory stimulation in return, their gestures are effective. In fact, "augmented" gestures can do even more than normal ones: it is enough to move your device to make the objects or information you are working with follow you. When wearing AR glasses, this could be as simple as turning your head or eyes. In this respect, AR seems to have fulfilled René Barjavel's prediction of virtual entities walking around our homes (cf. [2]). It has also realised the dream of Nikola Tesla, who in 1908 thought that, thanks to his discoveries, a businessman in New York would soon be able to dictate instructions that would materialise in London in the form of written words, and that we would soon be able to call anyone in the world, listen to music or the speech of a politician, and dispatch images to any corner of the globe [24: 165].

Although a man of science, Tesla was convinced that the wireless technology he helped create possessed a certain occult quality. Looking at the possibilities opened up by technologies like AR, it is not hard to see why: by multiplying our connections, giving us the ability to interact with inanimate objects and to move digital entities around with a gesture or a blink of the eye, AR seems to have finally found a way to obtain superpowers like telekinesis, telepathy, and teleportation.

6.2 The Automagic of Spatial Computing

Among the most important references for engineers working on these technologies in the 1980s and 1990s were studies by the US Air Force and blockbuster films such as *Top Gun* and *Terminator* (cf. [37]). The idea of finally acquiring superpowers is therefore implicit in the design of AR itself, as is the intention to use them to obtain unprecedented levels of efficiency. Allusions to the supernatural can still be found in many different materials dealing with AR (cf. [3, 9]), and are perfectly summed up by advertising mottos such as "Feels like magic."[3] Moreover, one of the leading companies in the AR field is called Magic Leap.

[3] "Lockheed Martin creates magic with HoloLens 2": https://www.youtube.com/watch?v=GRLyP7_cqNc; a similar motto can be found in the video "Introducing Apple Vision Pro": https://www.youtube.com/watch?v=TX9qSaGXFyg.

But what is constantly referred to as "superpowers," "supernatural" or even "magic" is in fact artificial intelligence (AI). The reason AR responds so well to us is that it has learned to track every action we take in our personal physical space. It is absolutely necessary for the machine to observe the user, with such precision that it does not create difficulties which discourage its use, such as motion sickness, and even gives us the impression that it can read our intentions. A milestone on this path is the three-dimensional head-mounted display designed by Ivan Sutherland and colleagues, so heavy and bulky that it had to be anchored to the ceiling, hanging menacingly over the wearer's head—hence the nickname "Sword of Damocles." Despite its obvious drawbacks, it was the first headset to track head movements, allowing digital objects and physical space to be interconnected seamlessly (cf. [43]).

Since then, AR devices have refined their tracking techniques. Indeed, augmenting reality means sensing it and extracting information from it; the most basic AR system requires at least the ability to follow and record users' movements in physical space [8: 660]. In this sense, a breakthrough year for the mass deployment of AR was 1999, when a camera, a mechanical gaze anchored to the surrounding environment of potentially any individual, was integrated into a mobile phone for the first time [46: 77].

Today, most smartphones are equipped with, at the very least, GPS, accelerometers and gyroscopes for tracking position and movement, as well as deep learning algorithms for text, object and speech recognition, transforming the ubiquitous computing envisioned by Weiser into an Artificial Intelligence of Things (cf. [5]). In particular, Convolutional Neural Networks (CNN), which are widely used in machine vision to "detect, analyse, and classify entities" [40: 80], are always implicit in AR systems, both portable and wearable, capable of "looking at" what is close by. CNNs are also used for eye tracking and gesture recognition, albeit the sphere of application of such classification algorithms is much wider (cf. [20, 23]).

When it comes to wearable AR devices, the process of human–machine pairing through algorithmic understanding of the environment is particularly evident. Some advertisements, such as that for the Xiaomi AR Glass Discovery Edition (a device that is not yet available to the general public) emphasise on the freedom granted to users: freedom of movement and, above all, freedom to manage the virtual world that the device reveals to them, wherever they want.[4] However, the Xiaomi AR Gesture Control system is not only called so because it allows you to control the gesture, but also because it exercises control *over* your gesture. The same goes for all other smart glasses: "The first time people don the new HoloLens 2 on their heads, the device automatically gets to know them: It measures everything from the precise shape of their hands to the exact distance between their eyes" [36]. The Microsoft HoloLens 2, currently the most widely used device in the professional sector, is in fact equipped with four visible light cameras for head tracking, two infrared cameras for eye tracking, a depth sensor, a five-channel microphone array and a frontal RGB camera for taking photos and videos, in addition to an accelerometer, a gyroscope and

[4] "Meet Xiaomi Wireless AR Glass Discovery Edition": https://www.youtube.com/watch?v=_B5Zgyq4EfM.

a magnetometer.[5] The Apple Vision Pro features two high-resolution main cameras, six world-facing tracking cameras, four eye tracking cameras, a depth camera, a LiDAR scanner, four inertial measurement units, a flicker sensor, an ambient light sensor, plus iris-based biometric authentication and Spatial Audio with dynamic head tracking.[6]

Classification algorithms are also necessary for the implementation of other types of AI, namely generative AI, because in order for the device to produce responsive digital objects, it must first analyse and catalogue what it "sees." In particular, Generative Adversarial Networks (GANs), first introduced in 2014, have begun to be used to create AR digital entities and filters (cf. [39]). By accessing public data repositories, GANs are able to create three-dimensional images that seamlessly adapt to the physical environment, enriching it with objects (such as a piece of furniture) or modifying the characteristics of existing elements (as is the case with make-up or ageing filters).

However, changing the way we appear in physical space thanks to AI is not limited to applying a hair colour filter. It can also mean creating a double of ourselves. In this respect, AR seems to be part of a general tendency towards the reification of the absent, making visible in a physical environment and available for interactive exchange what could not be close because it is distant in space or time. This phenomenon, also known as the "digital twin," allows users "to replicate and monitor existing physical objects and processes" (cf. [38]). Individuals themselves can be duplicated, so that they can use a real "digital doppelgänger" (cf. [4]), which allows them to appear like a ghost in someone else's space.

Customisation then reaches a radically new level, as does the idea of the "personal device," which is repeatedly reinvented and pushed to extremes by the big tech companies. Among the reasons why the Apple Vision Pro is praised for its ability to respect the unique qualities of its owner is a machine learning system that scans the face to create an avatar (digital "Persona," as they call it) which is as similar as possible in appearance. By tracking facial expressions in real time, it promises to preserve your personality even in virtual interactions. Also, at the pre-purchase stage, it is necessary to scan the user's face to ensure that the visor fits as comfortably as possible. Vision Pro therefore aims to synchronise with user's body in every respect,[7] even, if desired, with other Apple devices already in their possession, achieving an almost perfect illusion of transparency. Although the design of the device currently on the market is far from satisfactory,[8] the aim, true to Weiser's vision, is for the technological object to disappear, for the computer to be assimilated into the world.

---

[5] https://learn.microsoft.com/en-us/hololens/hololens2-hardware.

[6] https://www.apple.com/apple-vision-pro/specs/.

[7] For example, because Vision Pro cannot be worn with glasses, those who need glasses must provide the manufacturer with the details of their prescriptions so that graduated lenses can be installed on the device.

[8] "Apple Vision Pro review: magic, until it's not": https://www.youtube.com/watch?v=hdwaWxY11jQ.

In such a situation, Walter Benjamin's renowned theory, according to which the human body (both individual and collective) and the technical device are mutually innervated [41: 88], must be understood literally: AR is a "sticky" technology because it needs to integrate with the user's body and the environment in order to function properly. Certain organs, such as the skin, become core components of digital and networked technologies. For example, the prismatic lens integrated into Google Glass Enterprise Edition 2 was used to direct the light beam produced by the device directly onto the retina; to open the menu when wearing HoloLens 2, you have to raise your arm in front of the field of vision and click on the Microsoft icon that appears on the inside of your wrist. The tactile turn brought about by smartphones and tablets is being replaced by a more performative turn that engages the user from head to toe, since every part of the body acquires a gestural quality.

The idea that a moving body is a more effective one resonates strongly with classic studies, such as the work of Marcel Jousse, who as early as the 1950s recognised the value of using recording instruments, especially cameras, to capture the complexity of human gesture [21]. Even if they do not explicitly address the issue, AR designers recognise that bodily actions have epistemic value, as David Kirsh and Paul Maglio pointed out in 1994, demonstrating on scientific grounds that active manipulation and movement in space aid in the performance of cognitive tasks (cf. [22]). Eyes, hands, skin are thus incorporated as hardware elements of the device, along with the totality of bodily movements and behaviours that are fundamental to its functioning. Those parts of us that lend themselves as a support seem to be the seamless solution between animate and inanimate, a boundary where the distinction between human and non-human becomes blurred, so that a momentary integration between living body and technological body takes place [7: 145].

In his essay *Nature*, dating back to 1836, the philosopher Ralph Waldo Emerson described the ideal observer with a phrase that later became quite famous: "I become a transparent eyeball; I am nothing; I see all" [13: 39]. What Emerson may not have expected is that his description fits the machine better than the human: his metaphor seems to be an accurate analysis of ubiquitous and responsive digital technologies, and of AR devices in particular. The machines we have created for our purposes are watching us as we use them. There is thus a progressive confusion between an observing system and an observed system, which collapse into each other as the same measure of control can be exercised over the world and the user simultaneously [14: 154–155].

Rather than magic, what we are dealing with here is "automagic," something performed with such efficiency that it seems like sorcery.[9] Unlike magic, which held dominion in the world of approximation, i.e. before the introduction of precise measuring instruments, where alchemical formulae were used [25: 319], automagic is worked by algorithms aimed at optimising user performance with precision. The specific "wizardry" of automagic, especially in the case of AR, is to create the illusion of being in control from a distance of a set of operational gestures, whereas one is in fact under the control of something or someone else. This scenario essentially mirrors

[9] https://www.collinsdictionary.com/it/dizionario/inglese/automagic.

that typical of occult spellbinding, the bewitchment of another and their actions: AR users are under the spell of the technological apparatus.[10]

6.3 The Double Direction of Training

If you access the official website of Brilliant Labs, a start-up which produces smart glasses, you will be asked: "What if your glasses gave you AI superpowers?"[11] Some scholars have attempted to answer similar questions by highlighting the higher productivity fostered by AR, resulting from the high degree of gestural interaction and the positioning of digital documents in a physical workspace that can be accessed simultaneously in both proximity and telepresence. Indeed, one aspect underlined in all cases is the impetus towards collaboration within a group, even at a distance (cf. [27]). The art world also immediately looked with interest at this new technology, in an attempt to give the constant movement and interaction, induced by digital augmentation, a novel nuance to engagement, transforming them into mobilisation [35: 96]. As digital devices in general connect us to other people, places and times far away, the hyperlink seems to be the specific extension of contemporary creativity. In this respect, Ed Finn has spoken of an "augmented imagination" [15: 186] which is fundamentally collective in character. For this reason, and with specific reference to AR, there are those who do not hesitate to consider augmented imagination even posthuman (cf. [47]), contributing to marking an epochal transition for our species.

Of course, the emergence of AI "superpowers" also brought with it a host of criticisms. Even before one could speak of true artificial intelligence, in 1945, Vannevar Bush wrote:

> Such machines will have enormous appetites. One of them will take instructions and data from a whole roomful of girls armed with simple key board punches, and will deliver sheets of computed results every few minutes. There will always be plenty of things to compute in the detailed affairs of millions of people doing complicated things [6].

For the software we incorporate in our devices in order to recognise what we are framing with our mobile phone or smart glasses, a significant training process is required. Indeed, any technology involving AI has to draw on massive collections of data, which are continuously gathered, labelled and verified by underpaid (or completely unpaid) exhausting layers of labour. In recent years, scholars such as Luke Munn and Kate Crawford have explored this phenomenon in depth, highlighting the imperialist attitudes underlying the massive diffusion of AI. Today it is not just girls—as Bush suggested—, but the pool of these workers has simply diversified to include other groups of marginalised individuals, such as those from the Global South [31: 562], and more generally women, immigrants and members of

---

[10] Also, some of the symptoms reported by users, such as headaches or vomiting, are reminiscent of those considered to be manifestations of bewitchment and the evil eye (cf. [11, 44]).

[11] https://brilliant.xyz/.

oppressed communities (in the case of the United States, typically the Afro-American community). In short, "'dirty people' do the dirty work" [32: 298].

Automagic does not seem to imply a replacement or liberation of the human, but rather a meeting halfway. There is likely to be an ever-increasing co-dependency between individuals and mechanical systems, indispensable to each other in a process of mutual subjectivation [32: 293]. However, this mutual influence does not occur only in the training of the AI. If we look at the case of AR, it becomes clear that while we are training AI, users themselves are also being trained. And it is not necessarily the case that an increase in creativity is always the result. For example, in a professional environment, AR users may follow instructions to the point where they fail to recognise the unexpected, as might happen to surgeons using AR visors in the operating theatre (cf. [12]). Although smart glasses such as HoloLens 2 offer the possibility of "patenting" some original gestures to be stored in the toolbox, their use is strongly discouraged by the company as these would be difficult to export and teach to others and it would be a move away from what has been described as a more "intuitive" mode of interaction. Above all, it would mean moving away from standardising the gesture that the machine needs to interpret our behaviour: as Microsoft explicitly states, attention tracking is used to find out which remote collaborators you focus on the longest, to study how users generally relate to their action space in order to improve engagement with computerised devices based on interaction with physical objects; to monitor the work performance of wearers of the visor, who can also be subjected to tests that help them improve; to conduct market and interior design research.[12] The concern here is not with critical issues of individual privacy (not least because, in most cases, privacy concerns have already been addressed and solved by most AR applications: see for instance [16]). From the point of view of creating an intuitive and transparent interface, personal information is not in itself of interest, and so developers can easily dispense with it. Anything that can be traced back to a single person is worth far less than what stands for their humanity, namely the kind of data that makes it possible to understand how people in general move, how they behave, what they want. If the tool is designed to be as tailored as possible to the needs of each person, it is only to make the individual reveal what is generalisable about them. The gestures of a community as large as the users of a given technology can then be standardised along simplified lines, as has already happened with the case of the smartphone. With the Multi-Touch Gesture Dictionary (2006), Apple was the first company to propose the idea of patenting a dictionary of gestures for interacting with technology [17: 154], triggering a fierce competition that illustrates the importance of what is at stake.

Today, computers talk to us, turning code into meaningful information that we can see, hear and change in response to our gestures. But to do so, they must first convert the world into code. Machines do not think for us, but their intervention helps to change the patterns and qualities of thought. As we adapt the machine to understand us, we try to adapt ourselves to the machine. This step, like all transitions, is not neutral. In turn, as we use this technology, it teaches us a new, globally applicable

[12] https://learn.microsoft.com/it-it/windows/mixed-reality/.

standard for manipulating and interacting with the world. AR can therefore reconfigure our relationship with technology as well as the way we physically interact with objects and people.

6.4 Conclusion: Lands of Conquest

Who is the real magician in such a situation? The spell bound by this technology is effective because, while ubiquitous computing aims to integrate itself into our environment to the point of becoming invisible, its workings remain largely obscure to the majority of people who use it. In fact, any magic trick relies on our inability to understand the moves behind it: from this point of view, computers are indeed the most successful act of magic [33: 35]. AR is part of a more general tendency to see digital technology as the domain of "rational magic" [34: 226], belonging to the realm of the supernatural (if not the divine) and, as such, managed by specialists who alone are responsible for mediating with the rest of the world. Even in academia, many researchers argue that there is a definite association of the digital with the occult and magic (cf. [28, 42]). The ephemeral aspect of AR objects reinforces this tendency, leading one to interpret them as messengers on the threshold between the material and the immaterial [9: 465]. But it is possible to break the spell, starting from the observation that the functioning of AI inherently implies a symbiotic relationship with humans [33: 40].

The history of AI is a history of conquest: its operation involves the colonisation of both the lands that supply the minerals needed to build device components, and the imagination of users (cf. [10]). One of the battlegrounds, perhaps less talked about, is the physicality of users, in fact, AI is much more corporeal and human than one might think.

According to many philosophical and anthropological theories, since the dawn of our species we have needed technical objects to complete ourselves, so that humans can be described as essentially *homo faber* (cf. [19]). We use technology for almost every activity in our lives, from eating to sleeping, from loving to communicating. In this respect, AR seems to fulfil our technological destiny in the best possible way, as it aims to bring together all the tools we need in our very first tool: the body. At the same time, it realises in the most satisfying way so far the age-old question of magic: to control inanimate objects and interact with fantastical creatures. But even Weiser, a theorist and advocate of machine transparency, warned of the danger of putting a computer in every personal object: bringing private companies and governments into our own private spaces. As Karin Harrasser has pointed out in her reflections on AR smart glasses, it is crucial that we protect our technical nature and our thirst for technological novelty, while at the same time avoiding becoming "corporate cyborgs" [18: 85], whose bodies are a land of conquest, whose gestures are a source of profit for private companies.

References

1. Azuma, R.T.: A survey of augmented reality. Presence: Teleoperators Virtual Environ. **6**(4), 355–385 (1997). https://doi.org/10.1162/pres.1997.6.4.355
2. Barjavel, R.: Cinéma total. Essai sur les formes futures du cinema. Denoël, Paris (1944)
3. Bhola, I.: Magic: How Augmented Reality Will Change Our World. Independently published (2018)
4. Bode, L.: Digital doppelgängers. M/C Journal **8**(3) (2005). https://doi.org/10.5204/mcj.2369
5. Bunz, M., Janciute, L.: Artificial Intelligence and the Internet of Things: UK Policy Opportunities and Challenges. University of Westminster Press, London (2018)
6. Bush, V.: As we may think. Atl. Mon. **176**(1), 101–108 (1945)
7. Carbone, M., Lingua, G.: Screens'r'us: from bodies with prostheses to bodies as "quasi-prostheses"? In: Carbone, M., Lingua, G. (eds.) Toward an Anthropology of Screens, pp. 135–160. Palgrave Macmillan, Cham (2023)
8. Caudell, T., Mizell, D.W.: Augmented reality: an application of heads-up display technology to manual manufacturing processes. In: Proceedings of the Twenty-Fifth Hawaii International Conference on System Sciences, vol. 2, pp. 659–669. IEEE, Piscataway, NJ (1992)
9. Chaudhary, M.Y.: Augmented reality, artificial intelligence, and the re-enchantment of the world. Zygon **54**(2), 454–478 (2019). https://doi.org/10.1111/zygo.12521
10. Crawford, K.: Atlas of AI. Power, Politics, and the Planetary Costs of Artificial Intelligence. Yale University Press, New Haven (2022)
11. De Martino, E.: Sud e magia. Feltrinelli, Milano (1966)
12. Dixon, B.J., et al.: Surgeons blinded by enhanced navigation: the effect of augmented reality on attention. Surg. Endosc. **27**(2), 454–461 (2013). https://doi.org/10.1007/s00464-012-2457-3
13. Emerson, R.W.: Nature (1836). In: Ziff, L. (ed.) Selected Essays, pp. 35–82. Penguin, New York (1982)
14. Eugeni, R.: Capitale algoritmico. Cinque dispositivi post-mediali più uno. Morcelliana, Brescia (2021)
15. Finn, E.: What algorithms want. In: Imagination in the Age of Computing. The MIT Press, Cambridge-London (2017)
16. Geroimenko, V. (ed.): Augmented reality and artificial intelligence. In: The Fusion of Advanced Technologies. Springer, Cham (2023)
17. Grespi, B.: Figure del corpo. Gesto e immagine in movimento. Meltemi, Milano (2019)
18. Harrasser, K.: Corpi 2.0. Sulla dilatabilità tecnica dell'Uomo. goWare, Firenze (2018)
19. Ihde, D., Malafouris, L.: Homo faber revisited: postphenomenology and material engagement theory. Philos. Technol. **32**(2), 195–214 (2019). https://doi.org/10.1007/s13347-018-0321-7
20. Jiang, S., et al.: Emerging wearable interfaces and algorithms for hand gesture recognition: a survey. IEEE Rev. Biomed. Eng. **15**, 85–102 (2021). https://doi.org/10.1109/rbme.2021.3078190
21. Jousse, M.: L'Anthropologie du geste. Gallimard, Paris (1974)
22. Kirsh, D., Maglio, P.: On distinguishing epistemic from pragmatic action. Cogn. Sci. **18**(4), 513–549 (1994). https://doi.org/10.1207/s15516709cog1804_1
23. Klaib, A.F., et al.: Eye tracking algorithms, techniques, tools, and applications with an emphasis on machine learning and internet of things technologies. Expert Syst. Appl. **166**, 114037 (2021). https://doi.org/10.1016/j.eswa.2020.114037
24. Kluitenberg, E.: Connections machines. In: Kluitenberg, E. (ed.) Book of Imaginary Media: Excavating the Dream of the Ultimate Communication Medium, pp. 156–184. NAi Publishers, Rotterdam (2006)
25. Koyré, A.: Du monde de l'"à-peu-près" à l'univers de la precision (1948). In: Koyré, A. (ed.) Études d'Histoire de la pensée philosophique, pp. 311–329. Armand Colin, Paris (1961)
26. Liberati, N.: Augmented reality and ubiquitous computing: the hidden potentialities of augmented reality. AI & Soc. **31**(1), 17–28 (2016). https://doi.org/10.1007/s00146-014-0543-x
27. Lukosch, S., et al.: Collaboration in augmented reality. Comput. Support. Coop. Work **24**, 515–525 (2015). https://doi.org/10.1007/s10606-015-9239-0

28. Marenko, B.: Algorithm magic. Gilbert Simondon and techno-animism. In: Natale, S., Pasulka, D. (eds.) Believing in Bits: Digital Media and The Supernatural, pp. 213–228. Oxford University Press, New York (2019)
29. Milgram, P., Kishino, F.: A taxonomy of mixed reality visual displays. IEICE Trans. Inf. Syst. **77**(12), 1321–1329 (1994)
30. Montani, P.: Tre forme di creatività: tecnica, arte, politica. Cronopio, Napoli (2017)
31. Munn, L.: Digital labor, platforms, and AI. In: Werthner, H., et al. (eds.) Introduction to Digital Humanism. A Textbook, pp. 557–570. Springer, Cham (2023)
32. Munn, L.: Automation is a myth. In: Cinque, T., Vincent, J.B. (eds.) Materializing Digital Futures. Touch, Movement, Sound and Vision, pp. 289–304. Bloomsbury, London (2022)
33. Natale, S.: Deceitful media. In: Artificial Intelligence and Social Life after the Turing Test. Oxford University Press, New York (2021)
34. Novak, M.: Liquid architectures in cyberspace. In: Benedikt, M. (ed.) Cyberspace: First Steps, pp. 225–254. MIT Press, Cambridge, MA (1991)
35. Pirandello, S.: Fantastiche presenze. Note su estetica, arte contemporanea e realtà aumentata. Johan & Levi, Milano (2023)
36. Roach, J.: The making of the HoloLens 2: how advanced AI Built Microsoft's vision for ubiquitous computing. Microsoft Official Website. https://news.microsoft.com/source/features/innovation/hololens-2-shipping-to-customers/ (2019). Accessed 30 Apr 2025
37. Rosenberg, L.: How a parachute accident helped jump-start augmented reality. IEEE Spectrum. https://spectrum.ieee.org/history-of-augmented-reality (2022). Accessed 30 Apr 2025
38. Rudra, S.: What are digital twins and how can higher Ed use them? Digital Twin Technology Could Be the Next Step in Immersive Learning. EdTech. https://edtechmagazine.com/higher/article/2022/03/what-are-digital-twins-and-how-can-higher-ed-use-them (2022). Accessed 29 June 2024
39. Šoberl, D.: Mixed reality and deep learning: augmenting visual information using generative adversarial networks. In: Geroimenko, V. (ed.) Augmented Reality and Artificial Intelligence, pp. 3–29. Springer, Cham (2023)
40. Somaini, A.: Algorithmic images: artificial intelligence and visual culture. Grey Room **93**, 74–115 (2023). https://doi.org/10.1162/grey_a_00383
41. Somaini, A.: Innervazione e training. In: Pinotti, A. (ed.) Costellazioni. Le parole di Walter Benjamin, pp. 87–90. Einaudi, Torino (2018)
42. Susca, V.: Tecnomagia. Estasi, totem e incantesimi nella cultura digitale. Mimesis, Milano-Udine (2022)
43. Sutherland, I.: A head-mounted three dimensional display. In: Proceedings of the December 9–11, 1968, fall joint computer conference part I, pp. 757–764. Association for Computing Machinery, New York (1968)
44. Valletta, N.: Cicalata sul fascino volgarmente detto jettatura (1787). Imagaenaria, Napoli (2004)
45. Weiser, M.: The computer for the 21st century. Sci. Am. **265**(3), 94–105 (1991). https://doi.org/10.1016/B978-0-08-051574-8.50097-2
46. Wellner, G.: No longer a phone. The cell phone as an enabler of augmented reality. Transfers **3**(2), 70–88 (2013). https://doi.org/10.3167/trans.2013.030205
47. Wellner, G.: Posthuman imagination: from modernity to augmented reality. J. Posthuman Stud. **2**(1), 45–66 (2018). https://doi.org/10.5325/jpoststud.2.1.0045
48. Wellner, P., Mackay, W., Gold, R.: Computer augmented environments: back to the real world. Commun. ACM **36**(7), 24–26 (1993). https://doi.org/10.1145/159544.159555
49. Williams, A.S., Garcia, J., Ortega, F.: Understanding multimodal user gesture and speech behavior for object manipulation in augmented reality using elicitation. IEEE Trans. Visual Comput. Graphics **26**(12), 3479–3489 (2020). https://doi.org/10.1109/tvcg.2020.3023566

Sofia Pirandello (Ph.D., University of Milan 2023) is a post-doctoral researcher in Aesthetics at the University of Milan. Her research interests concern contemporary theories of imagination, philosophy of technology, media theory and contemporary art. She is the author of the books *Fantastiche presenze. Note su estetica, arte contemporanea e realtà aumentata* (2023) and *Emersioni. Realtà aumentata e poteri dell'immaginazione* (2025). She is currently working on the aesthetics and politics of augmented reality and other phantasmagoric devices and is also investigating the link between magic and contemporary technology.

Chapter 7
Algorithmic Economies of Light. The SARS-Cov-2 Micrographs as Algo-Images

Ruggero Eugeni

Abstract This chapter investigates algorithmic imaging dispositives, or *algo-images*, by situating them within the broader context of a political economy of light and its historical transformations. Initially, the chapter explores a detailed case study focusing on micrographic imaging technologies, encompassing both optical and electronic microscopy, utilised to visualise the morphology, behaviour, and material composition of the SARS-CoV-2 virus. The analysis elucidates how micrographic dispositives integrate two distinct image-production paradigms: one reliant upon manipulating electromagnetic energy flows, and another centred around digital data processing. Subsequently, the chapter argues that these paradigms correspond to two successive regimes within the political economy of light: the modern one characterised by *techno-images*, and the contemporary one dominated by *algo-images*. The conclusion highlights both divergences and continuities between these regimes, emphasising their common epistemological and practical aim: facilitating the cognitive and operational appropriation of reality. Ultimately, this analysis contributes to a deeper understanding of how contemporary visual technologies shape our relationship with the world through acts of appropriation and expropriation of the visible, which are presently carried out through data capture and management—both visual and invisual.

Keyword Algorithmic images · Microscopy imaging · Political economy of images · Techno-images · SARS-CoV-2 imaging

7.1 Islands in the Stream

The black and white image shows two whitish, rounded and bare islets seen from above. A crown of rocks surrounds each. All around, a milky sea troubled by mysterious currents. It is January 24, 2020. Less than a month has passed since December

R. Eugeni (✉)
Università Cattolica del Sacro Cuore, Milan, Italy
e-mail: ruggero.eugeni@unicatt.it

P. Conte et al. (eds.), *Algomedia. The Image at the Time of Artificial Intelligence*, Lecture Notes in Morphogenesis, https://doi.org/10.1007/978-3-032-08726-3_7

31, 2019, when the Chinese government notified the World Health Organisation of the spread of a new virus, provisionally labelled 2019-nCoV. The virus remained faceless for twenty-four days until scientists from the Chinese Centre for Disease Control and Prevention released the first image: the two crowned islands [52].

The first visualisation of the virus is the prologue to a story as complex as it is exciting: that of the use of micrographic images in virology's battle against SARS-Cov-2 [11, 31]. Visualising it is not enough to defeat the virus: you need to understand its mechanisms of movement, how it attacks human cells, and its internal conformation. These are all tasks in which micrographic images prove to be irreplaceable tools: not only means of information and dissemination but also practical *operational* objects [36].

In the first part of this chapter, I will follow the main stages of this story. In the course of the narrative, it will emerge that the images we are talking about present a peculiar characteristic: they are both *technological* (i.e. produced by management of electromagnetic flows aimed at engendering stable traces) and *algorithmic* (i.e. created by the processing and visualisation of dynamic data sets and data cubes). In the second part of the chapter, I will try to show that the two logics, although often intertwined, are distinct and that to grasp this distinction fully, we need to adopt the point of view of a broadened political economy: more precisely, the break that separates technological images from algorithmic ones refers to a different articulation of the *political economy of light*, which, from being a resource used to produce images, becomes an instrument for data extraction. But are these two economic regimes radically different? Or does the interweaving in micrographic technologies indicate an underground continuity between them?

7.2 Micro-imaging Warfare

7.2.1 How the Virus Looks like

The Chinese researchers arrived first in the frantic race to take a picture of the new virus. However, almost by chance, the Peter Doherty Institute for Infection and Immunity in Melbourne had obtained the same images a few days earlier but had not released them due to the quality of the image, which was deemed insufficient. Both groups worked with a *Transmission Electron Microscope* (TEM). This device does not use light rays to produce images but electrons. The reason is simple: as Ernst Abbe had demonstrated as early as 1873, the diffraction of rays related to the wavelength of light limits the resolution of *optical* microscopes to 200 nm (a nanometre is the billionth part of a metre). The primary means of circumventing this "diffraction barrier" (not the only one, as we shall see) has been *electron* microscopy and the resulting photomicrographic productions [21, 37].

Although the oscilloscope (the instrument for measuring electric current that displays electron flows on a phosphor screen) had been invented as early as the late nineteenth century, it was not until 1931 that a working group at the Technical University of Berlin experimented with the possibility of using such a device to make magnifications. Since the wavelength of electrons is much shorter than that of light, it is possible to break through the light diffraction barrier to reach a definition down to 0.2 nm. Subsequent developments, mainly due to the German company Siemens during the 1930s, perfected the *Transmission Electron Microscope* (TEM)—also thanks to the forced labour of prisoners in some concentration camps where Siemens built its production laboratories. In the TEM, a beam of electrons, once emitted from a source, is condensed through an electromagnetic field (a "lens," but the term is metaphorical) and then led to pass through the sample to be analysed. The resulting electronic print is "rarefied" by other electromagnetic fields and then projected onto a sensitive surface such as a phosphor screen.

At present, the screen has been replaced by a grid of sensors and electronic-digital transducers that transform the flow of electrons into a data set, which, after suitable refinement steps, can give rise to an image such as that of the two islets. In this respect, current research (also stimulated by the wave of analyses relating to SARS-CoV-2) is making enormous progress in applying machine or computer vision algorithms to the "raw" digital images produced by electron microscopes [47]. These algorithms perform several operations previously delegated to human agents [34]: they enhance image contrasts (*preprocessing*); identify the relevant areas and shapes by separating them from the "noise" of the context (*segmentation, feature extraction*); recognise the components thus isolated (*classification*: [7]), follow a particular shape within subsequently captured images to grasp its displacements and transformations (*tracking*); carry out automatic colouring (*staining*: [29]); artificially implement image resolution (*restoration/reconstruction*).

Scholars often link the origin of machine or computer vision to the introduction of the Perceptron, the first device capable of recognising visual shapes, in 1957 [13]. However, it began to be systematically applied to microscopy in the second half of the 1990s and underwent a sudden leap forward with the advent of machine learning algorithms, often labelled as Artificial Intelligence, in the second half of the 2010s [2]: scientists thus moved from algorithms facilitating hand-designed procedures and from "shallow learning" algorithms to "deep learning" ones [25, 26, 33, 39]. In this sense, the experience of the epidemic has been a decisive boosting factor, either through the use of algorithms inspired by human visual recognition, such as *convolutional networks* [3] or more recently through a new generation of algorithms inspired by linguistic understanding, such as *Large Language Models* [15].

In any case, TEM allows researchers to visualise the new virus (whose diameter measures between 80 and 120 nm) with sufficient definition. Thus, in less than a month, the virus was given a shape: on January 30, 2020, the US Center for Disease Control and Prevention (CDC) presented to the world Spiky Bob, an expressive CGI (Computer Graphic Imaging) image based on micrographic photographs taken with electron microscopes [20]. It is a proper mug shot, and as is often the case in crime fighting, the attribution of a well-defined face preceded the assignment of an exact name.

7.2.2 How the Virus Moves

In fact, it was only on February 11, 2020, that the virus received its official name: SARS-CoV-2. Two days later, on February 13, the Rocky Mountain Laboratories, part of the National Institute of Allergy and Infectious Diseases, or NIAID (an influential research centre headed by Dr Anthony Fauci, President Trump's pandemic crisis advisor), presented a series of entirely new images of the virus, much more realistic than the Chinese ones, with vivid colours and accurate lighting. The images were not only directed at scientists: they were disseminated by the NIAID on its Flickr account and, from there, widely relaunched by the media. It is evident that there was an intention to publicly promote the centre's image, metonymically aligning its visualisation capabilities with its ability to detect and combat the virus, thus justifying the substantial funding it receives.

The technological basis accounts for the vividness of these images: NIAID obtained its images using a different technology from TEM, the *Scanning Electron Microscope* (SEM). This device, too, is based on the use of electrons: it originated in the late 1930s as a refinement of TEM and has been marketed by DuPont since the 1950s under the name "Stereoscan." In this case, the *primary* beam of compressed electrons travels along the specimen line by line and is reflected by its surface; this process gives rise to a *secondary* flow carrying information about the surface topography. Originally, the "brush" of secondary electrons was amplified and visualised on a cathode ray tube with a phosphor screen. Today, this process involves converting the electrons into a grid of data, which is then processed by a series of algorithms. The result is less potent than TEM's (the maximum resolution is currently about 1 nm), but the images are more vivid, textural, and have a "tactile" feeling. They can also be digitally coloured: in an NIAID image, for instance, yellow ochre virions emerge from the body of a cell they have previously infected and colonised ([22], see also [4]).

The *second* direction of visual virus control thus shifts its focus from how the virus appears to how it acts. Other laboratories, however, considered images obtained employing SEM technology with some criticism (let us not forget the competition between research centres, given the funding that a good, adequately communicated discovery can secure). In particular, on March 5, 2020, a group of researchers from the Southern University of Science and Technology in Shenzhen presented the first virus images obtained through a *Cryogenic Electron Microscope* (Cryo-EM) [32].

Cryo-EM was introduced in the 1980s from the ascertainment that the electron beam from TEM sometimes damages the sample; hence, the idea of subjecting preparations to very rapid cooling to cryogenic temperatures (below about 180 °C). This idea earned its inventors Jacques Dubochet, Joachim Frank, and Richard Henderson the Nobel Prize in Chemistry in 2017. One of the advantages of this technique is that the virus is "frozen" in its context and possibly while it is acting; therefore, micrographs pass from a studio pose to "live" snapshots. Furthermore, since it is possible to freeze successive phases of the same action (e.g. the act of penetrating a cell membrane), it is also possible to recompose an animated sequence of that action

[24]. But there is more: the frozen sample can be "filmed" from different angles: scientists speak in this case of *Cryo-electron Tomography* (Cryo-ET); once fused, the resulting data can be visualised as a three-dimensional image, which researchers can rotate according to their needs [51], possibly with the aid of augmented or virtual reality viewers [43]. The algorithmic processing processes already observed for other electron microscopes are crucial in the case of the Cryo-ET, which "uses AI to automate the data analysis steps of particle picking, 3D map reconstruction, and local resolution determination" ([8]; see also [19, 42]). In other terms, even more than in the previous cases, Cryo-Em and even more Cryo-ET devices collect and process far more data within the data sets and data cubes than they display.

The results are remarkable. For instance, a group of researchers coordinated by Dr Gerhard Hummer of the Max Planck Institute for Biophysics in Frankfurt published a significant article in the October 2020 issue of the journal *Science* [48]: Cryo-ET led them to analyse "in situ" the movement of SARS-CoV-2's spikes (the S-proteins), and thus demonstrate the existence of three joints, which the researchers called "shoulder," "elbow" and "wrist": it is precisely the extreme agility of movement allowed by these appendages that makes SARS-CoV-2 so dangerous compared to other coronaviruses. The researchers showed all this in a sequence of images that reconstructed the virions' movements, just as Muybridge's photographs in the nineteenth century reconstructed the horse's gallop inside the zoopraxiscope.

7.2.3 *How the Virus Acts*

At this point, an unexpected development emerges. Optical microscopy, initially sidelined because of the diffraction barrier, makes a comeback. In fact, since the end of the nineteenth century and throughout the twentieth century, scientists have devised various techniques within light or optical microscopy to circumvent the diffraction barrier. A large family of methods is linked to the use of *fluorescent* reagents to mark the substances to be observed: the first microscope using this technique was developed by Oskar Heimstaedt in 1911; in 2014, Stefan Hell, a scientist of the Max Planck Institute, received Nobel Prize in Chemistry for his pioneering work in the field of ultra-high resolution fluorescence microscopy. In this case, the beam of electromagnetic components does not directly illuminate the specimen but activates these fluorescent reagents: the photons emitted by the sample are captured by the optical lens system, fixed on a surface and then transformed into an image. In other words, the process shifts from utilising reflected or refracted light to capturing *emitted* light.

Fluorescence allows optical microscopes to access forms of super-resolution microscopy [38]. Yet, it presents a problem: since it is activated in all focal planes of the specimen, it makes it impossible to distinguish its different levels. The solution is the Confocal Microscope, patented in 1955 by one of the future fathers of artificial intelligence, Marvin Minsky. In this case, two symmetrical and mutually aligned pinholes limit both the projection of the ray that excites the fluorescence (a reduced area at a time is illuminated) and the scope of the objective that allows it to be seen (a reduced area at a time is framed); hence, the sample is moved regularly while the photons gradually emitted are captured, amplified and deposited on an electronic receptive surface that stores them until the image is completed. If the use of Cryo-ET recalled the origins of cinema, in this case the analogy with the television apparatus is obvious; indeed, one of the ancestors of this technique is the "electric telescope" patented by Paul Gottlieb Nipkow in 1884, on which the first model of television device is also based. In recent models, the light beams that trigger fluorescence are replaced by laser beams: this is the case with Confocal Laser Scanning Microscope (CLSM); some advanced models, such as Stimulated Emission Depletion (STED) *Microscope,* can achieve very high resolutions, up to 40 nm. In this case, too, photons resulting from the process are captured by digital sensors, translated into data sets and then processed by algorithms that sharply improve "the signal-to-noise ratio, spatial resolution, temporal resolution and multi-colour capacity of live-cell imaging" [41: 443].

The confocal microscopes and fluorescence devices in general have various limitations: for instance, the emission of photons by fluorescent substances is time-limited—scientists speak of a limited "photon budget" of samples. However, there is one fundamental advantage: optical microscopes make it possible to observe phenomena "live" and thus to understand the unfolding of certain events in real-time—another feature that aligns fluorescence-based optical microscopy more closely with the logic of electronic and television imaging than with that of photographic or cinematographic representation.

Indeed, the productivity of such an imaging tool has not been long coming: for instance, Cortese et al. [9] published in December 2020 a paper resulting from observations conducted with a mixed technique of optical and electron microscopes: they clearly show how the virus induces a complete reorganisation of the infected cells—including endoplasmic reticulum, peroxisomes, mitochondria, the secretory apparatus and the cytoskeleton—to create ideal conditions for its replication.

7.2.4 How the Virus is Made of

In the meantime, a *third* direction of the micro-visual fight against SARS-CoV-2 was looming. Some laboratories were trying to explore and represent the molecular and atomic structure of the virus with increasing accuracy. In part, cryo-EM and cryo-ET make it possible to explore the protein and macromolecular structure of SARS-CoV-2. For instance, in October 2020, the company Nanographics, with the collaboration of Tsinghua University, provided a 3D illustration of SARS-CoV-2 obtained by Cryo-ET, revealing for the first time its structure and detailed architecture [51]. However, more advanced techniques exist, particularly *X-ray Crystallography* (XRC).

In 1912, Max von Laue recorded the diffraction of an X-ray beam by a copper sulphate crystal on a photographic plate; the discovery earned him the Nobel Prize in Physics in 1914. The following year, William Lawrence Bragg and his father,

William Henry Bragg, used the same device to analyse the atomic structure of crystals. Indeed, given the minimum wavelength of X-rays, XRC reaches a resolution of a few ängstroms (the ängstrom is the tenth part of a nanometer: the diameter of an atom ranges between 1.2 and 2.6 ängstroms). They, too, won a Nobel Prize for Physics in 1915. From the 1920s onwards, organic molecules also began to be analysed using the XRC thanks to the possibility of crystallising a sample before analysis (i.e., solidifying it to form a geometrically regular, non-amorphous structure of atoms, molecules or ions in all three spatial dimensions).

At present, XRC is the technique that allows the highest magnification of a sample: starting from the digital information collected and through its further computational processing, it is possible to reconstruct the average position of the atoms (and thus the density of the material) and their chemical bonds. A body of information that is extremely valuable for designing strategies to combat the virus: "This is like getting a high-resolution satellite image of your target. With that information you can best design your attack," comments Dr Gordon Joyce, who uses XRC at the Walter Reed Army Institute of Research (WRAIR), a US Army medical centre (the scientist does not deny the military nature of his institution: [46]).

Once again, the results have been significant, especially in identifying the virus's vulnerability points and the subsequent development of drugs that block its replication. Thanks to the joint work of Cryo-ET and XRC, it has been possible to ascertain the replication mechanism of the SARS-CoV-2 genome (the RNA-dependent RNA polymerase, or RdRp) and thus to develop the first drugs to inhibit this mechanism (in October 2020, the U. S. Food and Drug Administration approved the use of Remdesivir, a drug already used in the fight against the Ebola). Furthermore, XRC has enabled it to work at the protein level, particularly to identify new drugs capable of inhibiting protease, an enzyme indispensable for forming the virus's proteins [31].

And here, where the story of the defeat of SARS-CoV-2 begins, we can stop our account of its micro-visualisation.

7.3 Political Economies of the Visual

7.3.1 Techno-Images and Algo-Images

The history of the "micro-visual" struggle at SARS-CoV-19 opens to different considerations. For instance, following James Elkins [16: 117–155], we might observe how photomicrographic images compel us to rethink the concept of "representation;" or, we can read micrographic devices against the background of the broader category of all "those devices capable of making visible processes that are not in themselves visual—but rather chemical, thermal, magnetic, acoustic, or electrical—thanks to intersubjective performances" [5: 246]; or, we can even consider more specifically how they connect to the broader visual culture of the SARS-CoV-2 [27, 30, 40]. Without leaving these perspectives entirely aside, I will take a somewhat different approach, leading me to analyse in general terms the uses and role of images in contemporary living environments.

First, let us consider that micrographic imaging has allowed us to grasp, so to speak, "in vitro," the intertwining of two different logics of image production. On the one hand, images are produced and often visualised through the use of flows of electromagnetic energy (light, x-rays, electrons, etc.) which, following different treatments and paths of emission, propagation, condensation, expansion, rarefaction, multiplication, absorption, reflection, refraction, diffraction, come to leave a visual trace on certain surfaces (photographic plates, phosphor screens, electronic sensor grids). We can call this final object a *techno-image* [18]. From a media-archaeological point of view, it represents the oldest sediment, as it is the type of image that characterised modernity: the takeover of light to the electromagnetic spectrum (with Michael Faraday, in 1845) determined its detachment from the optical-geometric dynamics of the previous "classical" regime [50], and at the same time made it the instrument of the constitution of a new type of photographic, cinematographic, electronic image [10, 12, 17: 378–379, 35]. A new kind of micrographic image was also born in this context, organically embedded in the galaxy of modern images [28].

On the other hand, images are produced from a "capture" of electromagnetic fluxes that are transformed into dynamic data structures. These data sets or cubes constitute the image, although they are "invisual" [36]. Thus, the centre of gravity of the process has shifted: it no longer resides in the translation of the real into a stable iconic form, but rather in the automated processes of data extraction and computation, which operate largely independently of any form of visualization. This media-archaeological layer is more recent: it originated with machine vision processes at the end of the 1950s, so that we can call it computational [1] or algorithmic [44] imagery: in analogy with the term "techno-images," I will call them *algo-images*.

7.3.2 Political Iconomies

Second, and from a broader perspective, my story demonstrates that image production, whatever the logic behind it, is always radically connected to a complex but not random network of goals, interests, processes and social dynamics. In this sense, I have repeatedly emphasised the links between the micrographic image and "media" images: photography, cinema, television, and digital media; furthermore, I have observed how the production of the SARS-CoV-2 images was linked to economic investments, reputational returns, time management, the possibility or otherwise of extending the research to poorer countries, the re-use of military research centres, among other factors.

I propose to conceptualise this context as the environment within which the subjects inhabiting it experience and manage different sets of intertwined resources (including the images themselves), sharing their conditioning and affordances. In other words, this environment is characterised by the extraction or production of resources, their circulation and exchange, their transformation and commutation, and their disposal: all operations capable of modifying the shared environment within which they are conducted more or less in common. Some of these resources (in particular light, pictures, and more recently data) are immediately involved in the production and circulation of images and constitute what Peter Szendy [45] has called an Iconomy; however, they are at the same time intertwined with other sets of resources (raw materials, energy, labour, money, time, reputation, etc.) that need to be considered for a complete view of the phenomena involving images.

I define this perspective of analysis as an (enlarged) political economy approach [14]: in fact, I intend to take up some of the concepts and tools of analysis proper to political economy, i.e. the approach that analyses the strategic and purposive management of resources by given communities [6]; at the same time, however, I consider it appropriate to "de-economise" this approach, i.e. to remove it from a purely market and financial idea of resource exchanges, broadening its application to different fields such as social, ecological, psychological, technological ones. A glance at the story of microvisual research on SARS-CoV-2 immediately suggests that it would be incomprehensible if not framed within a political economy of knowledge and scientific research [49]. Furthermore, many of the details I have highlighted (from the use of forced labour by Siemens to technical expressions such as "photon budget") can be framed and re-read from this point of view.

In essence, the story I reported can also be read as a comparison between two political-economic circuits: the human one, based on cognitive, reputational, financial, technological and visual resources, and the virus one, based on purely biological resources. The two circuits operate on different spatial and temporal scales—and our entire narrative has traced how the human political economy gained access to the viral one. However, a common objective leads both to take control of a portion of the world. And we will have to return to this point in our conclusions.

7.4 Conclusion: The New Political Economy of Light

Let us come back to the distinction between techno-images and algo-images. We can observe that they correspond to two different regimes of the political economy of light, i.e. two different ways of using and exploiting the resource flows offered by the electromagnetic spectrum. On the one hand, the modern regime of techno-images establishes a predominantly *productive* economy by making light the instrument of automated image-making. On the other hand, the contemporary regime of algo-images uses light in a predominantly *extractive* key, as it operates the translation of visual traces into invisual data. In essence, the political economy of light has moved from being at the service of an economy of the visual to being attracted and finally incorporated into the broader galaxy of the data economy.

Yet, my narrative has shown that the two regimes are closely intertwined; it is, therefore, possible and necessary to flank the recognition of their differences with the identification of continuities. Once again, returning to micrographic devices is helpful, as they show that a common purpose determines the connection of techno-images and algo-images: "grasp the world as picture," according to Martin Heidegger's seminal expression [23: 67]. Techno-images and algo-images both aim at producing informational and operational models of the world, i.e. artificial reproductions that allow for an appropriation of reality both in cognitive terms (design, simulation, identification of patterns and recurrences, prediction of developments, etc.) and in pragmatic ones (transformation of states of affairs in the world based on operations performed with and on them). To pick up on what I was saying above, the political economy on a human scale uses images precisely how the virus uses proteins: that is, as tools for taking possession of and reorganising the world.

In conclusion, the transition from the modern to the contemporary political economy of light does not mark a rupture, but rather an intensification of a long-standing ambition: to transform the world into an image to master it. Algo-images do not *break* with the logic of techno-images—they *radicalise* it. By mobilising vast informational resources and automating cognitive and operational processes, they amplify the mechanisms of visibility and control inaugurated in the modern era. It is within this historical continuum that the rise of what [53] terms "surveillance capitalism" must be situated. Far from being a recent anomaly, it is the latest expression of a visual regime whose roots run deep in the modern project of government through light.

References

1. Anderson, S.F.: Technologies of Vision: The War Between Data and Images. The MIT Press, Cambridge, MA–London (2017)
2. Aswath, A., et al.: Segmentation in large-scale cellular electron microscopy with deep learning: a literature survey. Med. Image Anal. **89**(102920) (2023). https://doi.org/10.1016/j.media.2023.102920
3. Azuri, I., et al.: The role of convolutional neural networks in scanning probe microscopy: a review. Beilstein J. Nanotechnol. **12**, 878–901 (2021). https://doi.org/10.3762/bjnano.12.66
4. Brahim Belhaouari, D., et al.: The strengths of scanning electron microscopy in deciphering SARS-CoV-2 infectious cycle. Front. Microbiol. **11**(1), 2014 (2020). https://doi.org/10.3389/fmicb.2020.02014
5. Cappelletto, C., Galimberti, G.: Visually materializing nonvisual matter: challenges, paradoxes, and approaches. Reti, saperi, linguaggi. Ital. J. Cogn. Sci. **12**(2), 245–252 (2023). https://doi.org/10.12832/109345
6. Cardinale, I., Scazzieri, R. (eds.): The Palgrave Handbook of Political Economy. Palgrave Macmillan, Cham (2018)
7. Chen, M. (ed.): Computer Vision for Microscopy Image Analysis. Elsevier-Academic Press, London-San Diego (CA) (2021)
8. Chung, J.M., et al.: Artificial intelligence in cryo-electron microscopy. Life **12**, 1267 (2022). https://doi.org/10.3390/life12081267
9. Cortese, M., et al.: Integrative imaging reveals SARS-CoV-2-induced reshaping of subcellular morphologies. Cell Host Microbe **28**, 853–866 (2020). https://doi.org/10.1016/j.chom.2020.11.003
10. Cubitt, S.: The Practice of Light: in: A Genealogy of Visual Technologies from Prints to Pixels. The MIT Press, Cambridge, MA, London (2014)

11. Deroubaix, A., Kramvis, A.: Imaging techniques: essential tools for the study of SARSCoV-2 infection. Front. Cell. Infect. Microbiol. **12**, 794264 (2022). https://doi.org/10.3389/fcimb.2022.794264
12. Désile, P.: Généalogie de la lumière: Du panorama au cinema. L'Harmattan, Paris (2000)
13. Dobson, J.E.: The Birth of Computer Vision. University of Minnesota Press, Minneapolis-London (2023)
14. Eugeni, R.: Algorithmic Images: Post-media Dispositives and the New Political Economy of Light. Taylor & Francis, London (forthcoming)
15. Eisenstein, M.: Teasing images apart, cell by cell. Nature **623**, 1095–1097 (2023)
16. Elkins, J.: Six Stories from the End of Representation: Images in Painting, Photography, Astronomy, Microscopy, Particle Physics, and Quantum Mechanics, 1980–2000. Stanford University Press, Stanford (CA) (2008)
17. Elsaesser, T.: Film History as Media Archaeology: Tracking Digital Cinema. Amsterdam University Press, Amsterdam (2016)
18. Flusser, V.: Into the Universe of Technical Images (1985). University of Minnesota Press, Minneapolis-London (2011)
19. Galaz-Montoya, J.G.: The advent of preventive high-resolution structural histopathology by artificial-intelligence-powered cryogenic electron tomography. Front. Mol. Biosci. **11**, 1390858 (2024). https://doi.org/10.3389/fmolb.2024.1390858
20. Giaimo, C.: The Spiky Bob seen around the world. In: How CDC Medical Illustrators Created the Coronavirus Pandemic's Most Iconic Image. The New York Times, April 1, 2020. https://www.nytimes.com/2020/04/01/health/coronavirus-illustration-cdc.html
21. Hawkes, P.W., Spence, J.C.H. (eds.): Springer Handbooks of Microscopy, 2nd edn. Springer, Cham (2019). https://doi.org/10.1007/978-3-030-00069-1
22. Hawryluk, M.: The scientist behind some of the world's best Coronavirus image. Time, May 20, 2020. https://time.com/5839573/coronavirus-images/
23. Heidegger, M.: The age of the world picture (1938). In: Off the Beaten Track (1950), pp. 57–85. Cambridge University Press, Cambridge–New York (2002)
24. Hutcheon, S., Palmer, A.: Fighting the invisible enemy. Abc.net, April 28, 2020. https://www.abc.net.au/news/2020-04-28/putting-the-coronavirus-under-the-microscope/12158048?nw=0
25. Jones, M.L., Strange, A.: Artificial intelligence for image analysis in microscopy. Wiley Analytical Science Magazine, May 16, 2023. https://analyticalscience.wiley.com/content/article-do/artificial-intelligence-image-analysis-microscopy
26. Kalinin, S.V., et al.: Deep learning for electron and scanning probe microscopy: From materials design to atomic fabrication. MRS Bull. **47**, 931–939 (2022). https://doi.org/10.1557/s43577-022-00413-3
27. Keidl, P. D., et al. (eds.): Pandemic Media: Preliminary Notes Toward an Inventory. Meson Press, Lünenburg (2020)
28. Keller, C. (ed.): Brought to Light: Photography and the Invisible, 1840–1900. San Francisco Museum of Modern Art–Yale University Press, San Francisco–New Haven (2008)
29. Kreiss, L. et al.: Digital staining in optical microscopy using deep learning. A review. PhotoniX **4**(34) (2023). https://doi.org/10.1186/s43074-023-00113-4
30. Lancioni, T., Villa, F. (eds.): The visual culture of SARS-CoV-2. In: Special Issue of VCS Visual Culture Studies, vol. 2 (2021)
31. Leigh, K.E., Modis, Y.: Imaging and visualising SARS-CoV-2 in a new era for structural biology. Interface Focus **11**, 20210019 (2021). https://doi.org/10.1098/rsfs.2021.0019
32. Liu, C., et al.: The architecture of inactivated SARS-CoV-2 with postfusion spikes revealed by Cryo-EM and Cryo-ET. Structure **28**(11), 1218–1224 (2020). https://doi.org/10.1016/j.str.2020.10.001
33. Liu, Z., et al.: A survey on applications of deep learning in microscopy image analysis. Comput. Biol. Med. **134**, 104523 (2021). https://doi.org/10.1016/j.compbiomed.2021.104523
34. Merchant, F.A., Castleman, K.R. (eds.): Microscope Image Processing, 2nd edn. Elsevier–Academic Press, London–San Diego (CA) (2023)

35. Mikuriya, J.T.: A History of Light: The Idea of Photography. Bloomsbury Academic, London–New York (2017)
36. Parikka, J.: Operational Images. From the Visual to the Invisual. University of Minnesota Press, Minneapolis–London (2023)
37. Peres, M.R.: Laboratory Imaging and Photography. Best Practices for Photomicrography and More. Routledge, New York (2017)
38. Prakash, K., et al.: Super-resolution microscopy: a brief history and new avenues. Phil. Trans. R. Soc. A **380**, 20210110 (2022). https://doi.org/10.1098/rsta.2021.0110
39. Ragone, M., et al.: Deep learning modelling in microscopy imaging: a review of materials science applications. Prog. Mater. Sci. **138**, 101165 (2023). https://doi.org/10.1016/j.pmatsci.2023.101165
40. Ramírez-Blanco, J., Spampinato, F. (eds.): The Pandemic Visual Regime: Visuality and Performativity in the COVID-19 Crisis. Punctum Books, Goleta (CA) (2023)
41. Shroff, H., et al.: Live-cell imaging powered by computation. Nat. Rev. Mol. Cell Biol. **25**, 443–463 (2024). https://doi.org/10.1038/s41580-024-00702-6
42. Si, D., et al.: Artificial intelligence advances for de novo molecular structure modelling in cryo-electron microscopy. An advanced review. WIREs Comput. Mol. Sci. **12**, e1542 (2022). https://doi.org/10.1002/wcms.1542
43. Singh, R.P., et al.: Significant applications of virtual reality for COVID-19 pandemic. Diabetes Metab. Syndr. **14**, 661–664 (2020). https://doi.org/10.1016/j.dsx.2020.05.011
44. Somaini, A.: Algorithmic images: artificial intelligence and visual culture. Grey Room **93**, 74–115 (2023). https://doi.org/10.1162/grey_a_00383
45. Szendy, P.: The Supermarket of the Visible: Toward a General Economy of Images (2017). Fordham University Press, New York (2019)
46. Todd Lopez, C.: High-resolution virus pictures help researchers develop vaccines. US Department of Defense, March 17 (2020). https://www.defense.gov/News/Feature-Stories/Story/Article/2115411/high-resolution-virus-pictures-help-researchers-develop-vaccines/
47. Torralba, A., et al.: Foundations of Computer Vision. The MIT Press, Cambridge, MA–London (2024)
48. Turonová, B., et al.: In situ structural analysis of SARS-CoV-2 spike reveals flexibility mediated by three hinges. Science **370**(6513), 203–208 (2020). https://doi.org/10.1126/science.abd5223
49. Tyfield, D., et al. (eds.): The Routledge Handbook of the Political Economy of Science. Routledge, London (2017)
50. Watson, B.: Light: A Radiant History from Creation to the Quantum Age. Bloomsbury, New York–London (2016)
51. Yao, H., et al.: Molecular architecture of the SARS-CoV-2 virus. Cell **183**(3), 730–738 (2020). https://doi.org/10.1016/j.cell.2020.09.018
52. You, T., et al.: First pictures of China's killer coronavirus: microscopic images show the deadly new virus extracted from patients in Wuhan. MailOnline, January 27, 2020. https://www.dailymail.co.uk/news/article-7934091/Microscopic-images-capture-strains-new-coronavirus-extracted-patients.html
53. Zuboff, S.: The Age of Surveillance Capitalism: The Fight for a Human Future at the New Frontier of Power. Profile Books, London (2019)

Ruggero Eugeni is full professor of Media Semiotics at Università Cattolica del Sacro Cuore, Milan, where he also serves as Head of the Department of Communication and Performing Arts. He has held visiting professorships at Université Sorbonne Nouvelle—Paris 3, Sorbonne Université—Paris 4, and Goethe-Universität Frankfurt am Main. He has led several research projects, including *Subjective Experience and Estimation of Moving-Image Time* (2016–2020) and *Movement Experience Through Rhythmic Organization in Audiovisual Representational Texts* (2023–2025). His more recent publications include *Neurofilmology: Audiovisual Studies and the Challenge of Neurosciences* (with A. D'Aloia, 2014), *#Intelligence*, a special section of *NECSUS:*

European Journal of Media Studies (with P. Pisters, 2020), and *Algorithmic Images: Postmedia Dispositives and the New Political Economy of Light* (Taylor & Francis, forthcoming). Further information about his research and academic contributions can be found on his website www.ruggeroeugeni.com.

Chapter 8
From Windows to Databases: The Visual Politics of Facial Recognition

Samuel Solé

Abstract Drawing on visual culture and surveillance studies, this chapter provides readers with the fundamentals of facial recognition technologies (FRTs), their operating principles, ideological foundations, and socio-technical implications. In particular, it examines the paradigm shift induced by these technologies in the political and visual regime of images. First, the chapter explores FRTs' technical functioning, detailing the five-stage process from image acquisition to facial recognition. Next, it traces their historical development and deployment from early semi-automated systems to contemporary deep learning applications, highlighting how vested interests and positions shaped their implementation. In addition, it situates FRTs within a broader genealogy of face-reading techniques, examining how they perpetuate discriminatory operations that disproportionately affect already marginalised populations. Furthermore, it analyses how facial images have transitioned from instrumental to operational status in networked surveillance, critically altering the relation between face and identity. Finally, it discusses artistic counterstrategies and obfuscation techniques that challenge not only facial surveillance but also our basic understanding of the face itself.

Keywords Facial recognition · Biometric surveillance · Operational image · Anti-recognition mask · Visual culture studies

The author expresses his gratitude to Marie-Paule Cani, Professor of Computer Graphics at the École polytechnique, and Assia Wirth, Ph.D. candidate in Sociology at the ENS Paris-Saclay, for their valuable insights regarding the technical aspects of FRTs. His thanks also extend to Annette Sojic, Professor of Practice of French at Tulane University, for her careful proofreading, which greatly enhanced the linguistic quality of this chapter.

S. Solé (✉)
École normale supérieure, Paris, France
e-mail: samuel.sole@ens.psl.eu

Sorbonne Nouvelle University, Paris, France

P. Conte et al. (eds.), *Algomedia. The Image at the Time of Artificial Intelligence*, Lecture Notes in Morphogenesis, https://doi.org/10.1007/978-3-032-08726-3_8

8.1 Introduction

Facial recognition technologies (FRTs) generate considerable discussion and controversy in the public sphere, yet often remain poorly understood due to their wide variety of techniques, methods, and approaches. A precise definition is thus warranted. In their most elementary form, FRTs refer to biometric systems designed to identify individuals on the basis of their unique facial features. The process involves detecting a face in an image and comparing it to one or more faces previously identified in a database. While research into facial recognition dates back to the 1970s, its development and deployment have drastically accelerated since the 2000s, driven by the revival of neural networks and deep learning techniques, which had been sidelined for decades in favour of rule-based AI systems. Unlike other biometric systems such as fingerprinting or voice recognition, FRTs have the distinctive advantage of operating remotely, passively, and unobtrusively. This makes them particularly well suited to mass surveillance, but raises serious legal and ethical concerns about privacy, individual freedoms, and asymmetrical power relations.

In biometric surveillance, the face serves both as an identity marker and a data source to detect, identify, and monitor individuals. FRTs thus appear to be the privileged instrument of a "face politics." As Jenny Edkins argues, the face is permeated by a constant tension. On the one hand, it is still considered the "windows of the soul," which distinguishes us as human beings and grounds our ethical relation with others. On the other hand, FRTs break down this elusive and irreducible singularity into merely quantifiable and comparable data [12]. As Hans Belting emphasises, "the face as the only mirror of identity has become abstract and numerical, even when it serves merely as a storehouse of information" [4: 198]. From the perspective of visual culture studies, facial recognition raises a major issue, as it ushers in a paradigm shift in the visual regime of facial images. These are no longer intended to be visualised and interpreted by humans but automatically processed as massive datasets in machine-to-machine networks. How can we approach this political and visual economy of the face, where it becomes the site of both the expression of individual uniqueness and data production used for public surveillance?

8.2 How Does Facial Recognition Work?

Before examining their political and social implications, it is first necessary to consider the technical functioning of FRTs. The process usually comprises five distinct but interdependent stages: (1) acquisition, (2) detection, (3) normalisation, (4) extraction, and (5) recognition [22: 98]. The first stage involves capturing an image with a sensor such as a camera.[1] Images may present variations in lighting,

[1] FRTs involve various sensors centered around two criteria: the light spectrum (visible, infrared, or thermal) and the visualisation method (2D, 3D, or videos). Many methods exist, but most FRTs use 2D images in the visible spectrum due to their accessibility in existing databases, simplicity

camera angle and distance, facial orientation and expressions, as well as occlusive elements like beards, makeup, and glasses. These variations can affect system performance, which is highly dependent on the quality of data acquisition. The second stage involves detecting and locating one or more faces in the image. While early FRTs rely on rule-based AI, recent methods like deep neural networks enable systems to develop the most optimal solutions for face detection by themselves. This self-learning capability enables greater adaptability and efficiency by uncovering properties and correlations that have not been explicitly programmed by human operators.

Once the face has been detected, the system normalises the image, eliminating unwanted variations in framing, lighting, and contrast. The purpose is to produce standardised images for easy comparison. Spatial normalisation aligns facial features with predefined markers through geometric transformations such as scaling and rotation. Face extraction isolates the region containing the face and removes irrelevant elements such as the rest of the body and the background. Intensity normalisation adjusts grey levels to reduce variations in lighting and contrast. After this stage, the process can move on to feature extraction. In earlier methods, developers manually indicated all distinctive facial markers in the images. More sophisticated methods automatically encode relevant features as vector representations called "facial vectors." A facial vector is the unique digital imprint of an individual's face. All these vectors are projected into a multi-dimensional space where the closest vectors correspond to similar faces and the most distant vectors to different faces.

Deep learning methods have revolutionised facial recognition with algorithms that learn hierarchical representations of facial features on their own and show robust invariance to image variations [38]. The extraction process operates as a multi-layer recognition system, analogous to human visual perception. The system, initially devoid of any morphological knowledge, is exposed to a substantial corpus of images that enables it to gradually adjust its neural weights by identifying the essential features of facial morphology. Facial images pass through a series of filters in increasingly deeper layers of neurons. The first layers initially identify basic elements: contours, shadows, and textures. Intermediate layers assemble these elements into more distinctive structures: eyebrow arches, eye shapes, nose tip, and lip contours. Deeper layers integrate these elements into complex relational configurations: distance between eyes, proportion of forehead and cheeks, and facial symmetry. The final layers ultimately produce a multi-dimensional vector, which encodes abstract facial features.

The final stage of the process, facial recognition as such, establishes a correspondence between two facial vectors. The system compares the facial vector of an unknown individual with other facial vectors previously encoded and stored in a database. A similarity threshold is predefined to adjust the algorithm's error tolerance. If the similarity between two vectors exceeds the threshold, the system concludes that they are probably the same individual. FRTs have two distinct uses: verification

---

of computational processing, and possibilities for implementation in already deployed surveillance infrastructures such as CCTV systems [17: 126].

and identification. Verification is based on one-to-one correspondence. The system compares the image of an unknown face to one or more images of the same face to confirm an individual's identity (see smartphone user authentication like Apple's Face ID). By contrast, identification is based on one-to-many correspondence. The system compares the image of an unknown face to millions of already referenced faces. If the same individual appears in the database, their identity can be established. While verification answers the question "Are you really who you claim to be?", identification instead answers the question "Who are you among all these people?".

8.3 Development and Deployment of Facial Recognition

The history of FRTs should not be understood as a neutral and linear scientific progress, but rather as a complex intertwining of socio-technical forces, historical events, and political orientations. Their origin dates back to the 1960s and was intrinsically linked to the military-industrial agenda and national security concerns of the United States during the Cold War. Woodrow Wilson Bledsoe, co-founder of Panoramic Research, was commissioned by the CIA in 1963 to adapt his character recognition system to facial recognition. While innovative, his method was only semi-automatic: a human operator had to pinpoint relevant elements on facial images using a RAND Tablet. Following this manual input, a computer calculated the distances between the points and produced a unique biometric identifier allowing for quantitative comparison [6]. Only the data comparison was automated, with facial feature extraction still being handled manually.

What paved the way for autonomous facial recognition in the 1970s was the work of Takeo Kanade, then a doctoral student at Kyoto University. He developed a program capable of extracting facial properties without human intervention. The process began with analysing a greyscale front-face photograph from which only the contours were preserved. The algorithm relied on a generic face model to locate the positions of relevant facial markers. The computer then calculated sixteen distinct parameters—distances and angles between the markers—to create a unique digital vector for the face, which could then be compared with others [18]. This method achieved remarkable results for the time (75% correct identification) but suffered from significant limitations: the database was restricted to young Japanese men without facial hair and glasses, which limited the model's performance in recognising female or non-Japanese individuals. As we will see later, such methodological orientations and non-representative databases have contributed and still contribute to the perpetuation of social stereotypes and discriminatory operations.

The 1980s are known as the "AI winter," marked by technical limitations and investment reluctance, resulting in a slowdown in machine vision. As Kelly Gates explains, research moved toward less ambitious goals, favouring "optimized human-computer divisions of perceptual labor" [14: 32], where systems were designed to distribute recognition tasks between humans and computers, rather than pursuing fully automated solutions. It was not until the 1990s that facial recognition gained renewed interest with the introduction of the eigenfaces technique at the MIT Media Laboratory. Unlike early techniques based on discrete analysis of facial properties, this holistic approach treats the face as a global pattern whose relevant properties emerge statistically from the data itself [3: 40]. It generates an average face from the synthesis of a corpus of facial images and, using principal component analysis (PCA), it encodes individual variations of each face relative to this average face as vectors called eigenfaces. All these faces are then projected into a multi-dimensional space, in which their coordinates are determined by the weight of each eigenface in their composition. Facial recognition is performed by projecting an unknown face and determining the closest known face in the virtual space [36].

Building on these technical innovations, DARPA launched the FERET program in 1993 within the context of the war on drugs [14: 25–62]. This initiative aimed to provide a standardised database and rigorous evaluation protocols to test FRTs' performance in real-life environments. Beyond enabling scientific advances, the program legitimised FRTs in political and media discourses and played a critical role in their commercial uses in the United States. Three companies emerged as specialists in the field: Visionics Corporation, Miros Inc., and Viisage Technology, the latter having acquired the exploitation rights for the eigenfaces technique. The first real-life applications rapidly followed. Driver's licence offices in West Virginia and New Mexico implemented FRTs in 1997 to prevent the fraudulent acquisition of multiple licences under different identities. At the same time, the dramatic increase in prison population led to the FRTs' integration into the carceral system for inmate tracking and management across jurisdictions.

The 9/11 attacks provided political justifications for the accelerated deployment of FRTs, as media discourses spread the idea that facial recognition could have prevented this tragedy [14: 97–124]. The Smart CCTV experiment in Ybor City, launched three months earlier, illustrates this shift in public opinion: concerns about privacy protection and the risks of mass surveillance were rapidly superseded by security discourses advocating for FRTs' implementation in sensitive areas like borders and airports. Technical advances, such as real-time facial detection, responded to these new operational requirements.[2] The strategic alliance between the technology industry and the security sector resulted in a substantial homeland security economy. Increasingly prevalent themes and expressions like "asymmetric threats" and "unidentifiable enemies" reinforced the perception that FRTs were necessary for national security. Under the guise of technical neutrality, political and security practices implemented a "facialisation of terrorism," targeting specific ethnic and religious groups, mostly Arabs and Muslims, associated with profiles considered to be at risk.

---

[2] For example, the Viola-Jones algorithm [37] enables facial detection in real-time video. It is based on four key elements: Haar features that identify relevant contrasts in the image, the method of integral image to accelerate calculations, the AdaBoost algorithm that selects discriminative features, and a cascade of classifiers that quickly rejects regions of the image without faces.

Advances in deep learning in the 2000s and 2010s intensified the deployment of FRTs for commercial, military, and security purposes [1]. As early as 2010, Facebook implemented FRTs for automatic identification in images, processing over 350 million images daily, before developing DeepFace in 2014 with 97% accuracy, comparable to human performance. In 2017, Apple introduced Face ID on the iPhone X, the first consumer use of 3D identification with an error rate of 1/1,000,000. The FRTs market, estimated at $4.4 billion in 2019, is expected to exceed $10.9 billion by 2025. Meanwhile, facial surveillance increased in countries like China with the Skynet program aimed at identifying fugitives involved in financial crimes. The first judicial decisions were made, such as Bridges v. South Wales Police in 2020, when British courts ruled that South Wales Police's use of FRTs was unlawful due to inadequate legal safeguards, lack of proper data protection assessments, and failure to consider its discriminatory impact. More recently, Ukraine has used the Clearview AI software to identify deceased Russian soldiers and subsequently contact their families. This practice "might be interpreted as psychological violence towards the family members or even psychological warfare" [21: 115] and thus raises major concerns about the application of FRTs in military contexts.

As this non-exhaustive overview of facial recognition history demonstrates, priority areas of research and applications are not determined by technical possibilities alone, but shaped by many specific interests, whether governmental (public surveillance, national identification systems), military (access to restricted areas, tactical intelligence), commercial (biometric authentication, contactless payment, targeted advertising), and now commonplace consumer applications (images filters, smart TV, video games, and virtual reality). FRTs thus emerge at the intersection of multiple economic, political, social, and technical forces, including post-9/11 national security concerns, the profit-driven logic of corporate entities, and legislative frameworks that vary in restrictiveness according to national contexts. Citizen protests, court decisions, and ethical debates in recent years demonstrate that FRTs do not constitute an inevitable outcome of scientific progress but rather result from political decisions and vested interests that can be questioned, negotiated, and reoriented.

8.4 Face-Reading Techniques and Discriminatory Operations

While emerging in a specific socio-technical context, facial recognition cannot be understood without placing it within a wider genealogy of face-reading techniques, whose ideological foundations deserve to be questioned. FRTs perpetuate an epistemic stance whose origins can be traced back to physiognomy in the eighteenth century, a pseudoscience popularised by Johann Caspar Lavater, which postulated direct correspondences between facial morphology and moral personality. This idea has its roots in ancient Greece and was a major cultural practice in modern Europe, infiltrating all ideological, philosophical, and religious spheres of the period [30]. This hermeneutic framework was perpetuated by phrenology and craniometry in the nineteenth century, the latter being an anatomical measurement method aimed at determining the intelligence of human populations based on the bone structure of the skull. The methodological biases and racist assumptions of this method, however, have since been strongly criticised.

As it combines face-reading techniques and optical devices, facial recognition would be more precisely situated in the lineage of 19th-century anthropometry, mobilising the photographic apparatus, not as an unbiased instrument of representation, but as a power apparatus in specific institutional contexts such as justice, medicine, and science [35]. Unlike physiognomy, which relies on psychological interpretations, FRTs are more closely related to Galtonian and Bertillonian anthropometry, as they are based on a systemic and quantitative approach.[3] Notably, the techniques used by Francis Galton and Alphonse Bertillon corresponded to two distinct ideological and methodological paradigms, distinguished by Allan Sekula as "essentialist" and "nominalist" [32]. On the one hand, Galton's method, known as "composite portraits," consisted of superimposing several portraits of the same category of individuals—the insane, criminals, delinquents—to obtain the synthetic image of a social group. On the other hand, Bertillon's method allowed for measuring, describing, and classifying individual facial properties according to a precise nomenclature. Unlike Galton, who aimed to reduce faces to essentialised categories, Bertillon sought to make faces measurable, comparable, and identifiable.

Early anthropometric methods established a logic of face-oriented surveillance, police archiving, and database management that persists in contemporary facial recognition. One initial interpretation might align Bertillon's method with feature analysis approaches and Galton's method with holistic approaches: the former relies on the discretisation of facial images into separate elements, while the latter depends on the global synthesis of diverse facial images. Some scholars have aptly established a comparison between Galton's composite portraits and the eigenfaces technique, which rest on the same "reductive and statistical way of seeing" [19: 99]. They also share a striking visual similarity: ethereal faces with blurred contours where the eyes, nose, and mouth appear more distinctly than the rest, conveying that singular impression of a "statistical ghost." Yet it should be noted that the purposes of the two methods differ radically: while the composite portrait aims to make visible an imaginary collective essence, the average face in the eigenfaces technique is not an end in itself but a point of reference from which to derive individual variations.

From the perspective of modern biological anthropology, classifying faces according to racial or ethnic characteristics has no scientific grounding, since the majority of facial variations occur not between different populations but between individuals within the same population [15: 41]. By this principle, facial recognition

[3] FRTs aim solely to extract the facial features necessary for individual identification, not to infer information about the moral character of individuals. Related techniques, such as facial characterisation, are oriented toward attributing emotional states or behavioral intentions. While presenting similarities with physiognomic postulates, facial characterisation is distinct from facial recognition, strictly speaking.

is not primarily intended to distinguish between human groups but between individuals, thus aligning with an individualising logic (like Bertillon) instead of a typologising one (like Galton). This is not to say that FRTs are free from racial or ethnic biases: other purportedly unbiased biometric methods, such as fingerprinting, have historically contributed to the differential treatment of individuals and the stigmatisation of human populations based on race or ethnicity [10]. This holds true for more recent technologies, as Shoshana Amielle Magnet argues: "Biometric technologies, which claim to eliminate racial profiling through mechanical objectivity, simultaneously are explicitly based upon assumptions that categorize individuals into groups based on phenotypical markers of racialization and used in the implementation of programs specifically aimed at racial profiling" [25: 47].

Gender, race, and ethnicity biases are common in FRTs due to initial design choices and specific political applications.[4] As many scholars have demonstrated, biometric surveillance conveys the logics of racial segregation and whiteness normalisation, while maintaining historically unequal power relations [5, 7, 31]. A major study by Joy Buolamwini and Timnit Gebru measured these discriminatory operations by evaluating three commercial software products. The results reveal considerable accuracy disparities at the intersection between gender and race: dark-skinned women have the highest error rates (up to 35%), while light-skinned men benefit from the lowest error rates (less than 1%). Classifiers function less effectively on female faces than on male faces and on dark skin than on light skin [9]. Inaccuracies in facial recognition have serious consequences, with African Americans both more targeted by police surveillance and less accurately identified by FRTs. The case of Robert Williams, wrongfully arrested after an erroneous identification in Detroit in 2020, and more recently, Michael Oliver, Nijeer Parks, Randal Reid, and Porcha Woodruff in the United States, confirm that FRTs exacerbate the vulnerabilities of populations already disproportionately targeted by police harassment.

8.5 Operational Faces in Networked Surveillance

Given the clear disparities in facial recognition accuracy, many experts advocate for more diverse datasets and greater algorithmic transparency. While this is likely a necessary response, it remains fundamentally insufficient. A theoretically unbiased algorithm might seem socially acceptable, but it would preserve the process of mass surveillance and the biometric exploitation of faces as data production sites. This discursive position suggests that the problems associated with FRTs can be resolved through simple technical adjustments, without questioning their ideological foundations and socio-political implications. Critical discourse cannot be limited

[4] An accepted explanation for FRTs' failures is the use of non-representative databases, predominantly composed of white males. The algorithms are less trained to differentiate women and racialised persons, which consequently results in false positives or false negatives. Moreover, data acquisition and lighting conditions are often optimised for light skin, which can exacerbate this phenomenon.

to identifying the discriminatory operations of facial recognition but must extend to the underlying conception of the face that it presupposes. The question is not whether one type of face is less readable than another, but whether machines should be able to read any face at all, and what it does to a face to be machine-readable.

While reflecting preexisting stigmas and stereotypes, facial recognition contributes more broadly to the advent of what David Lyon has termed the "surveillance society" [24], a concept widely debated in surveillance studies over recent decades. Biometric technologies have profoundly reconfigured surveillance paradigms and practices, becoming increasingly "dispersed, pervasive, fluid, and invisible" [34: 1]. They are deployed jointly by state institutions, corporations, and various organisations, "closely articulated with some major modes of governance, through its capacity for social sorting" [23: 181], that is, the differentiated treatment of citizens based on their personal data. This new surveillance is organised into distributed networks of human and non-human agents, including computers, mobile devices, programs, sensors, and video cameras [26]. The opacity and invisibility of these surveillance devices make social protest difficult, while their existence is normalised in political and media discourses as a necessary prerequisite for national security [16: 90].

In this automated, distributed, networked surveillance, facial images pass through a pipeline of computational actions that significantly alter their visual regime. As Btihaj Ajana points out, all biometric technology aims to transform the body (or the face) "into an informational object, a readable text (or rather *palimpsest*) for statistical (re)measurements and data storage" [2: 90]. However, this text-body is clearly less and less readable by the human gaze in the age of machine vision.[5] As mentioned above, early FRTs rely on an analytical approach aimed at manually decomposing faces into discrete elements and geometric relations, in a way similar to traditional anthropometry. Nevertheless, holistic approaches and deep neural networks represent a major epistemic breakthrough, since facial processing is delegated entirely to machine vision. Algorithmic perception widely differs from our conception of what a face looks like. Machines do not "perceive" faces as coherent gestalts but as statistical distributions and multi-dimensional vectors that are completely beyond human perception.

What emerges here illustrates a paradigmatic shift in the status of facial images: from instrumentality toward operationality. "Instrumental" images are visualisation and calculation tools for representing the world and intervening in it, such as aerial images from the First World War [33]. Conversely, "operational" (or "operative") images, as defined by filmmaker and media theorist Harun Farocki, are non-visual images, generated, processed, and manipulated in machine-to-machine networks, without including human operators in the interpretive and decision-making process [13]. As Jussi Parikka observes, they imply "the primacy of action and function

[5] Vector representations generated by deep neural networks remain complex and resist direct, intuitive interpretation. Various visualisation techniques, however, have been developed to obtain partial glimpses into the perceptual and decisional mechanisms of algorithms. For example, feature activation maps help identify which spatial regions of an input image induce a significant response from neurons in the network.

instead of a picture to be seen and interpreted for meanings" [29: 12]. In FRTs, facial images become operational: they are not intended to be viewed and interpreted but to exercise control over bodies and spaces. The vision process is distributed in an integrated architecture of hardware and software, such that no image can be grasped as a fixed entity from a single viewpoint but as a residual and dispersed flow of data in multi-dimensional spaces. What matters here is not the image and its pictorial value, but the network of images and their transversal relations.

Drawing on the concept of operational images, Roland Meyer identifies a shift in the making and sharing of facial images. He notes the emergence of "operative portraits," which induce an inversion of the traditional portrait logic: "Rather than representing an individual whose autonomous individuality precedes the act of representation, the billions of digital images of faces circulating in social networks have become operative images, fuelling the ongoing production of digital identities" [27: 23]. Although the author focuses here on social media, the inversion of the image-identity relation also manifests itself in FRTs in a specific way. Instead of representing pregiven identities, facial images play an active role in the production and imposition of identities, assigning them arbitrary taxonomies related to gender, race, ethnicity, social class, or sexual orientation.[6] They are power apparatuses that perform a process of identification whereby identities are constructed as an ideal objective or an imaginary referent, and not as preexisting realities.

From a Levinasian perspective, facial recognition radically contravenes the "ethics of the face" and the "straightforwardness of the face-to-face." For Emmanuel Levinas, the face is not a simple object of perception but the epiphany of the Other, irreducible to categorisation: "The face is present in its refusal to be contained. In this sense it cannot be comprehended, that is, encompassed" [20: 194]. Conversely, FRTs are precisely designed to comprehend and encompass the face within organised and rigorous taxonomies. The bilateral face-to-face disappears in favour of unilateral surveillance, replacing the moral imperative "thou shalt not kill" with the techno-political injunction "thou shalt be identified." Whether they break faces down into discrete elements or apprehend them as statistical wholes, FRTs contradict the very assertion that "the face resists possession, resists my powers" [20: 197]. The face no longer resists powers but becomes a tool of power that produces and imposes identities, rather than manifesting an irreducible otherness. In Levinasian terms, "infinity" designates the ethical openness and radical exteriority manifested in the face of the Other, while "totality" refers to closed ideological systems aiming at assimilating otherness into sameness. Facial recognition thus materialises as a concrete implementation of this totalising position, which denies the infinity of the face and reduces the Other to the Same, difference to identity, infinity to totality.

6 A controversial Stanford study claimed to be able to identify sexual orientation from facial images with an accuracy rate supposedly higher than that of humans. This research illustrates how facial recognition can assign identity categories based on dubious statistical correlations and essentialist presuppositions about the bodily expression of sexuality.

8.6 Anti-recognition Masks and Obfuscation Strategies

In response to biometric surveillance, many artists and designers developed anti-recognition devices aimed at disrupting automated identification systems. Examples include patterned clothing, camouflage makeup, or distorting masks. They all rely on "obfuscation strategies," that is, the deliberate production of ambiguous, confusing, or misleading signals to diminish the reliability of data collection systems [8]. Despite the media hype surrounding these devices, they have sparked intense scholarly debate regarding their political effectiveness. For instance, Torin Monahan contends that "the works enact a play of individual avoidance, rather than asserting a collective right to look back and challenge authority." He further argues that "many of them reproduce discourses of universalism that elide difference as well as the marginalizing and discriminatory effects of surveillance" [28: 162]. Such critiques frame anti-recognition devices as manifestations of a frivolous and fashionable mode of resistance, which allegedly fail to propose effective, sustainable solutions for individuals who directly encounter algorithmic oppression in their lived experiences (Fig. 8.1).

However, Zach Blas's *Facial Weaponization Suite* (2012–2014) offers a compelling case for providing nuance to this argument. At first glance, Blas' colourful and vibrant masks could exemplify the aestheticisation of resistance pointed out by some critics. However, a closer examination of their production process and material properties reveals a more complex engagement with the politics of visibilisation and identification. The project comprises four series of coloured plastic masks, each representing an oppressed collective subject: pink masks for the queer community, black masks for African Americans, blue masks for feminist movements, and grey masks for migrants and refugees at the US-Mexican border. Crucially, the masks result from a collaborative process involving numerous participants from communities directly affected by biometric surveillance. Blas collected their facial data using a

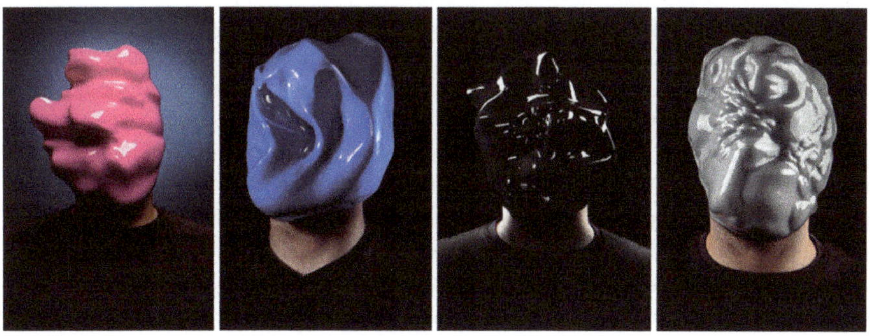

Fig. 8.1 From left to right: Fag Face Mask, October 20, 2012, Los Angeles, CA. Mask, May 31, 2013, San Diego, CA. Mask, November 20, 2013, New York, NY. Mask, May 19, 2014, Mexico City, Mexico. From Zach Blas, Facial Weaponization Suite, 2012-2014. Painted, vacuum-formed recycled polyethylene terephthalate. Courtesy of the artist

Kinect camera, then merged them into abstract masks using 3D modelling software, and eventually materialised them through 3D printing. The artist and participants subsequently organise collective actions and performances in which they appear publicly wearing these masks, thereby drawing attention to the dangers of biometric surveillance. This collective dimension counterbalances claims of "individual avoidance," revealing instead a communal resistance embedded in the visual composition and performative function of the masks.

Due to this amalgamation process, Blas's masks bear no resemblance to individual faces, but manifest as tumescent, undulating, amorphous surfaces, devoid of recognisable facial features. No eyes, nose, or mouth are discernible: only continuous, uninterrupted fields punctuated by random protrusions and depressions. In this way, they directly challenge the operating principles of FRTs by intentionally obscuring the organisation of the face upon which identification algorithms typically depend. But unlike other anti-recognition devices that directly interfere with algorithms through disruptive patterns, Blas's masks operate in a much simpler way by entirely covering the wearer's face. FRTs cannot operate, not because of sophisticated obfuscation techniques, but because there is simply no face to recognise. Blas's masks thus cannot be categorised as anti-recognition, as any other mask—from Guy Fawkes masks to Zapatista balaclavas—could serve the same function. What distinguishes them, however, is that they do not merely oppose facial recognition but engage in a critical reflection of its underlying principles through parodic inversion: they synthesise multiple faces and produce images that resist human legibility, thus mirroring and subverting the very same operations performed by FRTs.

One must engage with the Deleuzian conceptual framework to fully grasp the political and artistic significance of this operation of defacement. The material composition of Blas's masks—flowing contours, seamless surfaces, and organic fluidity—explicitly evokes Gilles Deleuze and Félix Guattari's theoretical distinction between "striated" and "smooth" spaces. Blas's masks correspond to the smooth space, that is, a topological field permeated with continuous variations, intensive movements, and tactile sensations. Their smoothness stands in direct opposition to the striated space, which is regulated by reference points, demarcation lines, and measurable distances. Although Deleuze and Guattari do not explicitly connect these concepts, one might argue that striated and smooth spaces correspond to the notional pair of "face" and "head." In Deleuzian terms, the face is a surface organised by metric coordinates and points of reference that enable categorisation, identification, and political domination: in other words, a striated space. By contrast, the head is an obvious example of smooth space: an unruly materiality that precedes signification, resists identification, and remains open to becomings. In this sense, I argue that Blas's masks deliberately transform the striated space of the face into the smooth space of the head. Echoing the function of the Greek prosopon, Blas's masks serve less to conceal than to reveal—not to conceal the face but to reveal the head.

For Deleuze and Guattari, the head precedes the face, and the transition from one to the other is enacted by a "faciality machine." This machine, semiotic rather than physical, designates the socio-political forces that transform the biological materiality of the head into the culturally encoded meaning of the face. "Concrete faces cannot be assumed to come ready-made. They are engendered by an *abstract machine of faciality*, which produces them at the same time as it gives the signifier its white wall and subjectivity its black hole" [11: 168]. This machine operates through a dual process: first, it produces faces by organising features into a recognisable system of signs, and second, it evaluates these faces against normative standards, accepting some while rejecting others that "do not conform, or seem suspicious" [11: 177]. We cannot overlook the striking homology between facial recognition and the faciality machine. Both systems, whether computational or conceptual, decompose faces into decipherable surfaces and recompose them into assignable identities. Both order a binary judgment according to an established norm, namely "the White-Man face," in relation to which "racism operates by the determination of degrees of deviance" [11: 178]. Despite being formulated decades before current biometric systems, the faciality machine somehow anticipates FRTs, which similarly discriminate against non-white faces that deviate from algorithmically encoded whiteness.

By undoing the face and returning to the head beneath, Blas's masks reverse the process of the faciality machine. They disrupt this power apparatus by dissolving facial features into unorganised, unquantifiable surfaces. Importantly, Blas not only neutralises facial recognition systems but more fundamentally challenges the underlying process of *facialization* that assigns arbitrary categories and identities to individuals based on facial features—operations that long precede biometric systems. Far from engaging in a merely playful resistance, Blas's masks constitute a genuine politics of collective becomings that undermines the semiotic forces that generate measurable and identifiable faces. By merging multiple faces into amorphous surfaces, Blas's masks embed a collective identity that transcends individual identification, thus concretising Deleuze and Guattari's political project, "to escape the face, to dismantle the face and facialisations, to become imperceptible, to become clandestine," by making "*faciality traits* themselves finally elude the organization of the face" [11: 171]. This explains why, unlike other anti-recognition devices that preserve the visibility of the wearer's face, Blas's masks completely cover and opacify it. The strategy is not about disrupting facial recognition but about disrupting its very condition of possibility, the face itself, as a power construct of the faciality machine.

8.7 Conclusion

This chapter has attempted to demonstrate how facial recognition marks an epistemic rupture in the political and visual regime of facial images. From instrumental images, intended for human visualisation and interpretation, they have become operational, circulating as massive data flows in automated, distributed, networked surveillance. In this context, facial images are coincidentally invisible to human vision and hypervisible to machine vision. Artistic counterstrategies, such as Blas's faceless masks, tend to reverse the underlying logic of this new mode of visibility. They render faces simultaneously invisible to human and machine visions, thus undermining all identification processes, whether human or machine-based. Instead of superficially contesting facial surveillance, they attack its very foundation: the face itself, understood as a power apparatus for identification and categorisation. The challenge is therefore not simply to "save face," nor to rehabilitate it as the site of individuality and humanity, but rather to reconfigure it as an uncertain and opaque territory, conducive to the formation of collective becomings. Against the politics of the face thus emerges a politics of defacement, opposed to the logic of facialisation that underlies facial surveillance.

References

1. Adjabi, I., Ouahabi, A., Benzaoui, A., Taleb-Ahmed, A.: Past, present, and future of face recognition: a review. Electronics 9(8), 1188 (2020)
2. Ajana, B.: Governing Through Biometrics: The Biopolitics of Identity. Palgrave Macmillan, London (2013)
3. Akbari, A.: Facial recognition technologies 101: technical insights. In: Matulionyte, R., Zalnieriute, M. (eds.) The Cambridge Handbook of Facial Recognition in the Modern State, pp. 29–43. Cambridge University Press, Cambridge (2024)
4. Belting, H.: Face and Mask: A Double History. Princeton University Press, Princeton (2017)
5. Benjamin, R.: Race After Technology: Abolitionist Tools for the New Jim Code. Polity Press, Cambridge (2019)
6. Bledsoe, W.W., Chan, H.: A Man-Machine Facial Recognition System: Some Preliminary Results, Technical Report PRI 19A. Panoramic Research Inc., Palo Alto (1965)
7. Browne, S.: Dark Matters: On the Surveillance of Blackness. Duke University Press, Durham, London (2015)
8. Brunton, F., Nissenbaum, H.: Obfuscation: A User's Guide for Privacy and Protest. MIT Press, Cambridge (2015)
9. Buolamwini, J., Gebru, T.: Gender shades: intersectional accuracy disparities in commercial gender classification. Proc. Mach. Learn. Res. 81, 77–91 (2018)
10. Cole, S.A.: Suspect Identities: A History of Fingerprinting and Criminal Identification. Harvard University Press, Cambridge (2001)
11. Deleuze, G., Guattari, F.: Year zero: faciality. In: A Thousand Plateaus: Capitalism and Schizophrenia, pp. 167–191. University of Minnesota Press, Minneapolis (1987)
12. Edkins, J.: Face Politics. Routledge, London, New York (2015)
13. Farocki, H.: Phantom images. Public 29, 12–22 (2004)
14. Gates, K.A.: Our Biometric Future: Facial Recognition Technology and the Culture of Surveillance. New York University Press, New York (2011)
15. Henneberg, M., Simpson, E., Stephan, C.: Human face in biological anthropology: craniometry, evolution and forensic identification. In: Katsikitis, M. (ed.) The Human Face: Measurement and Meaning, pp. 29–48. Springer, Boston (2003)
16. Hintz, A., Dencik, L., Wahl-Jorgensen, K.: Digital Citizenship in a Datafied Society. Polity Press, Cambridge (2019)
17. Jain, A.K., Ross, A.A., Nandakumar, K., Swearingen, T.: Face recognition. In: Introduction to Biometrics, pp. 119–173. Springer, Cham (2025)
18. Kanade, T.: Computer Recognition of Human Faces. Birkhäuser, Basel (1977)
19. Lee-Morrison, L.: Portraits of Automated Facial Recognition: On Machinic Ways of Seeing the Face. Transcript, Bielefeld (2019)
20. Levinas, E.: Totality and Infinity: An Essay on Exteriority. Duquesne University Press, Pittsburgh (2007)
21. Limante, A.: Faces of war: Russia's invasion of Ukraine and military use of facial recognition technology. In: Matulionyte, R., Zalnieriute, M. (eds.) The Cambridge Handbook of Facial Recognition in the Modern State, pp. 112–124. Cambridge University Press, Cambridge (2024)

22. Liu, C., Wechsler, H.: Face recognition. In: Wayman, J., Jain, A., Maltoni, D., Maio, D. (eds.) Biometric Systems: Technology, Design and Performance Evaluation, pp. 97–114. Springer, London (2005)
23. Lyon, D.: Surveillance Studies: An Overview. Polity Press, Cambridge (2007)
24. Lyon, D.: The Electronic Eye: The Rise of Surveillance Society. Polity Press, Cambridge (1994)
25. Magnet, S.A.: When Biometrics Fail: Gender, Race, and the Technology of Identity. Duke University Press, Durham-London (2011)
26. Marx, G.T.: Windows into the Soul: Surveillance and Society in an Age of High Technology. University of Chicago Press, Chicago (2016)
27. Meyer, R.: Operative portraits, or how our faces became big data. In: Geil, A., Jirsa, T. (eds.) Reconfiguring the Portrait, pp. 21–42. Edinburgh University Press, Edinburgh (2023)
28. Monahan, T.: The right to hide? Anti-surveillance camouflage and the aestheticization of resistance. Commun. Crit. Cult. Stud. 12(2), 159–178 (2015)
29. Parikka, J.: Operational Images: From the Visual to the Invisual. University of Minnesota Press, Minneapolis (2023)
30. Porter, M.: Windows of the Soul: Physiognomy in European Culture 1470–1780. Oxford University Press, Oxford (2005)
31. Pugliese, J.: Biometrics: Bodies, Technologies, Biopolitics. Routledge, New York (2010)
32. Sekula, A.: The body and the archive. October 39, 3–64 (1986)
33. Sekula, A.: The instrumental image: Steichen at war. Artforum 14(4), 26–35 (1975)
34. Simon, B.: The return of panopticism: supervision, subjection and the new surveillance. Surveill. Soc. 3(1), 1–20 (2005)
35. Tagg, J.: The Burden of Representation: Essays on Photographies and Histories. University of Minnesota Press, Minneapolis (1988)
36. Turk, M., Pentland, A.: Eigenfaces for recognition. J. Cogn. Neurosci. 3(1), 71–86 (1991)
37. Viola, P., Jones, M.: Rapid object detection using a boosted cascade of simple features. IEEE Comput. Vis. Pattern Recognit. 1, 511–518 (2001)
38. Wang, M., Deng, W.: Deep face recognition: a survey. Neurocomputing 429, 215–244 (2021)

Samuel Solé is a lecturer in Film and Media Studies at the Sorbonne Nouvelle University and a Ph.D. candidate in Art History and Theory at the École normale supérieure in Paris. Situated at the interdisciplinary nexus of art history, art theory, and visual culture studies, his current research investigates the resurgence and reconfiguration of the portrait genre in digital art and new media technologies. His multifaceted approach aims to contribute to scholarly discourse on the evolving relations between technical innovations and aesthetic practices in contemporary media environments.

Chapter 9
Prophetic Machines. Algorithmic Media as Late-Capitalism Divination

Anna Caterina Dalmasso

Abstract This essay addresses algorithmic media through the conceptual lens of divination, arguing that artificial intelligence today functions as a prophetic technology and can be understood as the divinatory practice of the fourth industrial revolution. It examines the predictive logic that defines algorithmic systems, not only in applications such as behavioural targeting, user recommendation, or decision-making, but in their fundamental operationality which consists in generating data-driven forecasts based on probabilistic models. The analysis situates this predictive logic within a broader media ecology structured by pre-emption, premediation, and feed-forward dynamics, showing how an anticipatory attitude shapes both technological infrastructures and affective-cultural orientations. It contends that the probabilistic modelling enacted by AI systems not only echoes long-standing practices of foresight but also underpins the foundations of modern science and informs contemporary neuroscientific conceptions of the mind and of human beings as forward-looking agents. By reframing AI as a predictive technology, the essay foregrounds the epistemological and political implications of algorithmic technocultures, raising critical concerns about how algorithmic foresight risks foreclosing futurity itself, and calling for the cultivation of fortune-telling competences.

Keyword Predictive media · Divination · Feed-forward · Premediation · Prophetic culture

The research that led to this essay was developed within the EXT—Coordinated Research Centre on Extended Reality, as part of the "Departments of Excellence 2023–2027" project, awarded to the Department of Philosophy "Piero Martinetti" at the University of Milan by the Italian Ministry of University and Research.

A. C. Dalmasso (✉)
Department of Philosophy "Piero Martinetti", University of Milan, Milan, Italy
e-mail: annacaterina.dalmasso@unimi.it

P. Conte et al. (eds.), *Algomedia. The Image at the Time of Artificial Intelligence*, Lecture Notes in Morphogenesis, https://doi.org/10.1007/978-3-032-08726-3_9

9.1 Divinatory Media

If we were to distil the essence of artificial intelligence (AI) to its core function, it could be described as the ability to make predictions. Whether supposed to recognise and interpret objects or to create new ones, neural networks do nothing but predict a desired output on the basis of the set of data on which they have been trained. By their very algorithmic structure, intelligent machines constantly exercise a capacity for prediction: what we are used to calling machine vision is in fact a form of pre-vision. Although algorithms simulate human perceptive and cognitive capabilities, their way of "seeing," "recognising," or "writing," is not at all similar to human apprehension and manipulation of the environment, but rather a form of probabilistic operationality, which is all the more compelling the more it is able to make accurate predictions. For instance, if a facial recognition system could not see a face, it could easily predict the probability of human-like features being included in a certain visual field, following specific parameters.

Probability plays a crucial role in AI across different stages [20, 32]: first and foremost, deep learning algorithms make predictions and decisions in a data-driven manner, that is, estimating an output through the analysis of vast amounts of data and the identification of meaningful recurring patterns and relationships. But AI does not make deterministic predictions—i.e. saying something will definitely happen —; instead, it uses probabilistic models to assess the uncertainty associated with its predictions and quantify the likelihood of its outcomes.

For example, an AI system would not be able to "draw" a cat, but it could generate a configuration of pixels with a certain probability of containing a cat. In other words, when an AI system produces an image, it does not actually do anything similar to what a human being would do in sketching out a figure. Instead, it makes a probabilistic calculation attempting to produce a grid of pixels which, with a good degree of approximation, will be perceived by humans as an image and, more precisely, as an image with such features and patterns likely to be interpreted as an image of something.[1] Thus, in the case of AI image generation by Generative Adversarial Networks (GANs), in which one neural network (the generator) attempts to create non-pre-existing images and the other (the discriminator) evaluates the outputs as being existing images or generated ones, the generator employs probabilistic methods to produce images that are more likely to "deceive" the discriminator—that is, to achieve the effect that it would not be able to tell whether they are generated or not and so deem them as being already part of the dataset.

Thus, when we formulate a question to elicit a text-to-image interface to generate something, we are prompting its ability to foresee. AI image generation should then be reframed as a form of *ekphrasis* [40], or, we could say, a *predictive ekphrasis*: just as in literature *ekphrasis* involves the detailed description of paintings, sculptures, and other visual artefacts, so that the reader may visualise it as vividly as possible, AI prompts function as accurate portrayals of images that have yet to come, as if the prompt were to instruct the algorithms' imagination. Hence, what appears to us as the

---

[1] About the shift from representation to prediction see Lev Manovich's contribution in this volume.

result of a poietic or mimetic process—the production of a digital object that features the physical appearance of a visual referent—is in fact the result of a probabilistic prognostication [28].

But we need to understand the ability of Generative Adversarial Networks to see into the future also beyond the mathematical probabilistic models which inherently inform their functioning. In fact, we could argue that this aptitude for prediction is not peculiar to algorithmic media but has long characterised contemporary mediality both in its processes and in its emotional-affective tonality. Indeed, the ability to make predictions, besides being the *modus operandi* of machines, could also be understood as an epistemic and anthropological constant informing the relationship between human beings and technologies. In the pages that follow, I will try to show, on the one hand, how this radical anticipatory or *prophetic* capacity of algorithmic media prolongs a trajectory that has already been unfolding in the current mediascape, and, on the other hand, I will propose to investigate AI as a contemporary form of futurology, so as to include it in a broader history of what I would call divinatory media or foreseeing media.

If machine learning is but a device of knowledge, a *nooscope*, as Pasquinelli and Joler [34] have suggested—from the Greek *skopein* "to examine, look" and *noos* "knowledge," that is an instrument to see and navigate the space of knowledge—, it is first of all a device of knowledge of the future, and, as the most recent stage in the trajectory of the mechanisation of reason, AI is carrying out the process of normalising divination as a form of knowledge.

Human beings have always tried to look into the future, to predict it, and even to influence the course of future events. Futurology or *prospective* [19, 30] is an ancient science, evidence of which can be traced in numerous and far-flung cultures. The ability to foresee happenings before they occur can be considered one of the oldest techniques developed by human civilisation, the ability to "read what has never been written" [4: 333] is among the earliest types of reading: reading before any writing, reading signs in entrails, in the stars, and even in the human body.

Technological innovation has always had a close connection with spectral entities and the paranormal [33, 38],[2] as science has always had to do with describing the invisible forces which regulate nature, resembling in this a form of divination. Thus, at least since the culture of attractions which characterised spectatorship during early modernity, media technologies have taken on the role of soothsayers, as attested by the presence of popular coin-operated fortune teller machines [22], which could be found from the late nineteenth century onwards in penny arcades and amusement parks, making clear the connection between automation and divination.

[2] For instance, abilities that now seem absolutely ordinary to us, such as believing in the existence of invisible waves and wireless communication at a distance, have for many centuries been categorised as belonging to the domain of the supernatural. Also, the very term we use to refer to the media is related to the term "medium" as the intermediary between the living and the dead in séances, the person believed to have the ability to communicate with the spirits [7]. Modern media practices share with pseudo-science and ghostly phenomena the goal of communicating at a distance with invisible entities and that of engaging in acts of foretelling.

But, as much as we human beings resort to the media to predict our future actions, the media themselves constantly predict our behaviour through the collection of data derived from our previous actions, from the traces we leave in our *onlife* movements—the virtual/online and physical/offline being inextricably related [16]—in the form of digital paths and biometric data collected by our portable and wearable edge devices. Our attitudes, decisions, and desires—from selecting which TV series to watch, road to take, or product to purchase, up to choosing a sexual partner—increasingly intersect with those of the machines. This predictive logic, which, as mentioned above, underpins the very functioning of AI, informs the operation of major digital platforms that are constitutively engaged in envisaging the behaviour of individuals, their choices, their tastes, as well as predicting complex (social, political, environmental, etc.) dynamics.

Thus, should machine learning be viewed as the modern equivalent of divination within late capitalism technoculture? The form of augury practised at the time of the fourth industrial revolution? To understand algorithmic media and to penetrate the ongoing structural recursivity between data-mining and human conduct, between prosthetic augmentation and sensible machines, it is of utmost importance to understand the characters of this hegemonic predictive attitude and the different forms of prefiguration of the real that underlie it, in order to grasp their genealogy and their anthropological implications. What is the connection between media and prediction? Does it pertain exclusively to technologically enhanced operations, or should it be regarded as an inherent component of human behaviour? And what is the difference between a simple forecast or prognostication and this radical capacity for anticipation brought forth by algorithmic media, which seems to be the key feature of twenty-first-century mediality?

9.2 The Future Belongs to Those Who Can See It[3]

The demand for knowledge about the future seems to be undiminished across times. Across different cultures and epochs, this drive had been met through a multitude of divination techniques, whether based on the observation of omens and the doings of gifted diviners, or involving specific rituals [9]. In this regard, the Romans made a distinction between so-called natural or inductive divinatory methods (*divinatio naturalis*) typical of inspired oracles and portents, as opposed to the artificial or deductive divination (*divinatio artificiosa*), relying on the observation of common events—such as the flight of birds, the entrails of animals, tea leaves, and so on—which could become signs of a supernatural order by the performance of ritual procedures that would then transform their status.

But, while divinatory practices bring us back to the earliest history of mankind and to attempts in various cultures to decipher the secrets of the universe by connecting

[3] The quote is from Barbara Kruger's work *Untitled (The Future Belongs to Those Who Can See It)* (1997).

the macrocosm to an observable microcosm, the age of modern science and the emergence of experimental scientific models were not exempt either from the fascination exercised by the power of prediction. Far from abandoning the idea that the future could be foreseeable, the modern era came to hold prediction as the most advanced form of knowledge. From the seventeenth century onwards, statistics, and in particular the more specific field of probability calculations, played an increasingly important role, focusing on the ability to anticipate future events and to define random variables and their distribution.

In a sense, the entire system of modern science can be viewed as a form of predictive knowledge. The concept of "prediction" lies at the very core of the logic pursued by the positivist model: la *méthode positive* which takes shape during the nineteenth century describes the goal of science as the capacity to predict future conditions, instead of simply researching for causes—according to Comte's formula "*Savoir pour prévoir et prévoir pour pouvoir*" [13: I, 3]. Arguing for the possibility that foreseeing the future could turn into a science, prediction, in its conjectural articulation, progressively came to assume an epistemological status, with a profound influence on various derivative theoretical models, including behaviourism in psychology and game theory in economics and social sciences. Thus, futurology, with its origins in magical thinking, has over the centuries acquired the traits of a science [21, 29], especially in connection with the cultural climate inaugurated by the Cold War and post-Second World War period [2]. As we cannot dissociate the history of science and technology from the previous attempts to dominate natural laws through magic and neglect the profound interconnections between magical thinking and scientific paradigm, we can observe how the same driving forces which animated divination have now come to sustain the course of action in strategic consultancy, the research of think tanks, and the entire field of future studies.

But, should prediction be understood, more broadly, as what constitutively structures our ability to know and learn? In the field of education, Pier Cesare Rivoltella has developed the hypothesis that what we call "learning" has essentially to do with "our ability to predict the behaviour of things and others around us, and even our own" [37: 5]. Therefore, in this sense, the function of education and any form of knowledge transmission could be understood as the operation aimed at making the learner imagine what consequences the manipulation of the variables that regulate the behaviour of a phenomenon might produce over time, and, conversely, learning itself would correspond precisely to the development of our ability to predict. In other words, prediction could be defined as the attempt to implement strategies to cope with an environment with a high degree of uncertainty or entropy, i.e. an environment that is sufficiently complex and ambiguous that it does not allow for a linear predictability of behaviour and phenomena.

Also, in the field of neuroscience prediction is held to be at the basis of how human beings get to know the world and to act in it not just because humans are forward-looking agents, constantly trying to anticipate the shape of future events, but insofar as the human brain would not be able to carry out complex tasks without being supported by a system of neural processing routines that underpin and unify perception and action. This is the thesis put forward by cognitive philosopher Andy Clark in his essay *Surfing Uncertainty*: "To deal rapidly and fluently with an uncertain and noisy world, brains like ours have become masters of prediction—surfing the waves of noisy and ambiguous sensory stimulation by, in effect, trying to stay just ahead of them" [11]. Clark posits an evolutionary origin for this anticipatory behaviour: if we had to constantly process information coming in from outside in order to interpret reality, we would have to expend an unsustainable amount of energy (see also [12, 25]). This attempt to save energy has turned our brains into "predictive engines," constantly trying to prefigure the structure and shape of the incoming sensory array: instead of actually processing it, to produce an inner conjecture, a mental anticipation of external reality, to be modified only when necessary, on the basis of the incoming sensory signal [18]. To put it in a nutshell, we learn about the world by constantly imagining it (predicting it instead of actually perceiving it), that is, by generating incoming sensory data for ourselves, from the top down. This manoeuvre revolves around the acquisition and unfolding of a "multilayer probabilistic generative model," as Clark describes it, a definition which, needless to say, fits perfectly with the description of how algorithms operate.

Are there actual similarities between (human) learning and (machine) processing? Or are we rather learning to think and shaping—once more—our understanding of the flesh-and-blood creatures we are on the model of machines? In fact, if we need to recognise that certain aspects make them more similar than we are inclined to believe—for instance the fact that automatism is part and parcel of human cognitive and behavioural processes, as much as it is a property of our technological prostheses—the predictive capacities of machines and of humans are not alike. As Clark suggests, having to reduce the activity of processing information, the human mind tends to imagine a single output and clings on to that in its guessing and anticipatory activity, while the machine, relying upon probabilistic calculation and lacking the ability to imagine, always prefigures a multiplicity of scenarios, which could not be handled by human cognitive capabilities, with the goal of assessing their likeliness and minimising the degree of uncertainty, while humans seem to be built to navigate through these.

9.3 From Feedback to Feed-Forward

But, if human beings have always been future-oriented organisms, how has their relationship to the future changed in relation to digital and algorithmic media? As André Leroi-Gourhan noted as early as 1964, with the advent of electronics, machines have become capable of exercising a form of perception and of implementing a system of gestures aimed at adjusting and reconfiguring their operational process: "the machine can orient, correct, or suspend its actions in response to messages it receives from its 'sensory' organs" [27: 251]. For the French anthropologist and palaeontologist, the distinctive capacity of twentieth-century media is their extraordinary ability to record and express human phenomena. However, since the recent development of AI, the learning process of machines is no longer based simply on the power of recording and computing while interacting with human beings, but rather on the ability to collect and reprocess the experiences produced by users in the form of data, so as to produce new content and information. Digital machines do not merely record and archive data; they connect with other machines, creating networks of complex interfaces. Information, then, is not simply processed and stored, but used by machines to adjust their operations and, at the same time, to predict user behaviour and even prompt it.

As the philosopher and media theorist Mark B. N. Hansen has pointed out, this combined mechanism of feedback and anticipation is what marks a fundamental discontinuity in the functioning of twenty-first-century media and our interactions with them [24]. By building their operations on user input–output, algorithms are not just receptacles of information but are able to influence user behaviour in real time: by constantly anticipating it, they affect it.

In this sense, according to Hansen, what distinguishes the way in which contemporary media operate is, first, that what is recorded is subliminal with respect to the experience that our body would be able to grasp at a spatio-temporal level and that, secondly, the data thus collected are oriented to modify the future actions of users. In addition to this, there is a third aspect, namely the fact that this collection of data by digital technologies is performed at the service of and in communication with other connected devices and interfaces [24: 40], in such a way as to produce a *retroactive effect* on the functioning of the same machines which perform this recording and collection of data, in order to improve and perfect their performance and anticipatory action itself. In other words, the efficacy of an operation or event is projected into the future, but in doing so, it also simultaneously influences the present in which it is generated—following a temporal paradox, dear to science fiction narrative.

This retroactive effect is precisely what characterises the form of anticipation of contemporary mediality, which Hansen expresses through the notion of *feed-forward*, a term that aims to define the way in which a technically enhanced sensibility generates a temporal path such as to modify our relationship to the environment and influence the way consciousness can experience its own operations [24: 30]. With the development of AI as a sensing technology and with the transformation of the human sensorium into a technologically-assisted sensibility [24: 25], we then observe the transition from a form of information-recording directed towards the past—that which characterised twentieth century media—to a form of systematic data-driven anticipation of the future, as a distinctive feature of twenty-first century media.[4]

In fact, the term *feed-forward* was first formulated by Richards [36] to define the reciprocal nature which already applied to the structure of a *feed-back* loop, in the domain of linguistic communication. Henceforth, his conception was taken up and had a decisive impact on the fledgling cybernetics community and on media theorists like McLuhan. When Richards introduced the term, he argued that feeding forward

[4] Pointing to the relationship to the future and its anticipation as the preponderant feature of the contemporary media system, Hansen thus formulates a dynamic that only a few years ago might still have seemed completely unheard of, but which has quickly become an increasingly shared awareness among our media businesses, particularly in the wake of the Cambridge Analytica scandal.

the context of one's principal message is the key to communicating one's intended message, which depends as much on the receiver's interpretation as it does on the sender's intention. The concept of *feed-forward* contrasts with the concept of simple *feed-back*, characteristic of the functioning of both early modern technologies and social communication: if the latter is *reactive* with regard to previous experience, the former is based on a *proactive* logic: it wants to know in advance of action. Instead of being finalised to the maintenance or internal functioning of a system, *feed-forward* mechanism is rather aimed at expanding its access and reach, and thus its *agency*, over the material elements of its environment.

In cybernetics, the concept of *feed-forward*, which is now key to deep learning and neural networks, describes a control mechanism in which the system anticipates changes and adjusts its actions accordingly *before those changes actually occur*. In contrast, simple "feedback" involves adjusting the machine's functioning on the basis of the results of past actions. Instead of reacting to errors after they occur, a system will try to prevent them from happening in the first place, so that it can adapt quickly and efficiently to changing conditions.

But, moving from cybernetics, we could extend the notion of feed-forward to describe the present state of technoculture and the relationship between humans and AI: a reciprocal feed-forward movement characterises the user-interface interaction since, as an AI systems collects data, so as to adjust its performance, the user also undergoes a training process, adapting, in turn, to the technology. Human beings and machines are tied by a two-way temporal relationship that gives rise to a *recursive feedback loop*: as much as intelligent machines have become the prostheses through which we encounter and understand the world and others, it is only through our sensible bodies that machines can experience the world and can develop and adapt to the environment according to the mechanisms of deep learning. The massive flow of data fuelled by our embodied-mediated experiences constitutes the invisible matter of the perception processes led by algorithms. In fact, our bodies increasingly function as prostheses of media, i.e. they act as organic prostheses of their technical prostheses, which in turn mediate our experience of the world and of other beings. As Vilèm Flusser suggested, in the human relationship to technology, every prosthesis always implies a process of *epithesis* as its reverse movement, since, as soon as the cane that I use becomes the extension of my arm or the pen with which I write the extension of my finger, in turn my organs need to adapt and modify to suit the functions and affordances unleashed by these extensions. Hence, the human body can be conceived of as an inward extension of its same instruments, in that it responds and adjusts to them [17: 165].

But, if a temporal relationship of co-implication inherently informs social communication and the relationship between humans and media, in recent decades we have witnessed a further metamorphosis: the strategy of radical anticipation of the future put in place by the media has globally inflected our way of apprehending the real—as it is inseparable from the functioning of algorithmic media and technologies.

9.4 Seeing and Foreseeing

Contemporary mediality is characterised by an affectivity of anticipation and a general state of anxiety. Digital media tend to keep audiences hooked by constantly whetting their appetites for future developments, updates, or events, fostering feelings of excitement, curiosity, or even apprehension and generating an emotional state where individuals are perpetually looking forward. Moreover, according to media theorist Richard Grusin [23], the media also constantly attempt to pre-emptively shape and influence future events, circulating narratives, images and scenarios, preparing the audience for various potential outcomes. This tendency first emerged in reaction to the events of 9/11 and their worldwide media coverage, which unfolded on a global scale, inaugurating the new Millennium, in the form of a collective shock.

In fact, at the end of the twentieth century, the approach of the media was characterised by the pursuit of ever greater immediacy [8], a veritable obsession with real time, i.e. with the possibility of witnessing live mediated sensational and surprising events. Grusin argues that the hypermedia event of 9/11, which made the people of the globalised world simultaneous spectators of the terrorist attacks, realised and fulfilled this desire for immediacy in such a way as to be deeply traumatic. Reacting to this shock, the temporal and affective posture of the media changed irreversibly: their goal became to prevent such a catastrophe from ever occurring again and exploding in such an uncontrolled manner. Since then, contemporary mediality has thus been characterised by a continuous and systematic attempt to prevent the future from emerging in the present without it having been somehow anticipated and prefigured, with the aim of preparing users and protecting them against yet another possible shock. Instead of communicating events, the media began to *prevent* them from happening: instead of *mediating* content, they began to *premediate* them, foreseeing every possible event as potentially traumatic. This stance ultimately results in a proliferation of anxiety, which finds its mirror in the emergence of the subcultures of preparedness [35].

This anticipation mechanism aims to make the future a sufficiently familiar place for it to be acceptable, so that what lies ahead cannot burst onto the world but only be received in the faded form of *déjà-vu*. By creating a system of continuous alerts and by directing the users' interactions through a pre-established system of interfaces, contemporary media both engender a permanent state of emergency and, at the same time, put in place a sensory training for any potentially traumatic threat. This strategy enacted by the media reflects the protective mechanism, so presciently envisaged by Walter Benjamin to describe the function exercised by early cinema, as capable of operating a "psychic vaccination" for the masses and preparing human sensibilities for the shocks of modern life [5: 38].

As Grusin points out, this tendency is paradigmatically reflected in the concept of pre-emption at the heart of the doctrine and rhetoric of the Bush-Cheney administration [23: 148], which characterised the politics of the war on terror that followed the attacks of 9/11 (see also [31]). Building on the failure of the logic of deterrence born in the context of the Cold War, pre-emption cannot be assimilated simply to a strategy of prevention, in that it does not operate by elaborating tactics to act on objective causes, but rather to convert an effect that has yet to occur into a potential or virtual cause. This is why pre-emption, inspired by a conditional logic, gives rise to a self-driving movement: potential threats can only be faced by potential strategies, which operate on a political and affective level, demanding control on every potential front. Indeed, even though the threat has not yet even formed, it remains indeterminate and undetectable, and still it is likely to emerge in the least expected way, without any warning. That is why what characterises pre-emption is not so much the prevention of threat as a "threat-o-genic" movement, the generation of threat itself [31].

It is worth noting that premediation differs from mere prediction, which already characterised the media even before 9/11: premediation does not so much aim at determining the future that is about to happen, but rather, as in the construction of virtual paths within a video game, at anticipating *all* possible scenarios which might arise [23: 46]. The media tend to prefigure and familiarise us with a multiplicity of virtual futures, so as to prepare us for *any* potentially imminent crisis.

In a way, the mechanism of broad-spectrum virtual anticipation that characterises premediation and pre-emption is inherent in the very logic of late capitalism or techno-capitalism as a system which, as Fredric Jameson has observed, manages to neutralise any opposition or alternative—including anti-capitalist drives—not through direct confrontation, but by absorbing and commodifying it. In this system, no genuine innovation can emerge since any deviation from the hegemonic *status quo* is constantly pre-empted and continuously reintegrated into it. Drawing on Paolo Virno's account of forethought as a central aspect of human agency, especially related to the linguistic relational capacity [44], Nigel Thrift has highlighted how late capitalism elicits new forms and habits of forethought, as cognitive and communicative labour [41, 42].

The *modus operandi* of intelligent machines is embedded in this process of systematic anticipation of the future, which combines the mechanism of feed-forward and the strategy of premediation, but it also entails a significant shift: this prophetic capacity is extended to every aspect of the way algorithmic media operate. In algorithmic mediality, "seeing and foreseeing," to paraphrase Paul Virilio, "tend to merge so closely that the actual can no longer be distinguished from the potential" [43: 4]. This point needs to be made more explicit, for we are not simply reaffirming the process of virtualisation of the real brought into focus by Baudrillard [3] with regard to hyperreality and simulacra, causing the collapse of reality and its reproduction. What is key here is the retroactive effect of the future on the present, such that it undermines the way we have traditionally thought about cause-effect relations: by envisioning what is by definition a portion of the invisible, pre-mediation places us in a temporality of *recursive anticipation*, in which it is no longer the cause, but rather the system of expectations revolving around the prefigurations of the future, that produces effects on the real. But, what happens then to the future, as such, when it is constantly anticipated and systematically foreshadowed?

9.5 The Augur and the Prompt Writer

Today, as we interact with contemporary devices, we are constantly receiving alerts, data feeds, and updates. We consult our screens to predict the time and route of our journeys, to check the weather, the date and time of a delivery, the arrival of trains, planes, and so on. Thanks to augmented reality, we carry out explorations and virtual location scouting in distant places, so as to pre-figure our future moves. Having become dowsers of wi-fi, our lives are increasingly supported by networks of connected objects. Very rarely do we make a decision or a plan without first turning to the reassuring touch of our smartphones, endowing them with the aura of divinatory devices. We consult our screens as prophetic surfaces, we *con-template* them, a term itself derived from the divinatory device of ornithomancy—that is the sacred rectangle traced by the augur called the *templum* [14]—as if they were crystal balls, as if by observing them we were receiving the pronouncement of an oracle. This prophetic power is a decisive component of the attraction contemporary media exert on us, of the fascination that binds us to these devices, which have become irreplaceable prostheses of human sensibility.

Embedded into the fabric of contemporary mediality, this forward-looking temporality and the anticipatory approach we have outlined now encompasses all aspects of individual and collective existence. This also dramatically contributes the realisation of a progressive erosion of the future. As with any phenomenon modified by its observer, the possibility of knowing what the future may bring also changes it. If the future is constantly pre-empted, it ceases to be, by definition, that which to us remains inaccessible, that which remains in an inexorable cone of shadow, perforce concealed, literally veiled from our eyes, shrouded in mystery. Since by the time the future reaches us, it has already been foreseen or prefigured, the future is then fundamentally deprived of its virtual character. Even before it has irrupted into reality, the future is already compromised, already undermined by the imminence of catastrophe or success, hijacked by a constant movement of anticipation that degrades it prematurely, prescribing its obsolescence—a “planned” obsolescence, as they say.

Hence, we tend to look at the future as if it were already potentially past, and therefore as a time already erased. As the philosopher Franco Berardi suggests, we have been witnessing a *slow erasure of the future* [6: 18]: the time to come can no longer be thought of as a dimension in which we can act, modifying the course of history. The future we face today is one that threatens our very existence as a species. Are we living in a future that is already in ruins, already in the process of collapsing [39]?

Thus, the media’s prophetic ability to premediate any possible future is already revealing its blind spots, namely the risk that recent developments in technology and mass communication end up obliterating our perspectives, reducing the potentially broad spectrum of futurability in loop-like self-referential paths. A sense of inevitability, making the future ultimately unattainable, is also the emotional tone that characterises what Mark Fisher has defined as “capitalist realism,” drawing on Jameson’s portrayal of postmodernism alluded to above: there is a dimension of inevitability that is inherent in the state of late capitalism as an economic system that proposes itself as the only possible way forward: it is easier to imagine the end of the world than the end of capitalism—a belief that can be condensed into Margaret Thatcher's propagandist slogan for her economic policy "There is no alternative" [15, 26].

It is on the basis of this feeling of inevitability and collapsed future, and by living in the shadow of a climate catastrophe, that we turn today to intelligent machines and trust the predictive power of algorithmic governmentality. In fact anyone who has ever had their palm read—even for mere amusement— knows that fortune tellers *always tell the truth.* For this, they need not be genuinely endowed with supernatural skills: their talent lies in being able to glean enough information about us from our gestures, words and manners, including nuances of meaning that we ourselves are unaware of, so that in predicting the future they merely render to us our own message, which we tend to accept as true precisely because it comes across as absolutely familiar.

AI wields this kind of fortune-telling power. We feed machines with information that we ourselves have assembled for them, imbued with layered connotative nuances that even we are not fully aware of, so that in them we find yet another version of ourselves, as if looking into a distorting mirror. Hence, the potential contribution of algorithmic interfaces to the advancement of human knowledge is unprecedented, but also unprecedented is the risk they carry of jeopardising it by preventing any actual diversion from the already known. As they constantly prevent it from breaking through into reality, instead of helping us to know the future, they elaborate an accurate description of what we already know or think, what we have already seen and so on. As has been argued [1], this is likely to engender an autophagous mechanism, that is, a self-consuming loop compromising the possibility of achieving truly new results. As is the case with soothsaying, on the one hand, the answer is already in our questions, and, on the other hand, every prediction inevitably turns into a self-fulfilling prophecy, i.e. a prediction which comes true because it has been formulated, insofar as the subsequent belief drives the behaviour of the individuals concerned.

Thus, the newly-born art of prompt engineering requires from us the development of divinatory skills, it urges us to pursue and promote a prophetic culture [10]. In other words, the prompter's capacity lies not so much in their ability to match the machine's way of operating, but rather in their capacity to hijack the code: diverting the functionality of machines, making algorithms hallucinate, as contemporary artists claim, is the only truly productive way to discover something new, in the gaps between what we expect and what they unexpectedly bring forth, so revealing what for us remains latent or unconscious.

To study the new regime of truth *inaugurated*—literally, as we speak of the divinatory ritual of augury—also means to study how data are diffracted and distorted by the lenses of AI [34]. So, instead of being subject to their inbuilt aberrations, we would become able to leverage AI to generate new knowledge. That layer of meaning which we involuntarily communicate and of which we are not yet aware is the true invisible reality that AI is able to envision, not just to prevent or predict the future, but to outline a diagnosis or symptomatology of our present and visualise the figures these complex lenses inevitably refract in their crystal ball.

References

1. Alemohammad, S., et al.: Self-consuming generative models go mad. *arXiv Preprint* arXiv: 2307.01850 (2023)
2. Andersson, J.: The Future of the World: Futurology, Futurists, and the Struggle for the Post-Cold War Imagination. Oxford University Press, Oxford (2018)
3. Baudrillard, J.: Simulacra and simulations (1981). In: Crime and Media, pp. 69–85. Routledge, London (2019)
4. Benjamin, W.: On the mimetic faculty. In: Reflections: Essays, Aphorisms, Autobiographical Writings, pp. 333–336. Schocken Books, New York (1978)
5. Benjamin, W.: The Work of Art in the Age of Its Technological Reproducibility and Other Writings on Media. Harvard University Press, Cambridge, MA (2008)
6. Berardi, F.: After the Future. AK Press, Oakland-Edinburgh (2009)
7. Berton, M.: Le médium (au) cinéma: Le spiritisme à l'écran. Georg, Geneva (2021)
8. Bolter, J.D., Grusin., R.: Remediation: Understanding New Media. MIT Press, Cambridge, MA (1999)
9. Bouché-Leclercq, A.: Histoire de la divination dans l'Antiquité. Millon, Grenoble (2003)
10. Campagna, F.: Prophetic Culture. Recreation For Adolescents. Bloomsbury, London (2021)
11. Clark, A.: Surfing Uncertainty: Prediction, Action, and the Embodied Mind. Oxford University Press, Oxford (2015)
12. Clark, A.: The Experience Machine: How Our Minds Predict and Shape Reality. Random House, New York (2024)
13. Comte, A.: A Discourse on the Positive Spirit (1844). Reeves, London (1903)
14. Dalmasso, A.C.: Cadre et templum: Une archéologie des limites de l'image. La Part de l'Œil 33, 133–151 (2020)
15. Fisher M.: Capitalist Realism: Is There No Alternative? Zero Books (2009)
16. Floridi, L. (ed.): The Onlife Manifesto: Being Human in a Hyperconnected Era. Springer, Cham (2015)
17. Flusser, V.: Gestures (1991). University of Minnesota Press, Minneapolis (2014)
18. Friston, K.J., Parr, T., Pezzulo, G. (eds.): Active Inference: The Free Energy Principle in Mind, Brain, and Behavior. MIT Press, Cambridge, MA (2022)
19. Gaudin, Th.: La Prospective. Presses Universitaires de France, Paris (2005)
20. Ghahramani, Z.: Probabilistic machine learning and artificial intelligence. Nature 521, 452–459 (2015)
21. Gidley, J.M.: The Future: A Very Short Introduction. Oxford University Press, Oxford (2017)
22. Grespi, B.: The power of the mechanism: cinema and the machinery of desire. In: Malavasi, L., Tongiani, S. (eds.) Technophobia and Technophilia in the Media, Art, and Visual Culture, pp. 161–178. Aracne, Roma (2020)
23. Grusin, R.: Premediation: Affect and Mediality After 9/11. Palgrave Macmillan, New York (2010)
24. Hansen, M.B.N.: Feed-Forward: On the Future of Twenty-First-Century Media. University of Chicago Press, Chicago (2015)
25. Hohwy, J.: The Predictive Mind. Oxford University Press, Oxford (2013)
26. Jameson, F.: Archaeologies of the Future: The Desire Called Utopia and Other Science Fictions. Verso, New York (2005)
27. Leroi-Gourhan, A.: Gesture and Speech (1964). MIT Press, Cambridge, MA (1993)
28. Marzo, J.L.: Las videntes: Imágenes en la era de la predición. Arcadia, Barcelona (2021)
29. Masini, E.: Why Future Studies? Grey Seal, London (1993)
30. Massé, P.: De la prospective: Textes fondamentaux de la prospective française, 1955–1966. L'Harmattan, Paris (2007)
31. Massumi, B.: Ontopower: War, Powers, and the State of Perception. Duke University Press, Durham, NC (2015)
32. Murphy, K.P.: Probabilistic Machine Learning: An Introduction. MIT Press, Cambridge, MA (2022)

33. Natale, S., Walsh Pasulka, D.: Believing in Bits: Digital Media and the Supernatural. Oxford University Press, Oxford (2019)
34. Pasquinelli, M., Joler, V.: The Nooscope manifested: artificial intelligence as instrument of knowledge extractivism. nooscope.ai (2020). Accessed 9 May 2025
35. Pellizzoni, L.: The time of emergency: on the governmental logic of preparedness. Sociol 536 Ital AIS J Sociol 16, 39–54 (2020)
36. Richards, I.A.: The secret of feedforward. Saturday Rev. 14–17 (1968)
37. Rivoltella, P.C.: La previsione: Neuroscienze, apprendimento, didattica. La Scuola, Milan (2014)
38. Sconce, J.: Haunted Media: Electronic Presence from Telegraphy to Television. Duke University Press, Durham-London (2000)
39. Sofia, Z.: Contested zones: futurity and technological art. In: Malloy, J. (ed.) Women, Art, and Technology, pp. 86–101. MIT Press, Cambridge, MA (2003)
40. Somaini, A.: On the photographic status of images produced by Generative Adversarial Networks (GANs). Philosophy of Photography 13(1), 153–164 (2022)
41. Thrift, N.: Knowing Capitalism. SAGE, Thousand Oaks, CA (2005)
42. Thrift, N.: Non-Representational Theory: Space, Politics, Affect. Routledge, New York (2007)
43. Virilio, P.: War and Cinema: The Logistics of Perception. Verso, New York (1984)
44. Virno, P.: A Grammar of the Multitude. For an Analysis of Contemporary Forms of Life. Semiotext(e), Los Angeles (2004)

Anna Caterina Dalmasso is tenure-track assistant professor in Aesthetics at the Department of Philosophy "Piero Martinetti" at the University of Milan, where she teaches "Media Theory and AI". She is a member of the research centers "EXT. Extended Realities" and "HCAI. Human-Centered Artificial Intelligence." She is also an associate researcher at the Centre Prospéro at UCLouvain—Brussels, where she was a Marie Curie Cofund Fellow (2017–2019), and a member of the research center *Vivre par(mi) les écrans* at the University "Jean Moulin" Lyon 3. A specialist of Merleau-Ponty's philosophy, her work combines phenomenology, aesthetics, film and media studies, media archaeology, and visual culture. Her current research focuses on the aesthetics of immersive environments and the status of virtual and algorithmic images.

Part III
Algorithms and the Arts

Chapter 10
Ebrah K'dabri. Agency and Authorship in AI-Enabled Artistic Creation

Domenico Quaranta

Abstract A mere tool, or a non-human collaborator: this is how a sizable portion of artists working with artificial intelligence (AI) describe the role it takes in their practice. However intimidated by a cumbersome, presumed "creativity of the machine," they do not see their authorship as limited. Yet, what if we were to reverse this perspective? What role could we consider the artist-user to play from the standpoint of the machine? By interacting with an AI system, a user consumes a commercial product, provides additional training data, fine-tunes the software, and expands its reach and appeal to other users and to investors. From this perspective, it is not the user but the machine that assumes the dominant position. Is it possible to find a mediation between these two seemingly irreconcilable forms of the relationship? Who, in the human-AI relationship, does the using, and who is it that is used? Who is in control? And should a reconciliation not prove to be possible, what then is the role of art and artists interfacing with AI? By exploring a number of case studies, this text identifies the path of art through critical exploration, while eschewing mystification and unveiling the biases that inform the current form of AI.

Keywords Authorship · Medium · Black Box · Creativity · Bias

10.1 Ebrah K'dabri

"With AI's increasing sophistication, the role of the prompt writer has become central. Each image generated needs a directive to initiate computation. These prompts involve a blend of intuition, imagination, and understanding of the AI model's behaviour" [29]. With these words, Canadian artist Jon Rafman accompanies the online presentation of his video *Counterfeit Poast* (2023) on the *DIS.art* web platform. With an extensive and multitiered use of artificial intelligence (AI) systems, the work takes us through a series of character studies, each one the protagonist of a micro-story told in the first person, by someone confessing or seeking advice on an

D. Quaranta (✉)
Accademia Di Belle Arti Di Brera, Milan, Italy
e-mail: info@domenicoquaranta.com

P. Conte et al. (eds.), *Algomedia. The Image at the Time of Artificial Intelligence*, Lecture Notes in Morphogenesis, https://doi.org/10.1007/978-3-032-08726-3_10

online forum, in the style of Reddit's creepypasta: absurd, visionary, weird stories that easily pull us into a nightmarish spiral, always on the edge between reality and fiction, between the prosaic and mystification. AI is remarkably well suited to take us into that uncanny space inhabited by Rafman's stories, in which every fragment of reality appears "counterfeited" by the filter of memory and hallucination. The unimaginable acquires an uncomfortable concreteness, at once painterly and post-photographic, nourished as it is by the details of hundreds of stolen images; the commonplace, in turn, always has something alien about it, while retaining a familiarity of its own.

According to Rafman, recent developments in AI research have, on a technical level, made AI tools accessible to anybody, but they have also introduced another layer of sophistication: the one tied to language control. Conscientious use of the prompt, according to Rafman, requires not just a good level of confidence with the language but also a "deep familiarity" with the machine. Like a conductor, the prompt writer directs a set of computational processes, "transforming human imagination into machine-readable instructions," and like a detective, proceeds by trial and error; like a poet, the "artist-promptist" must have the gift of synthesis, and like a mystic or a magician, he or she must be able to activate the creative power of language: "The ancient Hebrew phrase *ebrah k'dabri*, colloquially known as *abracadabra*, translates to *I create like the word*."

Significantly, when writing about the creative use of the prompt, Rafman is not focusing, as seen in emerging manuals on prompt engineering, on professionalism and skills; his conductor-detective-poet-mystic outlines a more complex, elusive figure and practice. One might say that you have to be an Artist in order to be a "prompt artist." Throughout the essay, AI is sometimes described as a mere tool, sometimes as a collaborator, but ultimately, it's the artist who utters the abracadabra, who spells the word that creates.

The centrality of the role of the artist is put into question near the end of the text, when Rafman addresses the topic of AI biases. These, according to Rafman, do not just amount to ideologies and worldviews inadvertently filtering through AI systems from datasets and keywords, but they also include the alleged "moral" limits embedded in the products by their manufacturers, which limit their users' freedom of creative exploration. This turns the AI system into a deforming lens, which does not seem immune to its own agency.

A collaborator, a technological facilitation, or a distorted and biased lens? The unresolved contradictions Rafman encounters in his attempt to illustrate his relationship with AI actually nourish his work, leading it to acquire the complexity and subtlety that makes it not into the virtuosic outcome of a skilful use of prompts, but rather into that of a struggle, which at times takes the docile and consensual form of collaboration, and at other times, the more antagonistic form of overpowering, one that always involves awareness of the ambiguity of the *other*.

10.2 An Active Medium

> Then and now, the computer is a conversation partner—one that consistently offers surprising results we cannot imagine on our own.
>
> Valyi-Nagy 2024 [32]

Far from being exhausted, the debate on the role of the machine and the exact terms of the human–machine relationship in the creative process has gained surprising vitality in recent years. The reasons for this rekindling of the discussion are many, and it is worth reviewing them briefly, lest we run the risk of attaching undue importance to those of more recent manifestation: the astonishing level of creative autonomy achieved, from 2023 onward,[1] by large language model (LLM) based technologies such as ChatGPT and diffusion models such as DALL•E and Midjourney for image generation, and the subsequent explosion of public debates about AI, partly spontaneous, partly masterfully and artificially accelerated by statements, made in various contexts by astute entrepreneurs or by recognised protagonists of AI research, about the existential risk that could follow from its uncontrolled development.[2]

As is well known, the perceived simplicity in the use of these models, and the awesomeness of their results, made them immediately popular among a very large pool of users, causing ChatGPT to become one of the apps with the fastest-growing user bases in the history of digital technologies.[3] Predictably, this immediate and widespread popularity, and very rapid adoption curve, has had the side effect of narrowing the public perception of AI to these technologies, zeroing in on their historical and phenomenological complexity in every sphere of discussion, including the arts: after November 2022, to talk about AI means to talk about generative models that produce text, images, sound, and, more recently, video in response to a textual prompt.

Before this date, the revival of thinking about the role of the machine in the creative process could be traced to two different, albeit nearly simultaneous, processes: the invention of generative adversarial networks (GANs) in machine learning research, and their rapid adoption in the arts; and the growing popularity, fuelled by the NFT economy,[4] of blockchain-based generative art, which has spearheaded the late or posthumous rediscovery of some of the Computer Art pioneers of the 1960s and 1970s, as well as a wide range of practices that developed in the first two decades of the 2000s straddling the visual arts, visual design, and creative coding around programming environments such as Processing.

First presented by U.S. researcher Ian Goodfellow in a 2014 paper [13], GANs are a class of machine learning methods in which two neural networks are competitively trained in the context of a zero-sum game. The outputs produced by GANs are

[1] We adopt here the symbolic date of the launch of ChatGPT, which, released online on November 30, 2022, by January 2023 had already surpassed the impressive number of 100 million active users.

[2] See in particular [3, 15]. See also [28].

[3] See Humlum [17].

[4] On the NFT boom, see Quaranta [27].

both progressive—with a range that can go as far as convincing photorealism—and surprising in the eyes of a human viewer. Their malleability made them far more appealing than the "Inception" model implemented shortly before by Google Deep Dream, developed in 2014 [31] and made publicly available in 2015, in which a single neural network trained on a dataset is forced to recognise faces and patterns in given images, producing a kind of algorithmic pareidolia that inserts human, dog or cat faces at different points in the image.

GANs require from their users a technical ability to intervene both at the level of the algorithm and of the dataset on which they are trained, and in their variety of results, they feature a certain aesthetic homogeneity. But neither did the technical difficulties nor this obvious "designation of origin" prevent, in the second half of the 2010s, a wide diffusion of GANs in art and visual design. At the culmination of this trend, on October 25, 2018, Christie's auction house hammered out for $432,500 the *Portrait of Edmond Belamy*, a picture generated by a GAN trained by the French collective Obvious on a dataset of portraits painted between the fourteenth and twentieth centuries.

It should be noted here that the adjective "generative," used systematically since the introduction of GANs into the debate surrounding art and AI, is far from a novel term in digital art territory. Codified in a now classic paper [12], the term stands for all those artistic practices in which the artist does not directly produce the visual outputs of the process, but programs—we might say, paraphrasing Casey Reas, who in turn paraphrases Sol LeWitt—the machine (the code, the algorithm, the system) that creates the art.[5] In generative art, computer code has the role that the score has in music: it does not constitute the final artifact, but produces it at the moment it is activated, executed or performed [30]. "Generative" thus refers to all those practices—analogue or digital—in which the work manifests itself as the enactment of a set of coded instructions: from Sol LeWitt's *Wall Drawings* to Fluxus' performances and the early examples of Computer Art by the Algorists' generation. It is in this sphere that first emerged the idea, later taken up in the work with generative AIs, that the machine is not a mere executor; it is here that the machine was first conceived as an active agent of collaboration, as a "conversation partner," to which the code leaves—by negotiating with the machine's limits or by introducing random factors—a more or less substantial margin for intervention and surprise. As artist and engineer A. Michael Noll wrote in a seminal 1967 essay, the computer "is an active medium with which the artist can interact on a new level [...] The artistic potentialities of such a creative medium as a collaborator with an artist are truly exciting and challenging" ([25]: 93–94).

A few years later, German artist Manfred Mohr explicitly wrote about the computer as a partner in his work [22]. And later on, in a short essay: "The computer is not conscious of what it is doing and can only execute orders from outside: from us! That means: a computer itself cannot create or invent something... Abraham Moles once said: 'La machine ne pense pas, elle nous fait penser!'" [23].

[5] "The idea becomes a machine that makes the art" [18].

Statements like these recur in the texts of those early pioneers, and they set up the debate in a way that still seems valid in our day and age. As in the 1960s and 1970s, today's machines do not think, cannot create autonomously, and have no consciousness of what they do. Yet, more than any other medium, the computer opens up a realm of possibilities that makes its use in the creative process something more than the instrumental use of a mere technological extension, qualifying it rather as an active medium, a creative partner that makes us think and that "helps to realize what is most subjective, what is most profound in man" [24].

10.3 "Created with" or "Created by"?

But if answers have indeed already been given, and if the machine has not leap forward toward self-awareness, creative autonomy, or autonomous thinking, why do artists and theorists feel the need to reformulate, specify or contextualise the terms of their and our collaboration with the machine?

The answer is simple. Every step taken toward greater automation seems to threaten the creative autonomy of the artist. As we have seen, generative art is based on a deep dialogue between the artist and the machine, mediated by programming code. However, the level of control over the operation of the machine by artists such as Noll or Molnar, who programmed the movements of a plotter in BASIC, is not the same as that of a contemporary artist, who programs in an extremely sophisticated computer environment, made up of layers of software stacked pyramidally, and incorporating automation at various levels of depth. Similarly, if with GANs, the collaboration is still mediated by code and the artist can intervene in the process by customizing the training algorithms and datasets, with large generative models, the dialogue with the machine is mediated by a textual prompt in natural language, and the machine itself is a complex system, pre-trained on datasets that aggregate millions of contents from the Internet, optimized through continuous interaction with an invisible and widespread community of human trainers, and able to make "decisions" autonomously. As it's been repeatedly stated, no medium is neutral, but in digital media, neutrality is inversely proportional to the level of sophistication, which introduces new and invisible levels of bias and agency.

Such a degree of automation cannot fail to fascinate, and it produces several spillovers at the level of communication, language, and even in the judicial sphere. In 2018, the aforementioned *Portrait of Edmond Belamy* was presented at auction as a work of art created by an algorithm: "not the product of a human mind," but an artefact "created by an artificial intelligence" [8]. A fragment of the algorithm was even used by Obvious to "sign" the portrait, in the conventional position in which the artist's signature is put on a painting. This way, Obvious enacted a conscious and strategic transfer of authorship and agency in order to raise an argument "that algorithms are able to emulate creativity." Yet, we shouldn't forget that it was Obvious that collected the dataset of paintings on which to train the GAN; it was Obvious that selected the outputs; it was Obvious that constructed the narrative of the fictitious

Belamy family; and, finally, it was Obvious's agency that drove the project, using the classic portrait form and the certifying function of an auction to demonstrate that algorithms can emulate creativity.

As regards language, one should note the growing tendency to fall back on formulations and expressions that attribute an unprecedented degree of authorship and agency to AI. A news title like "AI won an art contest, and artists are furious" [21] comes as one out of many examples of how generative AI is discussed in journalism. Additional shadows of ambiguity on the issue of authorship are cast by the English expression "AI-generated art," and by the frequent slippage between the preposition "with," implying instrumental use or collaboration, and "by," implying passive action. On a broader level, the very term "artificial intelligence" seems questionable: the use of the term "intelligence" inevitably leads to a subjectification of the machine, on which much of the AI narrative has converged for decades. The linguistic process of humanising AI extends to a whole range of other actions attributed to it, with or without consciousness, such as dreaming or hallucinating: metaphors that are fascinating but dangerous in their implicit animism and anthropomorphism.

Finally, at the legal level, the situation is no less paradoxical or complicated. On Aug. 18, 2023, U.S. District Judge Beryl Howell ruled that copyright did not apply to an image generated with AI, described in the ruling as "a work created absent any human involvement" [16]. The claim for copyright attribution had been made by Stephen Thaler, an American computer scientist who in 2016 used an AI system of his own making to generate a vaguely impressionistic landscape image entitled *A Recent Entrance to Paradise*. The paradoxical nature of the ruling stems from the fact that the plaintiff acknowledged that the work lacked human authorship but demanded that the machine's authorship be recognised for legal purposes, thereby transferring copyright rights to the AI owner itself. The court rejected the possibility of awarding copyright in the absence of "human authorship," effectively recognising the AI full authorship of the work. The ruling threatens to set a dangerous precedent, not so much for not recognising human authorship but for attributing full authorship to a non-human agent. In a similar case that anticipated the Thaler ruling by a few months, some images generated with Midjourney to illustrate a graphic novel were not considered a product of human authorship either. The key difference is that while Thaler claimed copyright as the owner of the AI system, Kris Kashtanova made the claim as a user of a popular TTI (text-to-image) model, not unlike Rafman. For now, the refusal to apply copyright to AI-generated images at least protects users from having ownership of (their?) creations attributed to the companies that own the models used—a dystopic but not totally unrealistic scenario [14].

10.4 Posthumanist Agential Assemblage

> Working with AI imagery is much like your dog dragging you down the street and a friendly stranger asking you: "Who's walking who?"
>
> Laurie Simmons, in Chen [5]

Despite the aforementioned attempts to fully attribute the authorship of AI-generated content to the machine, and in the face of concerns raised by common sense, if we listen to the words of artists, the prevailing models seem to remain, however, those of instrumental use of and collaboration with the machine, which often alternate, schizophrenically, in the statements of the artists themselves. German artist Mario Klingemann goes as far as to compare the machine to a slave, a Golem that we have built for our intellectual or emotional entertainment ([26]: 237–238), yet, in other passages in the same interview ([26]: 239–240), Klingemann stresses the need to team up with the machine, and highlights its autonomous ability to capture something of the human condition, elsewhere [5], he speaks of collaboration, explaining that it's in the limitations of technology—its inability to capture "the nuances of human perception, emotion, and intuition"—that a space for action opens up for the artist. British artist and scientist Mike Tyka, co-founder of Google's Artists and Machine Intelligence program, also speaks of a creative process that is based on the exploration of the tool, in which, however, artistic choices are manifested in a context where one never has total control ([26]: 258). And Italian engineer and artist Mauro Martino makes it clear that AI is not a simple expressive medium, but one made of uncertainty and indeterminacy, forcing upon the artwork "[its] own aesthetics that we have to adapt to" ([26]: 303–304).

Interviewed by New York's MoMA on the occasion of the launch of *Unsupervised* (2021), a project that uses the museum's entire collection as a dataset to train a GAN that produces new visual outputs, Refik Anadol also speaks of collaboration with the machine, albeit attaching decisive importance to the moments of control and choice by the artist: the supervision of parameters, the moment when the decision is made to stop training, and the moment when outputs are selected from countless options [1]. Similarly, artist and programmer Casey Reas explains [1] his interest in AI with his fascination with latent space—the space of possibility from which images emerge in TTI models—but specifies elsewhere [5]: "I always train my own datasets rather than working with the larger models that are pre-trained. I think training datasets is where a lot of the agency comes in."

We could go on and on, but it should now be quite clear how, in human–machine collaboration, control—in order to stay in the hands of the artist—must continuously be reclaimed, through the coercive force of language (the prompt, the code, the algorithm) and intervention in the training process. In this regard, the words of British artist Anna Ridler, which mirror an experimental and unprecedented creative practice, seem more interesting—and potentially able to guide us toward a way out of the tool/collaborator dilemma. Ridler does not make use of pre-trained models like Midjourney, nor does she train her GANs on pre-existing datasets, but she personally creates, using photography or other media, the images used to train the neural networks. In *Fall of the House of Usher* (2017), Ridler makes a 12-min video by training a GAN on a dataset of more than 200 hand-drawn ink-on-paper drawings, based on frames from the silent film made in 1929 from Edgar Allan Poe's short story. Instructed by this preliminary training, the GAN generates images that Ridler redraws manually, creating a second "refinement" dataset that is submitted to the machine. The final video is the result of this circular feedback loop between algorithmic generation and manual creation, a "natural-artificial system" ([2]: 113) in which the machine learns anomalously (incorporating drawing hesitations and handprints into the generative process), and the artist learns something about herself by watching the machine's outputs. Ridler explains: "It is my work and at the same time it is not—it is recognisable as mine, but it is not something I would have been able to make myself. [...] I don't see GAN as a tool [...] but I don't consider it a real creative partner either. I am not sure what it really is" ([2]: 126).

Commenting on this passage, Italian philosopher Roberto Diodato proposes to reread the human–machine exchange in terms of "non-instrumental collaboration," or better yet, of "relationship" ([10]: 526). The concept of relationship is fundamental to addressing the idea of authorship from the perspective of, precisely, relationality, and it is taken up by Austrian researcher Martin Zeilinger in his formulation of the concept of "posthumanist agential assemblage," defined as "a way to push AI—and AI art—beyond the humanist subject of the singular, unified artist, their individualized voice, and their original and uniquely spirited creative expression." Considering the human–computer collaboration as a unified creative subject, the posthumanist creative assemblage allows "the productive entanglements between different human and non-human elements [to] find expression in manifold redistributions of agency across all participating elements" ([33]: 45).

If it is true, in the words of artist and theorist James Bridle, that creativity "is an emergent property of encounters, experiences, materials, and relationships—only some of them identifiably 'human', and many of them non-human and entirely constructed" [4], then, the usefulness of the concept of posthumanist agential assemblage lies not so much in questioning the notion of authorship. It lies, instead, in emphasizing, even at the terminological level, that the creative act now takes place at the crossroads between different agencies, interests, and systemic (environmental and social) influences, and that it must therefore be redistributed among multiple agents, human and non-human. This redistribution of agency is particularly evident in the non-instrumental relationship with AI, but it is not exclusive to it, in fact affecting any process of creation.[6]

One problem affecting Zeilinger's concept of the posthumanist agential assemblage, however, is that he seems to understand it as applicable to the artwork as an automated process, rather than to the artist, and that, by accepting the shallow idea of the "creativity of the machine,"[7] he ends up confusing "creative agency" with "artistic agency," as can be seen when he considers one of his key case studies: the installation-performance *All We'd Ever Need Is One Another* (2018) by Canadian artist Adam Basanta. The installation involves a process that is fully automated and pre-coded: two desktop scanners resting on their sides with the scanning surfaces facing each other are controlled by a computer script that randomly changes the

[6] The need for a posthuman notion of creativity is also emphasized by Joanna Zylinska, who derives it from Vilém Flusser's reflection on the relationship between human intentionality and the photographic apparatus. See Zylinska [34]: 53–54.

[7] For a criticism of the concept of the "creativity of the machine," see [19, 20, 34].

software interface settings and initiates the scanning process. Each image generated is then analysed by a series of deep learning algorithms trained on a database of contemporary artwork: when an image has at least an 83% match to an existing image, it is "validated as art" and uploaded to a dedicated website and social media. In Basanta's words, "the installation acts as a golem-like assemblage, continuously and mindlessly self-producing without regard for human spectators."[8] According to Zeilinger:

> Conceptually and practically (if not legally), the setup of the installation disqualifies Basanta as the default creator of any of its image outputs [...] *All We'd Ever Need Is One Another* constitutes a posthumanist agential assemblage in which the human artist is participant rather than creator. Basanta's role is decentered, and his capacity to function as author is significantly limited ([33]: 104).

This statement completely misunderstands the role of the artist in the work, reducing the project to the images it generated, as opposed to the process that generated them. The authorial function here manifests itself not in the production of images, which is fully automated, but in the orchestration of the various agencies (mechanical, environmental, and human) that come into play throughout the project. This is not to re-attribute to Basanta the "humanistic subjectivity of the artist," but rather to recognise in himself, and not in the work/process he has activated—the posthumanist agential assemblage—the orchestrator who coordinates the various agencies without reducing them to mere instruments. It also means to distinguish the "creative agency" of the machine that generates the images from the "artistic agency" of those who trace these images back to an artistic projectuality that provides them with the framework within which their transition from "creative output of the machine" to works of art becomes possible. It means, finally, defining a conceptual basis on which to reconcile the possibility of an authorial subject (however hybrid and posthuman) with the algorithmic generation of the image.

10.5 Strategic and Tactical Uses of AI

All along this path, machine agency has been sketched somewhat simplistically, as that unpredictable and uncontrollable ability of the latent space to generate something new and surprising, discounting all attempts to drive this generative ability through dataset construction, parameter modulation, algorithm customization, or writing a good prompt.

If the machine's agency were limited to this, Klingemann's slave metaphor would not be so inappropriate: an unpaid collaborator who satisfies us when she—as it is usually gendered—does her duty but even more so when she makes mistakes, when she is hallucinating, when she freaks out; a collaborator who perhaps requires, as D'Isa notes ([11]: 31), an "empathic exchange" to give us what we want, but who receives nothing from us in return, at least on the surface.

---

[8] Cf. https://adambasanta.com/allwedeverneed.

In reality, the exchange is not just empathic, and it is far from free. The AI models in use today are the products of gigantic companies, which have aggregated immense capital with the myth of automation and are co-opting even larger capital flows with the threat of Artificial General Intelligence (AGI). Algorithms and datasets incorporate human-inherited cognitive biases that spill over, more or less transparently and legibly, into whatever is produced with them [6]. LLM-based online models also unilaterally implement censor filters and community rules that strongly condition what can be generated with them. Their staggering growth depends on the rapacious appropriation of everything available online, from the countless resources in the public domain to the equally numerous resources "protected" by different types of licenses; and early attempts at public regulation and control of AI, instead of balancing this elitist nature, seem to consolidate it by attempting to impose limits on both inputs (the training materials) and outputs, through the application of watermarking to generated images. Anyone who uses TTI and LLM models seems to benefit from the consequences of this qualitative growth, but only their owners benefit from them economically; if these models are slaves to anyone, we users are not the slavers.

In the essay accompanying their anatomical map of an AI system, Kate Crawford and Vladan Joler famously describe the Echo user as a chimaera, "simultaneously a consumer, a resource, a worker, and a product" [6]. This model can be easily remapped onto the user of any AI system, including generative models. We are consumers because we use, for free or for a fee, a service that provides us with a range of extraordinary features. But we are also resources because our inputs are mined, collected, and classified, becoming part of the immense database that feeds the AI system. We provide (unpaid) labour because our prompts and feedback become useful resources in an ongoing process of system refinement. And we are, in the end, products, because our participation broadens the user base of the service, and our generating, sharing, and discussing increases the popularity of these models, becoming a commodity to investors.

The elements listed here make the human–machine relationship suddenly more complicated. If the machine is to the human, or to the artist, a collaborator, from the machine's point of view, the human is trapped in a complex relationship of exploitation and extraction, one of disproportionately small benefit compared to what is extorted, in terms of labour, data, and value. We should not forget, moreover, that in the recent AI hype, AI's ability to generate cultural content, its alleged ability to compete with humans on the terrain of creation, have been instrumental to showing that we have entered a new phase in the history of AI. Simply put, art—the idea of art, the machine's supposed ability to create—has become the ultimate Turing test, and has been used and abused to promote AI.

If no medium is neutral, and if—as Crawford and Joler also note—this multiple identity of the user manifests in other technological systems as well, the relational imbalance in favour of the system presents itself, with AI, on an absolutely unprecedented scale, which in our opinion imposes, especially on the user-artist, to take a responsible stance. It is here that Zeilinger's thought comes in handy again, when he borrows from Michel de Certeau's well-known essay, *The Practice of Everyday Life*

(1984), the distinction between tactics and strategies. In De Certeau's formulation, strategies serve administrative and managerial agendas, drawing on the control structures inherent to the system, often for the purpose of containing critical or divergent drives. They are goal-oriented and functional to the acceptance and maintenance of the system. Conversely, tactics are recalcitrant, critical and oppositional, seeking shortcuts and ways out of the ordered grids within which institutions seek to harness their uses and behaviours ([9]: 29–42). Particularly illustrative is the example of the city and the street: there, the strategies of governments and institutions are manifested in the urban planning of the city, but those who walk through the city move tactically, resisting the urban map, pandering to their own needs and preferences.

In his context of interest, Zeilinger finds a strategic approach to AI in works such as Obvious, which consolidate rather than dismantle anthropocentric views of authorship, and a tactical use of AI in works that, like Basanta's installation, exploit its potential ability to subvert traditional notions of agency, creativity and ownership ([33]: 50–55).

The distinction suggested by Zeilinger proves particularly functional for the discourse proposed here and lends itself to an extension of both territories. The strategic area, therefore, includes all those artists who accept artificial intelligence, and the evolutionary direction proposed by the companies that develop it, as a given; those who set out to creatively explore the possibilities introduced by the medium, and who extol and celebrate its outcomes. Not surprisingly, we can locate in this area the works of those artists who have benefited from the research and development programs of companies investing in AI, such as Google's Artists and Machine Intelligence Program (AMI), co-founded by Mike Tyka. AMI has hosted and supported, among others, Refik Anadol, author of installations that, in their spectacularity and imitation of Renaissance altarpiece tropes, elicit an attitude of religious respect and uncritical awe before the mysterious algorithms that govern their animations.

In the tactical area, on the other hand, are all those works and artists who field reflections and/or practices that are critical of the current development of AI, of the roles to which the user is relegated, and of the mainstream narratives built around technology and its seemingly unstoppable development; those who claim for themselves a position in the debate, and the ability to influence its reception in the present and its development in the future. Tactical are the works of those artists who resist being reduced to consumers, resources, workers or products: works such as Trevor Paglen and Kate Crawford's *Imagenet Roulette*, an online project that in 2019 powerfully raised the issue of racial and gender bias embedded in one of the most widely used datasets in machine learning, influencing the public debate [7]. Yet, returning to De Certeau's street metaphor, tactics need not be barricade activism, a no-holds-barred urban guerrilla warfare: they can also manifest themselves in seemingly more innocuous and lightweight forms—a walk, a psychogeographical drift, an unpredictable detour, a tag left on a wall. Thus, Ridler's handmade datasets are also tactical, forcing the algorithm into a one-on-one confrontation between artist and machine, and circumventing the problems of large model bias as well as appropriation and exploitation of unpaid labour. Tactical are those works that thematize, or bring into visibility, the scale and types of invisible human labour that is put to work to maintain the myth of automation; those that bring to light the relationship between the emergence of AI in the generation of cultural artifacts and the production of fakes; those that investigate machine vision for the purpose of bringing to life images that resist machine vision. And tactical, finally, are those works that—as in Rafman's case—manage to optimise the generative capacities of the "word that creates," bypassing the machine's censorial limits and biases to force it to produce images that reflect the artist's own vision, style and imagery.

Art that uses or deals with AI strategically is an ornament of power. It is as decorative, superfluous and condescending as regime art, or as the academic art of the Salons in the nineteenth century. It cannot be said to be useless because it is functional in generating consensus, astonishment and uncritical abstention around an issue on which there is an urgent need to activate a debate, to nurture a plurality of views, to keep the critical thinking alive.[9] Art that uses or addresses AI tactically responds to these urgencies, and participates in a movement of reappropriation from below of a technological apparatus whose evolution is not inevitable, but whose involution, if left in the hands of strategists, is certain.

References

1. Anadol, R., et al.: Modern dream: how Refik Anadol is using machine learning and NFTs to interpret MoMA's collection. MoMA Magazine, November 15 (2021). www.moma.org/magazine/articles/658
2. Barale, A. (ed.): Arte e intelligenza artificiale. Be My GAN. Jaca Book, Milan (2020)
3. Bengio, Y., et al.: Pause Giant AI Experiments: An Open Letter. Future of Life Institute, March 22 (2023). https://futureoflife.org/open-letter/pause-giant-ai-experiments/
4. Bridle, J.: Is creativity over? Wepresent, February 20 (2023). https://wepresent.wetransfer.com/stories/james-bridle-on-creativity-and-ai-collaboration
5. Chen, M.: 'A Collaborator Who Understood My Vision': Laurie Simmons, Casey Reas, and Mario Klingemann on How AI is Fueling Creativity. Artnet News, December 20 (2023). https://news.artnet.com/art-world/artists-ai-creativity-2340982
6. Crawford, K., Joler, V.: Anatomy of an AI System. The Amazon Echo as an Anatomical Map of Human Labor, Data and Planetary Resources (2018). https://anatomyof.ai/
7. Crawford, K., Paglen, T.: Excavating AI. The Politics of Images in Machine Learning Training Sets. September 19 (2019). https://excavating.ai
8. Christie's: Is Artificial Intelligence Set to Become Art's Next Medium? Christies.com, December 12 (2018). www.christies.com/en/stories/a-collaboration-between-two-artists-one-human-one-a-machine-0cd01f4e232f4279a525a446d60d4cd1
9. De Certeau, M.: The Practice of Everyday Life. University of California Press, Berkeley–Los Angeles–London (1984)
10. Diodato, R.: The dream of a machine. An example of the chiasmatic image. Paradigmi **3**, 517–528 (2023). https://doi.org/10.30460/112739
11. D'Isa, F.: La rivoluzione algoritmica delle immagini. Arte e intelligenza artificiale. Luca Sossella, Bologna (2024)
12. Galanter, P.: What Is Generative Art? Complexity Theory as a Context for Art Theory. Available at https://philipgalanter.com/downloads/ga2003_what_is_genart.pdf (2003)

[9] A scathing critique of this type of art is also found in the work of Zylinska [34]: 72–76.

13. Goodfellow, I., et al.: Generative Adversarial Nets. arXiv:1406.2661. https://doi.org/10.48550/arXiv.1406.2661 (2014)
14. Hals, T., Brittain, B.: Insight: Humans vs. Machines: The Fight to Copyright AI Art. Reuters, April 1 (2023). www.reuters.com/default/humans-vs-machines-fight-copyright-ai-art-2023-04-01/
15. Hinton, G., et al.: Statement on AI Risk. CAIS, May 30 (2023). www.safe.ai/statement-on-ai-risk
16. Howell, B.A.: Stephen Thaler v. Shira Perlmutter. Available at: https://fingfx.thomsonreuters.com/gfx/legaldocs/lbvgooeoqvq/AI%20COPYRIGHT%20LAWSUIT%20thalerdecision.pdf (2023)
17. Humlum, A., Vestergaard, E.: The Adoption of ChatGPT. Becker Friedman Institute for Economics Working Paper No. 2024–50, University of Chicago, April 25 (2024). https://doi.org/10.2139/ssrn.4807516
18. LeWitt, S.: Paragraphs on conceptual art. Artforum **5**(10) (1967). www.artforum.com/features/paragraphs-on-conceptual-art-211354/
19. Manovich, L.: Defining AI arts: three proposals. In: AI and Dialog of Cultures, exh. cat., Hermitage Museum, Saint-Petersburg (2019). http://manovich.net/index.php/projects/defining-ai-arts-three-proposals
20. Manovich, L.: Seven arguments about AI images and generative media (2023). In: Manovich, L., Arielli, E. (eds.) Artificial Aesthetics. A Critical Guide to AI, Media and Design (2021–). http://manovich.net/index.php/projects/artificial-aesthetics
21. Metz, R.: AI Won an Art Contest, and Artists Are Furious. CNN Business, September 3 (2022). https://edition.cnn.com/2022/09/03/tech/ai-art-fair-winner-controversy/index.html
22. Mohr, M.: Artist's statement. In: Gaudibert, P. (ed.) Manfred Mohr Computer Graphics – Une Esthétique Programmée, exh. cat. ARC–Musée d'Art Moderne de la Ville de Paris, Paris (1971)
23. Mohr, M.: The machine seen as an extension of us. Computing Europe, February 28 (1974)
24. Molnar, V.: Léonard de Vinci s'il eût eu un ordinateur. +-0 (Plus-Moins-Zéro) 52 (1989). http://www.veramolnar.com/blog/wp-content/uploads/VM1981_leonard.pdf
25. Noll, A.M.: The digital computer as a creative medium. IEEE Spectrum **4**(10), 89–95 (1967)
26. Pedrazzi, R.: Futuri possibili. Scenari d'arte e intelligenza artificiale. Jaka Books, Milan (2021)
27. Quaranta, D.: Surfing with Satoshi. Art, Blockchain and NFTs. Postmedia Books, Milan (2022)
28. Quaranta, D.: Demystifying artificial intelligence. Flash Art **361**, 87–96 (2023)
29. Rafman, J.: Ebrah K'dabri. DIS, April (2024). https://dis.art/jon-rafman-ebrah-kdabri
30. Reas, C.: The Work. Github, April 30 (2016). https://github.com/REAS/studio/wiki/The-Work
31. Szegedy, C., et al.: Going deeper with convolutions. In: Computing Research Repository (2014). https://doi.org/10.48550/arXiv.1409.4842
32. Valyi-Nagy, Z.: Between the lines. The art of Vera Molnar. Artforum **62**(9) (2024)
33. Zeilinger, M.: Tactical Entanglements: AI Art, Creative Agency, and the Limits of Intellectual Property. Meson Press, Lüneburg (2021)
34. Zylinska, J.: AI Art. Machine Visions and Warped Dreams. Open Humanities Press, London (2020)

Domenico Quaranta is a contemporary art critic, curator, and professor at Accademia di Belle Arti di Brera in Milan (http://domenicoquaranta.com). His research focuses on the mutual interconnections between technology, culture, and society. His writings have been published in magazines, journals, newspapers, books, and exhibition catalogues. He is the author of *Beyond New Media Art* (2013) and *Surfing with Satoshi: Art, Blockchain, and NFTs* (2022), among other works, and has edited several volumes. Since 2005, he has curated exhibitions for both national and international institutions, including: MNAC, Bucharest; iMAL, Bruxelles; LABoral, Gijon; HeK, Basel; 319 Scholes, New York; Kunsthaus Langenthal; + MSUM, Ljubljana; Palazzo delle Esposizioni, Rome; Centre d'Art Contemporain, Genève; MGLC, Ljubljana; GAMeC, Bergamo; OGR Turin.

Chapter 11
Crypto Art and NFTs: A Semiotic Inquiry

Marion Colas-Blaise

Abstract The first purpose of this article is to show that *non-fungible tokens* (NFTs), i.e. cryptographic tokens created by the uploading of an artwork on, for instance, the decentralised Ethereum blockchain that uses native cryptocurrency, accept verification tools for authenticity, provenance, scarcity and preservation. The paper turns to the opposition between autographic and allographic systems developed by Goodman (*Languages of Art: An Approach to a Theory of Symbols*, The Bobbs-Merrill Company Inc, Indianapolis, 1968). Then, attention is focused on the reception practices, when the transfer of a certificate of ownership of the digital artwork entails specific ways of collecting art. Finally, the question of the work of art is addressed. The art criteria are redefined, in the light of art movements of the twentieth century.

Keywords Non-fungible tokens (NFTs) · Autographic vs. Allographic · Authenticity · Reception practices · Work of art

11.1 Introduction

To what extent is the "digital revolution" [31] bringing radical changes to the arts? In particular is crypto art[1] a simple side effect of the blockchain technology, a distributed ledger technology often associated with cryptocurrencies such as Bitcoin, or does it reveal its wide-ranging impact? Is it enough then to say that some artists are

[1] Crypto art, also called cryptoart or digital art, is a form of digital art that uses cryptography to create unique pieces that can be authenticated and sold. Every work of art is linked to a NFT (non-fungible token), as for the use of cryptocurrencies. NFTs thus are a type of crypto art, unique tokens being minted thanks to a blockchain. They represent a digital asset. NFTs is a larger category that includes images, photos, videos, audio files and comic strips, games and collectible cards. So it is remarkable that all the generic classifications, all the distinctions between the arts, "minor" or otherwise, are being thrown overboard: NFTs can also cover sporting goods. The boundaries between the artistic, the technological and the playful are tested.

M. Colas-Blaise (✉)
University of Luxembourg, Luxembourg, Luxembourg
e-mail: marion.colas@uni.lu

P. Conte et al. (eds.), *Algomedia. The Image at the Time of Artificial Intelligence*, Lecture Notes in Morphogenesis, https://doi.org/10.1007/978-3-032-08726-3_11

"cryptofinanciers," the value of the art works being first of all a monetary one? Do new artistic forms and innovative artistic practices emerge that follow in the footsteps of former artistic movements and reshape them?

The blockchain technology and digital "time-stamping" have their roots in a journal paper by Stuart Haber, a cryptographer, and Scott Stornetta, a physicist [22]. It is highly significant that from the beginning on, blockchains may apply to texts and to art. Much more, "any string of bits can be time-stamped, including digital audio recordings, photographs, and full-motion videos" [18, 22]. Crypto art remained a "niche artistic movement," before the market boomed in the late 2020, which is confirmed by sales of crypto artworks in auction houses for record-breaking amounts of dollars [18].

An artistic *non-fungible token* (NFT) is a cryptographic token created by a transaction, i.e. the uploading of an artwork on, for instance, the decentralised Ethereum blockchain that uses native cryptocurrency. Uniquely linked to a work of art and testifying to ownership and authenticity, the token is moved to the artist's cryptographic wallet. The artwork may be distributed over the nodes of the InterPlanetary File System (IPFS) network,[2] i.e. a file-sharing and storage protocol. Named with a code, it is waiting to be sold, exchanged, et cetera, via auctions [19]. It may also be gifted, or even destroyed.

From a semiotic point of view, the association of an artwork with a NFT may be a strong reaction against some more controversial aspects of digital art: among these, the fact that a digital artwork can easily be reproduced and manipulated. On the one hand, Fake Art is said to be not only counterfactual (that which has not been achieved although it could have been), but also a counterfeit (a fraudulent copy) [8]. On the other hand, when there are multiple versions that differ from the original work of art, it becomes difficult to identify the *princeps* version and the notion of copy should be questioned [21]. To this we can add that digital art is, by definition, ephemeral. The urgent question is how to preserve and secure it. Finally, how does this affect creative practices and the way collectors handle different market variables?

First, the paper seeks to show that NFTs accept verification tools for authenticity, provenance, scarcity and preservation. The paper turns to *Languages of Art* by Goodman [21]: does the opposition between autographic and allographic systems allow us to better understand crypto art?

Secondly, we focus on the reception practices: what does it mean to transfer a certificate of ownership of the digital artwork? Are there new ways of collecting art?

Thirdly, when does a NFT become a work of art? To what extent do the reception practices differ from the traditional "aesthetic encounters" [28] that a spectator may experience? We will try to redefine the art criteria, wondering whether crypto art draws its origins from art movements of the twentieth century.

---

[2] The InterPlanetary File System (IPFS) is a peer-to-peer network that stores files and data and shares them in a distributed file system.

11.2 The Artistic NFTs and Their Specific Features: Authenticity, Provenance, Preservation

Deep generative networks are widely used for bridging the gap between a low-dimensional latent space and a high-dimensional data space. Creative abilities are however often limited by the objective of "perfection," when the output data seek to be similar to the input data extracted from the examples used in training. In 2020, Berns and Colton have stated that generative models still are best at producing "more of the same." Models with millions of parameters identifying the patterns in data sets as accurately as possible may simply memorise and reproduce the examples given in the corpora mobilised at the beginning of the generative process. Despite the generation loss due to the double movement of compressing and decompressing data by the encoder and the decoder, overfitting hinders learning systems from being more creatively autonomous. However, random generation and experimentation come to the fore in early 2022, with the arrival of new deep learning technologies belonging to the field of generative AI, such as Midjourney and DALL•E 3. Images are created not only from images, as with Generative Adversarial Networks (GANs),[3] but also from textual prompts.

Hence, in order to avoid standard usages and to generate surprising artefacts, computational creativity[4] *a minima* relies on evolutionary algorithms, on the mixing of training sets, on divergence techniques that focus on "sub-optimal" values. They attest to the desire to go beyond the regularities in the data and thus facilitate the evolution of generative models towards autonomous systems. The determining factors are a distribution of possibilities beyond any predictable combination and the reduction to a minimum of what Berns and Colton [4] call "clip values," i.e. constant average values. This is in favour of data value components that are above or below a specific threshold. In their article "Seeding Diversity into AI Art", Zammit et al. [37] mobilise a database of 81,444 artistic images of various styles. They envisage replacing 5% of an individual's genes associated with randomly chosen segments with random integers; the result is a mutation coefficient that can create perceptible disturbances in the image without making it non-identifiable.

Crypto art diverges from the original objective of mere reproduction conforming to the patterns of the training corpus. Thanks to milestone architectures in AI Arts such as GANs, the artists put forward the criteria of experimentation, novelty, rather than targeting only typicality. However, are NFTs protected from reproduction? More importantly, are they safeguarded against manipulation, transformation, and various forms of reinterpretation—what Colas-Blaise and Tore [11] refer to as being "re-enounced," that is, captured by new semiotic processes? Let us consider digital art in general: thanks to the interface design [32], the internet user may increase contrast, change tones, enlarge ore reduce dimensions and introduce filters. Characteristic features may be added, replaced, subtracted, in combination with local operations

---

[3] GANs were introduced during the mid-2010s. The aim is less to analyse and classify images than to modify them or to create entirely new images.

[4] Computational Creativity is known as an emerging branch of artificial intelligence studies [34].

such as rotation, zooming, cut-out, exposure. The result is a large number of variants whose origins cannot always be precisely traced. The variants can be considered as successive, but they can also be captured simultaneously. In that case, they are denied any anchorage in chronology.

In this context of re-generation, in the double sense of either reproducing (nearly) the same or transforming images (in French: *ré-génération*[5]), to what extent does the blockchain bring radical changes? Does it activate the search for new artistic strategies? As mentioned above, this technology emerged from a paper by Haber and Stornetta [22], who wonder how to "time-stamp a digital document." From a semiotic perspective, the epistemological background is most interesting: the question of truth [36]—how can we trust a digital file that may have been manipulated?—is even more important nowadays, at a time when we are confronted with fake news every day. Are we able to distinguish between an original, a fraudulent copy or a replica, according to [21]? The replica may be unlimited. The origin may be "an eddy in the stream of becoming" [3]. Then, how can we keep the record, in order not to be overwhelmed by increasing amounts of unstructured data?

Blockchain in crypto art is responsible for "digital scarcity" [18]. The work of art is no longer caught in a blending transformation process. Each item is fixed before user modification or distortion. The blockchain technology artificially limits the range of possible renderings. Crypto art is no longer part of a broad range of renderings of the same data, nor is it governed by the principle of internal variation. It lacks the inherent reproducibility of contemporary digital art—not so much because it represents a new art form, but rather because its artistic practices differ. These differences stem primarily from storage on a blockchain, decentralisation, and a new possibility for artists to maintain control over their works, as will be demonstrated below.

Crypto art hosted by a blockchain represents something unique. When the crypto art is reproduced, one might say that the work is not a replica of a replica of a replica…,[6] but a copy distinguishable from the original by the date.

As the blockchain technology stops the dynamic process of replication, one may suggest that NFTs are autographic, according to Goodman's [21] distinction between autographic and allographic systems.[7] Among others, what sets digital art in general apart from traditional art is the fact that historicity is initially masked in an "insolent presentism that defies the passage of time in an assertion that disregards veracity" (Bachimont 2022; my translation). But Bachimont immediately adds that a "digital diplomacy" under the banner of *digital forensics* can "establish the authenticity of content," making it possible to "trust content" (Bachimont 2022; my translation). The history of production is taken into account. This applies to digital art in general and

[5] In French, we can distinguish between the "re-génération" (generate once more) and the "ré-génération" (creative process).

[6] Although digital works can be copied exactly, there is no degree of similarity necessary for a replica.

[7] The non-crypto digital art is also authentic to some extent, assuming that the digital code holds the status of the original. When the code is changed, the meaning of the visual piece is reconstructed anew.

to crypto art in particular. In the latter case, the verification tools for provenance are especially effective. Metadata like title, description and tags of artworks are stored on networks like IPFS and may easily been downloaded [18].

Still, like all digital art, crypto art is coded before being displayed and the "copy" is "genuine" as long as the notation or score is reproduced exactly, i.e. is accurate ("orthographic correctness"). If allographic artforms such as text are amenable to replication, it is because they consist of sequences of symbols, of constitutive properties that are distinguishable. In the latent space, the conversion of the image consists of turning the features into discrete data, i.e. into numbers (embeddings) as well as into formal, structural relationships. At the root of the generative process in AI lies the reduction of features of images, the "contingent" ones being dismissed, down to a "digital" diagram,[8] before semantic and syntactic density is added at a second stage. The act of combination and composition in a high-dimensional data space is followed by the act of semio-rhetorical [12] "performing," especially through the visualisation of the artifact.

We argue that digital art, be it crypto art or not, is allographic *and* autographic. The authenticity may be said "complex" [10]. The complexity then varies in degrees according to different weightings. Current non-crypto digital art is more allographic than autographic, first of all because it is reproducible, provided a basic "symbolic notation" [21][9] is respected, the reverse being true for crypto art. In the latter case, authenticity and authentication are highlighted and they play a major role on the art market.

Finally, the storage of a NFT on a blockchain helps preserving and securing the digital artwork. Traditionally, digital art is ephemeral for two main reasons: as mentioned above, it is easily manipulable and distorted. Crypto art, on the contrary, is fixed. Secondly, in order to be fully "realised" [15],[10] digital art needs to be visualised via a type of screen, the visualisation techniques contributing to the production of meaning each time again. The artwork thus may disappear from the screen, be replaced by a different artwork, come back and announce other artworks, and disappear once more, according to an "analeptic" (referring to the above) and "cataleptic" (anticipating the following), "hypertextual," logic [9]. As we have seen, crypto art gains in stability.

Let's put our hypotheses to the test with crypto artworks by Kevin Abosch. In particular with the NFT series *Sun Signals* (2021), which includes 1010 unique generative artworks. Adopting an ecological perspective that is sensitive to climate change and the carbon footprint, Abosch wants to celebrate solar cycles and radiation on Earth. The solar signals are generated "off-grid" using computer servers powered by solar energy. Released on October 10th 2021, the series was made available

---

[8] According to Goodman [21], the purely digital diagram constitutes a completely differentiated system, syntactically and semantically, with lines and points functioning in a notational language.

[9] For Goodman [21], a symbol system is called "notational" notably if the symbols are non-compact.

[10] Fontanille [15]: 289–290 distinguishes four modes of existence of entities: the virtual, the potentialised (a discursive form is lodged in the background, awaiting a new manifestation), the actualised and the realised (when the discursive form meets a reality, such as a material one).

on OpenSea marketplace on the Ethereum blockchain. The artist uses encryption algorithms (or the encryption of documents using algorithms)[11] in order to reduce the vulnerability of data. The blockchain prevents any changes thanks to interconnected blocks and painting. It is unique in that it sends a transaction, in the form of a block of encrypted data, to computers all over the world. Each node hosts a copy of the database containing the transaction history. Finally, the images and metadata are stored on Arweave's Permaweb, as part of a wider project, "1111," designed on the basis of a decentralised autonomous organisation (DAO) and a "green" satellite project, "1111 KOSMOS." The works of art are "dropped into space," before being captured on earth.

Once again, we see that artistic NFTs differ from non-crypto digital art works primarily by means of *artistic practices* that are creative in a broader sense. In the case of Abosch, *creative gestures* include storage on a blockchain not only to experiment new ways of raising awareness of ecological issues, but also to secure the works.

These artistic practices require further study.

11.3 From Transferring the Certificate of Ownership to Collecting

First, let us adopt the spectator's point of view. Since the invention of blockchain certificates of ownership in 2014, the presence of this cryptographic asset on a blockchain constitutes a digital equivalent of a certificate of authenticity for a unique, highly distinctive, identifiable and non-exchangeable object. The fact that the NFT can be viewed in a definitive form establishes a relationship with the owner that is reminiscent of the purchase of a "physical" work of art. This is another reason to say that, unlike repeatable, reproducible and manipulable digital productions, NFTs are more autographic and less allographic. Thanks to the metaverse, the work of art destined to be attributed to an owner-collector is credited with a market value. The conventional art market is rejected.

But does decentralisation, which allows so-called copies to be hosted on millions of computers simultaneously, call into question the uniqueness of the NFT? One solution would be to say that the NFT is spread over different computers, which the very nature of the NFT contradicts from the outset: the NFT is non-fungible, i.e. it cannot be divided or split. It seems more satisfying to say that the NFT is located in different places without fundamentally changing neither its "qualitative identity"—one cannot say that "contingent" features like the thickness of the lines are dismissed—nor its "quantitative" or "numeric" identity. The certification through a smart contract on the blockchain grants the uniqueness of the NFT despite its being potentially stored on different computers.

[11] Encryption is the process of protecting information or data through mathematical models: only the parties who have the key are able to access it.

Some changes are inevitable. According to Genette ([20]: 261),

The "unique" physical objects in which these works [artworks] consist are [...] not unique only from the point of view of numeric identity; qualitatively, they are unique only in the purely theoretical immutability of the instant; in the duration of their persistence, they are temporally plural, through the effect of endless more or less perceptible mutations. This is called growing old [or] the universal fact that a (specific) identity never ceases to modify itself, spontaneously or by intervention, and that the life of works is not a restful one (my translation).

Every "aesthetic encounter" with a work of art results in a new interpretation, which makes its meaning evolve, although the "numeric identity" remains the same. At the same time, the "temporal plurality" is here combined with a "local plurality." We suggest that these changes do not affect the uniqueness of a NFT artwork, which is a one-of-a-kind collectible. Even distributed on several nodes of the network, it keeps the same name.

What does it mean to own a NFT artwork? Basically, storing such a work of art on a blockchain is akin to transferring the ownership to some collector. On the crypto art market, pieces by digital artists including Beeple, for example, are sold by crypto art galleries such as MarketPlace and Nifty Gateway as well as by Christie's and Sotheby's. As for non-digital art, one may assume that *mutatis mutandis* the market value is attributed in "concentric" circles, from peers to experts, to dealers and collectors [24]. To what extent is the "general public" involved? As for non-digital artworks, it is possible that the owner does not own the intellectual property rights, i.e. the copyright and the right to use the work for commercial purposes. It all depends on the terms of the smart contract. Furthermore, The *Gazette Drouot* (2017) underlines the principle of the independence of a work and its support: according to article 29 by the French law of 11 March 1957, then L. 111-3 by the French law of 13 July 1992, "the intangible property defined by article L. 111-1 is independent of the ownership of the material object" (my translation). Strictly speaking, the owner of a work of art owns the material support of this work: a frame, a canvas, materials such as bronze or marble. He does not own intangible artistic property. Rather, it is the object in which the "work of the mind" can materialise.

This leads us into the debate on the materiality versus immateriality of digital documents. On one hand, the digital document is often described as virtual or immaterial. It is considered to be the result of a process of dematerialisation. Materialisation would slow down circulation and hinder access to immediacy [30]. We may then focus on a genealogy that stems from conceptual art [19]. On the other hand, Bachimont (1996)[12] argues for a material hermeneutic, considering that (i) computer science is a science of nature: "computation must be considered as a 'natural' object, that is, as a 'material' object in the sense that it belongs to what nature produces" and that (ii) "all knowledge rests on the mediation of a material inscription" (my translation). Hence the problem of "artifacture": "if computation is a material object, we can consider it as a dynamic material medium of inscription, and therefore we

[12] Doctoral Thesis titled *Herméneutique matérielle et Artéfacture: des machines qui pensent aux machines qui donnent à penser. Critique du formalisme en intelligence artificielle*, May 24, 1996.

can consider computer artifacts as dynamic material media of inscription in order to study the constitution of the knowledge they authorise" (Bachimont 1996; my translation). The materiality of the artefact thus enables a flow of thought to be "permanent." Inscription is both "transcendental," in that it is not only a consequence of, but also a "precondition for experience and the emergence of knowledge," and it is "empirical," in that it is "constituted by our material and concrete experience" [1], my translation). Although the digital schema is not dependent on the medium in which it is modelled, although schematisation frees itself from the environment associated with the machine, the "implementation" and the visualisation on a screen *reveal* so-called "immaterial" information. Crozat et al. [12] distinguish (i) the ideal object as calculability, (ii) the material existence through implementation, (iii) the practical consistency through perceptible manifestation and (iv) the interaction at the semio-rhetorical level. Fontanille [16], for his part, contrasts two modes of existence, internal and external, of the material support and the formal support: first, the "material support of inscription" (the electronic file), which is "electrical, physical, digital and imperceptible" (2005: 189), is opposed to the formal support of computer coding secondly, the external material medium of the screen, etc., is distinct from the formal medium of the "screen page." Hence the distinction between different levels or strata, as shown by Moutat ([29]: 60–63): the levels correspond to material support $_{n}$ (the screen), material support $_{n\text{-}1}$ (navigator), material support $_{n\text{-}2}$ (compiler) and material support $_{n\text{-}3}$ (assembler). Finally, the materiality of crypto art is linked to its "perceptible manifestation."

In that sense, NFTs require insightful resolutions to a longstanding debate. We might conclude that the owner of crypto art *does not own* the formal support: "The formal support is the structure that accommodates the inscriptions, the set of topological rules of orientation, of dimension, proportion and segmentation, in particular, that constrain and give meaning to the inscribed characters" [16]: 186, my translation). The gesture of creation remains beyond his reach. Unless we broaden the notion of the creative gesture to include storage on the blockchain, as will be shown below, and reception gestures for which the owner is partially responsible. Such a change of focus encourages further inquiry into the fundamental principles governing the *access* to NFTs.

Indeed, the originality of NFTs hosted by a blockchain also lies in a particular way of accessing the artwork. The owner-collector is given a cryptographic wallet containing the keys to access the digital assets. Only a unique "boot phrase" or "recovery phrase" gives access to the wallet. Limited access (often with pre-registration) therefore has to be negotiated, as well as display, which is *limited* in time. The important thing is that the methods of display (they provide access to the wallet) are unique and differ from the methods of production and acquisition of the artworks.

What does it mean being a collector of NFTs? The market booms since Christie's March 2021 sale of *Everydays: the First 5000 Days* (2021) by Mike Winkelmann, aka Beeple, for USD 69.3 million. Owners collecting NFTs do not rely so much on criteria such as the intrinsic artistic value of the artwork, the genres and the selected themes (for instance, collect only landscape paintings or portraits relating to specific periods, art movements, etc.), but on reasons that may differ from the ones governing the traditional non-digital art market. So German collector Peter Niemann, who, together with Gunda, founded exhibition *Haus N Athen*, a non for profit private space, in 2017, bought works from Nagel Draxler, arguing that he wanted to understand "how the emergent technology works," how the NFT market is evolving, and "for some fun."

Thus, one of the most central questions concerns the emergence or the non-emergence of a new "artistic paradigm" ([23]: 52). From the very outset, it is essential to note with Heinich that a paradigm does not simply encompass the chronological dimension nor the dimension of classification dear to aesthetics. It also includes the whole range of discourses on a socio-cultural, economic or political environment, and the ways in which works are circulated, disseminated and received. Does crypto art benefit from the "digital revolution," which has given rise to a new form of life based on connectivity? As with non-digital contemporary art, some traditional expectations are not met ([23]: 63): that of immediate visibility: as mentioned above, the display of the work needs to be negotiated. Regarding subjectivity, in his paper « Le computationnel et le sujet» (2015), Salanskis notes that "the computational era is not necessarily in harmony and phase with the ethos of subjectivity" (my translation). One may add that, in contemporary non-digital art as well as in digital art, there is no perfect control over the creative process: the unsupervised process of generating surprising artefacts includes randomness. At the same time, through its galleries and collectors, crypto art not only centralises and democratises the art market, but also allows for more individual control [19]. Crypto art can also be considered expressive in that it does not simply reproduce existing forms of meaning identically, in line with pre-established conventions, even if the criterion of expressiveness is probably not the first one to be adopted by collectors of NFTs. Even if the "disruptive" nature of the crypto work of art is valued, there is every reason to believe that its "numeric identity" and the scarcity of the NFT are often outweighing its "qualitative identity." The criterion of uniqueness applies to its singularity: creativity is highlighted by the Computational Creativity movement, even if, in the case of NFTs, the notion of creativity needs to be broadened, as will be shown below. It also relates to the "material" mode of existence of the work of art.

11.4 What Qualifies a Work Associated with a NFT to Be an Artwork?

Once more, let us consider *Sun Signal* #596, by Kevin Abosch (Figs. 11.1 and 11.2).

The specialist in Visual Semiotics will be sensitive to the protuberance (simulation of a three-dimensionality) of a red disc, which is apparently projecting towards the viewer. He will identify it as the figure of the sun, of the "red planet," as Marin [27] notes with regard to the painting *Ad marginem* by Paul Klee (1930). According to Bordron [5, 6], analyses can be carried out at two different levels: that of forms, using

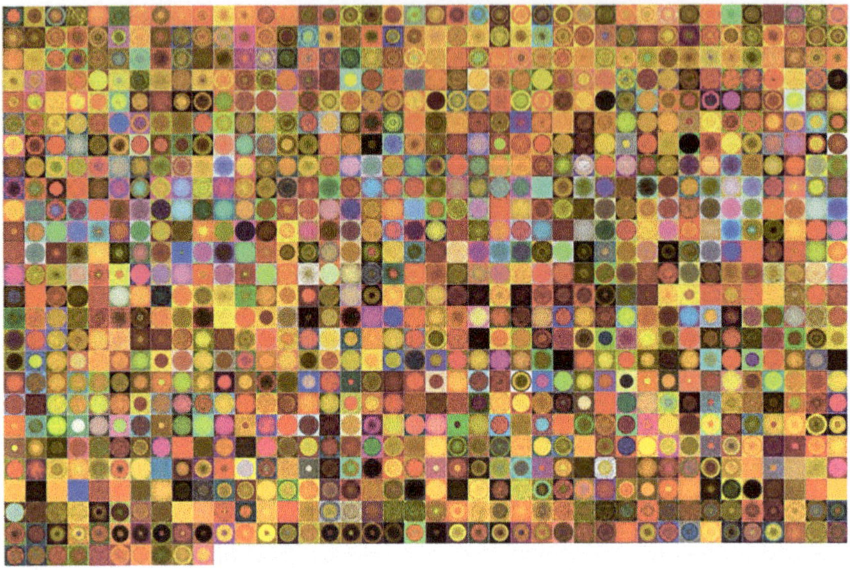

Fig. 11.1 Kevin Abosch, *Sun Signals,* 2021

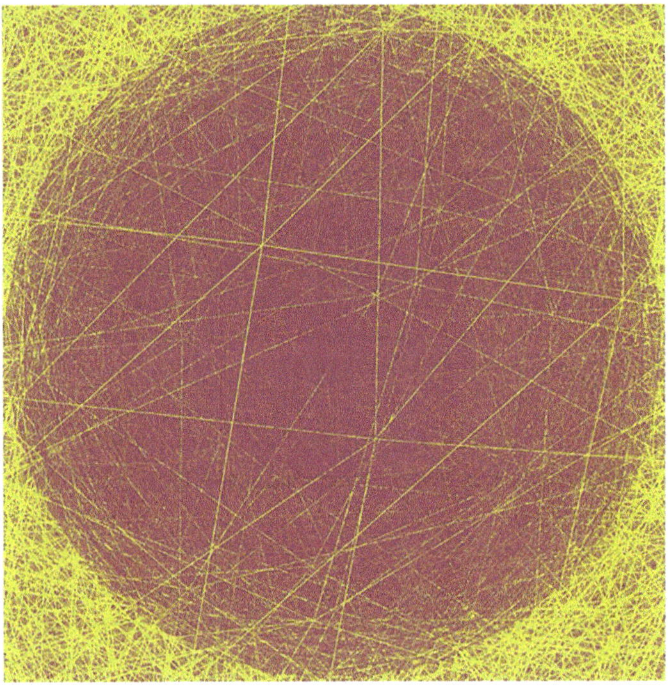

Fig. 11.2 Kevin Abosch, *Sun Signals*, #596, 2021; courtesy of the artist

a visual language with rules and conventions, and that of the underlying converging and diverging forces of a composition or morphology, prior to the control of the symbolic language. As they venture onto the surface of the red disc itself, the yellow lines coming from the peripheral parts are responsible for the cohesion of the whole, central part and margins, without the underlying tensions being completely resolved.

This underlines the ecological issues at stake in Kevin Abosch's work, which, let's not forget, includes *Sun Signals* as part of the wider project "1111 Kosmos." No doubt that he wants to represent the world through more or less dense networks of fine lines, multiplying and intersecting, which symbolise the innumerable connections of every living being, human and non-human, with other living beings. Geometric shapes, patterns and (ir)regularities may emerge, essentially at the center of the image. The density of the gridded space increases as we move from the center to the periphery.

But that is not all. Like the NFT series titled *Preludes* by Trevor Paglen, a generative art project of 250 outputs, which is supposed to recall algorithmically generated pieces by composers such as John Cage and Iannis Xenakis, besides Haydn and Brahms who encoded messages into their music,[13] Abosch seeks to convey a sense of mystery. *Preludes* is part of a gamified project that includes "speculative reality work": the acquisition of the NFT allows you to decode and enter the line "Pm fvb hyl ylhkpun aopz, fvb ohcl ahrlu fvby mpyza zalw puav h shynly dvysk." Once decoded, the message says: "If you are reading this, you have taken your first step into a larger world." The code leads the owner of the NFT to claiming his prize, a vinyl LP. The latter contains further puzzles, exemplified by a seemingly meaningless sequence of numbers: "65 76 69 82 84 32," which is read by a computer voice.[14] We are told on the net that the owner of the NFT is involved in a "new art genre that combines elements from Alternate Reality Games (ARGs), with aspects of historical 'unfiction', experimental narrative, and cooperative problem solving." A hint is given of how to play the game: collaboratively, in a well-organised group.

It would appear that the aesthetic quality of crypto art, as defined by more traditional art criteria, is not always relevant. Attention needs to be focused on the *art practice* as a whole, which includes not only the generation of digital artworks, but also their storage on a blockchain as well as the wide variety of ways in which they are acquired and "interpreted" (including the solving of enigmas). This multifaceted practice, strongly rooted in the cultural, social and even political context, invites us to renew the notion of the creative gesture, in production and reception. If the aesthetic

[13] Trevor Paglen comments on his practice: "[...] I've always been interested in the numerous methods humans have developed to hide secrets and the ingenious ways others have created to expose them. For the PRELUDES works, I challenged myself to think about how one could hide secrets using the particularities of an entirely on-chain medium. While the blockchain's underlying technology, of course, relies on the same types of cryptography used by intelligence agencies, secure banking systems, and other entities, the question of how to hide something in a medium where everything is invisible poses a particular type of problem. I went back to my early training as a composer to find an answer"; https://beta.rarible.com/token/0xea698596b6009a622c3ed00dd5a8b5d1cae4fc36:5000044.

[14] https://news.artnet.com/market/trevor-paglen-new-nft-project-cyclops-2300762.

form is distinguished from the work of art that is exposed, sold and circulated (process of "artistisation"),[15] one may suggest that more than ever, the production and the reception of the NFT crypto art as object are decoupled from aesthetic creation in a narrow, traditional sense. What is validated and positively evaluated is an original way of "artistising" (circulating, storing, archiving), the form of life that underpins it, rather than the result displayed. The act itself can become a work of art. It is the artistic idea that takes precedence over the aesthetic qualities as such. This shift from the aesthetic to the artistic merely reinforces a phenomenon typical of the contemporary art paradigm as described by Heinich ([23]: 53): it falls under the heading of the "peri-artistic," requiring, as Verlaine ([35]: 462) notes, that "professional intermediaries, art dealers and critics considerably modify their means of action." As for contemporary art, the "construction of strategies of address" (inscription in the field of art) is emphasised [26].

It is all the more significant that some NFTs are on display in crypto museums. If the goal of the National Cryptologic Museum, which is located next to the headquarters of the National Security Agency at Fort George G. Meade, Maryland, is to educate visitors about the role cryptology plays in shaping history, unique cryptology artifacts are collected, preserved and showcased in the Museum of Crypto Art (M○C△). Including 251 artworks by 251 crypto artists since December 2020, it uses a multi-stakeholder decentralised platform dedicated to art curation and exhibition in order to "challenge," to "create conflict," to "provoke".[16] Its mission statement reveals the profound issues at stake in such an undertaking, which go far beyond any purely trading or gaming experience. Indeed, the Museum addresses those who "dare to believe in a better future that prioritised sovereignty, market access, and freedom of expression in the arts." As visual aesthetics are essential for communicating these "ideals," the museum seeks to combine the process of "artistisation" with the process of "aesthetisation." Together they may give rise to a new "socio-economic paradigm," whose aim is to deeply connect artists and collectors through a "cohesive structure that can shape [a] communal identity, culture, and recognise what is fundamental," so that crypto art may "permeate everyday experience." Collaboration through sharing the collections is incentivised. The public is invited to "explore the beautiful and soul-quenching need to create and recreate." The M○C△ Genesis Collection allows the active viewer to delve into the depths of the Metaverse. Representing "the earliest etchings on the blockchain," the artworks are considered as the "digital cave paintings of our transhumanist narrative." Without any real surprise, we learn that people involved are called "evangelists".[17]

Should it be seen as more than an artistic project, as a political commitment, for example to the environment? Does the crypto art museum take on a real mission, his philosophical background having religious overtones? We are faced with a project which aims to be societal, with an "ideological tribute" that relies on "fair distribution

[15] The artistic is often subordinated to the aesthetic, i.e. to an appreciation (Dickie 1992), but it may also be dissociated, for instance in the case of Duchamp's *Fountain* [13].

[16] https://museumofcryptoart.com.

[17] https://museumofcryptoart.com.

models." It wishes to propose an unprecedented form of life, that is to say, according to Fontanille's definition of the form of life [17], the congruence of values, narrative roles and passions with a syntagmatic made up of actions, demands, collaborations and attitudes attesting, in our case, deeply-grounded beliefs.

More generally speaking, in the domain of crypto art, there are alternatives to a purely financial goal, with market values being replaced by existential ones. These ambitions are not fundamentally foreign to artists such as Paglen, who, by proposing a "treasure hunt" involving the solving of riddles, merely confirms the political and cultural commitment demonstrated by other works. Combining the social sciences with journalism and art, Paglen looks at the societal consequences of artificial intelligence technologies, in particular their use by governments for surveillance purposes.

11.5 Conclusion

To conclude, let us ask ourselves whether crypto art, which can be seen as subversive, has any links with twentieth-century art movements, whose ambitions it seeks to take to the extreme. The new forms of digital art, since the early 2010's, are often referred to as post-conceptual art.[18] Artists are then related to generative art that relies on autonomous systems of algorithms. Does post-conceptualism reinvent Pop Art? In both cases, the merchandising of artworks plays a major role. Yet, unlike unlimited Pop Art, crypto art stored on a blockchain seeks scarcity, authenticity, uniqueness and value-preservation, as mentioned above. What about Conceptual Art? At first glance, crypto art and conceptual art seem to be at odds with each other, the latter blocking any form of commercialisation by preferring unfinished works that cannot be sold. What they both have in common, however, is the strong socio-political commitment demonstrated by Joseph Beuys's social sculpture, for example. If conceptual art attaches an artistic value to thought processes and production methods, we could try to find this constant in crypto art as well: as if all the artistic NFTs were valued not for their intrinsic aesthetic value, but as vehicles for a key idea. Which once again raises the question of the "dematerialisation" of crypto art, on the model of conceptual art. As far as they are not physical copies, which again leads us to re-examine the nature of the autographic regime, they may be considered "dematerialised" manifestations of a basic concept, unlike, in that case, the object chair, the lexicographic description and the photo of a chair, whose association, in *One and Three Chairs* by Kosuth, may suggest the concept of an ideal and invisible chair.

At the same time, we argued that crypto artworks are interwoven with a digital form of materiality. When they are implemented and visualised, when they are stored

[18] The term "post-conceptual" refers to the California Institute of Art, under the influence of John Baldessari. If the work of the artist dates back to the early 1970's, the term was brought forth by the British philosopher and critic Peter Osborne, who strived to describe art objects that do not convey a particular meaning.

on a blockchain, when they circulate on the art market, when they are sold and "interpreted" by their owner (for instance through solving riddles), they are *art objects* rather than purely formal, perhaps immaterial, algorithmic combinations. Materiality is also a question of "activating material relations" [7], i.e. embodied relations with the owner. As for conceptual artworks, planning beforehand is involved in the whole process of creation. Unlike conceptual art, the execution is not only a perfunctory affair.

Perhaps we could say that crypto art profoundly renews the German idea of the *Gesamtkunstwerk* (of the "total work of art"). In this particular case, the focus is not on the interplay of different art forms, but on the creative process in a broad sense, encompassing not only the generation of the digital image, but also its transformation into crypto art associated with a NFT, its storage on a blockchain, and the various commercial practices, not forgetting the reception practices that contribute, in their own way, to making the work of art meaningful. From this point of view, "artification" [25][19] comes to the rescue not so much of aesthetisation as of artistisation (for example, through the exhibition in a museum). The "global crypto art" requires a sequence of more or less immediately creative actions. At the same time, each singular art NFT bears the mark of all the environmental, social, political and cultural influences as well as of the choice of different media (such as the art market and the crypto museum), with their more or less decentralised distribution circuits.

References

1. Bachimont, B.: Le numérique comme support de la connaissance: entre matérialisation et interprétation. In: Gueudet, G., Trouche, L. (eds.) Ressources vives: Le travail documentaire des professeurs en mathématiques, pp. 75–90. PUR, Rennes (2010)
2. Bachimont, B.: Donner du sens aux données: les ruses du numérique. Les disciplines du document face à la métis du calcul. Interfaces numériques **11**(2) (2020). https://doi.org/10.25965/interfaces-numeriques.4838
3. Benjamin, W.: The Origin of German Tragic Drama (1928). NLB, London (1977)
4. Berns, S., Colton, S.: Bridging generative deep learning and computational creativity. In: Proceedings of the 11th International Conference on Computational Creativity (ICCC '20) (2020)
5. Bordron, J.-F.: L'iconicité et ses images. PUF, Paris (2011)
6. Bordron, J.-F.: Dynamiques des images. Signata **10** (2019). https://doi.org/10.4000/signata.2267. Accessed 15 Feb 2025
7. Bruno, G.: Surface: Matters of Aesthetics, Materiality and Media. University of Chicago Press, Chicago (2014)
8. Caliandro, S.: Fake Art: entre le contrefait et le contrefactuel. Interfaces numériques **11**(2) (2022). https://doi.org/10.25965/interfaces-numeriques.4889
9. Colas-Blaise, M.: L'énonciation: Évolutions, passages, ouvertures. Presses universitaires de Liège, Liège (2023)
10. Colas-Blaise, M.: La fiction au risque de l'art numérique. Actes sémiotiques **131** (2024). https://doi.org/10.25965/as.8651. Accessed 15 Feb 2025

[19] Cf. also a different use of "artifact": the generation of a computational artifact, in a broader sense.

11. Colas-Blaise, M., Tore, G.M.: "Re-". Répétition et reproduction dans les arts et les médias. Mimésis, Milan (2021)
12. Crozat, S., Bachimont, B., Cailleau, I., Bouchardon, S., Gaillard, L.: Éléments pour une théorie opérationnelle de l'écriture numérique. Document numérique **14**(3), 9–33 (2011)
13. Danto, A.C.: The Transfiguration of the Commonplace. Harvard University Press, Boston (1989)
14. Dickie, G.: Defining art. Am. Philos. Q. **6**, 253–256 (1969)
15. Fontanille, J.: Sémiotique du discours (1999). Pulim, Limoges (2003)
16. Fontanille, J.: Du support matériel au support formel. In: Arabyan, M., Klock-Fontanille, I. (eds.) L'écriture entre support et surface, pp. 183–200. L'Harmattan, Paris (2005)
17. Fontanille, J.: Formes de vie. Presses universitaires de Liège, Liège (2015)
18. Franceschet, M.: The Sentiment of Crypto Art. CHR 2021: Computational Humanities Research Conference, Amsterdam (2021)
19. Franceschet, M., et al.: Crypto art: a decentralized view. Leonardo **54**(4), 402–405 (2019)
20. Genette, G.: L'œuvre de l'art (1994). Seuil, Paris (2010)
21. Goodman, N.: Languages of Art: An Approach to a Theory of Symbols. The Bobbs-Merrill Company Inc., Indianapolis (1968)
22. Haber, S., Stornetta, S.: How to time-stamp a digital document. J. Cryptol. **3**(2), 99–111 (1991)
23. Heinich, N.: Le paradigme de l'art contemporain: Structures d'une révolution artistique. Gallimard, Paris (2014)
24. Heinich, N.: Qui décide de la valeur d'une œuvre d'art contemporain? Nectart **2**, 93–102 (2016/1)
25. Heinich, N., Schapiro, R. (eds.): De l'artification: Enquêtes sur le passage à l'art. Editions de l'École des Hautes Études en sciences sociales, Paris (2012)
26. Lenain, T.: Le geste auctorial. Revue Philosophique de Louvain **119**(3), 42–70 (2022)
27. Marin, L.: De la représentation. Gallimard, Paris (1994)
28. Morizot, B., Zhong Mengual, E.: Esthétique de la rencontre. Seuil, Paris (2018)
29. Moutat, A.: Écriture et matérialité numérique: de la substance aux formes de l'expression. Semen **49**, 51–67 (2021)
30. Negroponte, N.: L'homme numérique. Robert Laffont, Paris (1995)
31. Rabot, C., Chartier, R.: De l'imprimé au numérique: une révolution de l'ordre des discours. Biens symboliques **7** (2020). https://doi.org/10.4000/bssg.478. Accessed 15 Jan 2025
32. Reyes-Garcia, E.: La datavisualisation comme image-interface. I2D—Information, données & documents **52**, 38–39 (2015/2). https://doi.org/10.3917/i2d.152.0038. Accessed 15 Jan 2025
33. Salanskis, J.-M.: Le computationnel et le sujet. In: Larsonneur, C., Regnauld, A., Cassou-Noguès, P., et al. (eds.) Le sujet digital, pp. 82–98. Presses universitaires de Paris Nanterre, Paris (2015)
34. Veale, T., Cardoso, F.A.: Computational Creativity: The Philosophy and Engineering of Autonomously Creative Systems. Springer Nature, Cham (2019)
35. Verlaine, J.: Les Galeries d'art contemporain à Paris: Une histoire culturelle du marché de l'art, 1944–1970. Publications de la Sorbonne, Paris (2012)
36. Whitaker, A.: Art and blockchain. Artivate: J. Entrepreneurship Arts **8**(2), 21–46 (2019)
37. Zammit, M., Liapis, A., Yannakakis, G.N.: Seeding diversity into AI art. In: Proceedings of the 31st International Conference on Computational Creativity (2022)

Marion Colas-Blaise is emeritus at the University of Luxembourg. She was the Head of French Studies Bachelor in European Cultures and of the Institute for Roman Studies, Media and Art. She has authored and co-authored (collective) books and scientific papers in the fields of literary and visual semiotics, enunciative linguistics, stylistics, pragmatics, and Discourse Analysis. Her work develops at the intersection of different disciplines (language sciences, art theory, philosophy, sociology, anthropology). She is co-director of the scientific journal of semiotics *Signata* and she is a member of the editorial board of the collection “Sigilla”. Among her recent publications: *L'énonciation. Évolutions, passages, ouvertures* (2023); *Matérialités du discours rapporté* (2024, co-ed.); *“Re-.” Répétition et reproduction dans les arts et les médias* (2021, co-ed. G.M. Tore); *La communication à l'épreuve du geste numérique, MEI*, 47 (2019, co-ed. P. Basso-Fossali, M.G. Dondero); *Synesthésies sonores* (2018, co-ed. V. Estay Stange); *L'énonciation aujourd'hui. Un concept clé des sciences du langage* (2016, co-ed. L. Perrin, G.M. Tore).

Chapter 12
Exhibiting Contemporary Artworks Co-produced with Artificial Intelligence

Aluminé Rosso

Abstract This article analyses the exhibition of AI co-produced artworks in art museums from a dual perspective, focusing on their historical, aesthetic, and institutional implications. The first part of the study, grounded in theories of contemporary art and curatorial studies, examines how AI technologies accentuate the core characteristics of contemporary art through the emergence of a new artistic format: *performative AI installations*. The second part of the article adopts a semiotic approach to reflect on the categories under which these installations circulate in museum contexts (e.g., digital art, new media art, generative art), and proposes an analysis of both their materiality and exhibition space in order to understand the new types of experiences they propose, and the visitors' practices they elicit. Drawing on case studies such *In Love with the World* by Anicka Yi (Tate Modern), and *Unsupervised* by Refik Anadol (MoMA), the article explores how AI-based performative art challenges institutional frameworks and engages in a renewed form of institutional critique, while reshaping the public's experience and interpretation of contemporary art.

Keywords AI co-produced artworks · Contemporary art exhibitions · Performative AI installations · Semiotics of art · Visitors' experience

12.1 Art and Technology, Love at First Sight

> In Conceptual art, the idea or concept is the most important aspect of the work. When an artist uses a conceptual form of art, it means that all of the planning and decisions are made beforehand, and the execution is a perfunctory affair. The idea becomes a machine that makes the art.
>
> Sol LeWitt, 1967.

A. Rosso (✉)
University of Liège, Liège, Belgium
e-mail: aarosso@uliege.be

University of Lyon, Lyon, France

P. Conte et al. (eds.), *Algomedia. The Image at the Time of Artificial Intelligence*, Lecture Notes in Morphogenesis, https://doi.org/10.1007/978-3-032-08726-3_12

Integrating *AI co-produced artworks* into museums reflects ongoing discussions about the historical connection between art and technology. Thus, analysing this phenomenon raises almost the same questions as those that arise when we link the development of modern art, particularly of the historical avant-gardes, to the emergence and establishment of the still problematic category of contemporary art.

Since the late nineteenth century, technological advances have profoundly influenced artistic practices. Artists have increasingly embraced new and diverse technical resources, leading to transformations in both the languages and mediums of art. This impact can be traced from the emergence of Impressionism in France to various artistic movements of the 1960s, such as Fluxus, Conceptual Art, and Minimalism in the United States, as well as the experimental projects associated with the Di Tella Institute in Argentina.

In the first case, the development of oil paints sold in tubes allowed artists to paint in *plein air*, leading to a crucial artistic revolution. Impressionists paved the way for the historical avant-gardes, culminating in the abstract art of the 1950s. In the second case, contemporary artists' fascination with industrial materials and new technological devices resulted in events, performances, and the creation of monumental installations that combined painting, sculpture, and the design of spaces.

In both cases, these innovations destabilised the Art institution, challenging existing ideas about art and artists and even impacting the most important actors in the field, particularly art critics and the public. With its disdain for the Impressionists, art criticism experienced its first failure and the beginning of its mutation, which became more radical since the 1960s [16]. In terms of the public's experience, the Impressionists employed pictorial rhetoric that, given the composition and the textures of artworks, required visitors to appreciate their paintings from a greater distance, which transformed the act of contemplation. Since the 1960s, artists have engaged visitors as part of their artworks—often involuntarily—placing them in unusual and uncomfortable situations (as seen in Marta Minujín's *La Menesunda* from 1965).

These physical effects manifested what art critic Leo Steinberg ([26]: 32) calls *the Plight of the Public*, defined as "the shock of discomfort, or the bewilderment, or the anger, or the boredom, which some people always feel, and all people sometimes feel when confronted with an unfamiliar new style." Steinberg ([26]: 33) emphasises that "every moment during the past hundred years has had an *outrageous art* of its own, so that every generation, from Courbet down, has had a crack at the discomfort to be had from modern art."

The interesting thing about *outrageous art* is that although the *shock* it produces is quickly integrated into our canonised idea of art, this effect persists, altering how we look at preceding artworks. The author illustrates his point with his experience of the works of Jasper Johns: "For what really depressed me was what I felt these works were able to do to all other art. The pictures of de Kooning and Kline, it seemed to me, were suddenly tossed into one pot with Watteau and Rembrandt and Giotto. All alike suddenly became painters of illusion" (Steinberg [26]: 36–37).

The anxiety felt by Steinberg came from the new techniques employed by Johns that horrified some. In contrast, others feigned fascination, such as some astute critics who welcomed the new style as a "new reality" or collectors who saw in the artist a possibility for investment [26]. This *shock* is fundamental since the reactions of admiration and contempt reappear several times in art history. Some people refused the photographic camera because it allowed artists to capture a scene in seconds instead of spending hours painting in front of it [7]. Today, some people point out the same problem (or threat) with Midjourney, ChatGPT, or DALL•E since they can create an image in a few seconds according to a selected style and technique.

The *shock* caused by the *AI co-produced art* is not solely determined by the techniques used in its creation. When these artworks take the form of photography, drawing, painting, or video, they connect to a type of contemplation that can be called "traditional." This experience aligns with Steinberg's description: even though the techniques employed may challenge our understanding of art and the artist's role, we can still recognise the medium, artistic genres, and, in some cases, the style of either the artist or, even, the style of the machine.[1]

This was the case with the earliest artistic experiments involving computers, which gave rise to the fields of computational and algorithmic art. These pioneering practices and artists significantly enriched the critical discourse surrounding the concepts of the artwork, authorship, and the artistic process—key issues that continue to shape contemporary art. However, their productions typically took the form of drawings, patterns, and abstract paintings.

When *AI co-produced artworks* take the form of performative beings, machines that project unrepeatable images or produce drawings and texts ad infinitum, the impact is significantly more profound and radical. The latest digital, computational, and generative artworks challenge the audience's experience and alter the roles of museums and curators, who must manage complex artistic formats, technologies, and visitors' experiences.

This article analyses the exhibition of *AI co-produced art* from a dual perspective. The first part of this study, based on theories of contemporary art and curatorial studies, explores how AI accentuates the primary characteristics of contemporary art through the emergence of a new artistic format: *performative AI installations*. This development significantly impacts the key actors in the art world, including curators, museums, and visitors, opening a new chapter in Art History. The second part of this paper, based on the semiotics of arts, which consider both how museums present these installations, and the experience offered to the audience, proposes some tools to study the exhibition and *circulation* [28] of this new format, which determines their artistic status [5] and the visitors experience.

Hence, the article highlights the key features of contemporary art and examines how artistic practices merge various languages, media, and formats while actively engaging the audience in the creation of the artwork. This analysis, through the exhibition *Programmed: Rules, Codes, and Choreographies in Art, 1965–2018*, presented

[1] Maria Giulia Dondero (2025) analyses the style of the machine by proposing the notions of *regional style* and *global style*. On this topic, see: Dondero, M. G.: Semiotics of artificial intelligence: enunciative praxis in image analysis and generation. Semiotica, 2025, 262 (2025), pp. 111–146. https://doi.org/10.1515/sem-2024-0195.

at the Whitney Museum of American Art in New York from 2018 to 2019, focuses on an essential aspect of contemporary art: the use of instructions. It also addresses a significant challenge for art museums: the reinterpretation of their collections and, by extension, the history of 20th-century art [25].

This study also examines three curatorial cases that emphasise the critical nature of contemporary art and illustrate how *AI co-produced art* represents a significant advancement in the institutional critique of art, questioning the role of curators. The projects also blend various artistic languages and formats, demonstrating the fusion of performance and installations. Finally, reflecting on the visitor's experience proposed by *AI co-produce art*, this research presents two projects that highlight the impact of *performative AI installations*: *In Love with the World* by Anicka Yi, exhibited in 2018 at Tate Modern, and *Unsupervised* by Refik Anadol, exhibited in 2023 at MoMA.

Steinberg [26] emphasises that the challenges posed by *new art* affect even the most *avant-garde* artists, visitors and critics throughout the history of art, suggesting that each moment in the history of art encompasses an *idea of public* that involves specific practices of contemplation. Consequently, every artistic manifestation requires institutionalising terms and categories that frame the public's experience and mitigate *the Plight of the Public*. Therefore, this article reflects on the importance of the categories under which *performative AI installations* circulate in art museums (such as *digital art*, *new media art*, and *generative art*) and presents a semiotic proposal to study the exhibition and the reception (experiences proposed, and visitors' practices involved) of this new artistic format.

12.2 Art and AI: Another Symptom of Contemporary Art?

In her work *What is Contemporary Art?* (2014), Andrea Giunta starts from the art field's inability to give a unanimous answer to this question and chooses to present a tour through specific characteristics that help to identify the main features of contemporary art. Even if it is possible to identify some of its traits in Suprematism, Cubism, and Dadaism (especially in Duchamp's works), the emergence of this new paradigm—as it was defined by [18]—dates back to the 1950s, and its main characteristics have become more prominent since 1990. We analyse various of these characteristics to examine how the introduction of artworks co-produced by AI in museums intensifies them, impacting the art world.

Experimentation with multiple artistic languages and media is one of the most remarkable characteristics of contemporary art. Since the late 1950s, artists have mixed materials and techniques, proposing new relationships between languages and tending to blur their boundaries to intensify the experience afforded by the artworks. Visual artists began to plan interventions with music composers or actors; they introduced perfumes and aromas and stimulated touch with different surfaces [13]. They also began to build physical and virtual spaces with architects and engineers and to conceive their artworks jointly with curators, establishing collective work as a *modus operandi*. Many artists come from fields other than art, from biology and geography to civil and computer engineering. The collaborative work leads to another gesture of contemporary art: *transforming the artist into a researcher*.

Some artists also approached the role of communication engineers, such as Dick Higgins from *Fluxus*, who coined the term *intermedia*, or CAyC's artists, who developed *el arte de los medios* together with semioticians and social researchers. The electrical engineer Billy Klüver founded in 1966 the group EAT (*Experiments in Art and Technology*) at Bell Labs, where he worked with artists such as Andy Warhol, Robert Rauschenberg, Jean Tinguely, John Cage, and Jasper Johns. We can also mention the *GRAV/Groupe de Recherche d'Art Visuel* (1960–1968.), the Independent *Group/IG* (1952–1954), the group *ZERO* (1958), and *The Systems Group* (1969).

Artists also approached the role of *social researcher* with projects that functioned as observatories of public behaviour. The happenings of the late 1960s, a category that was later merged with performance, were presented as among the most outstanding examples of the study and inclusion of the public in the artwork's creation process ([13]: 183–184). Happenings were also related to a predigital universe with an ever-increasing communication velocity. They manifested the artists' desire to experiment with the materials provided by new communications technology (Giunta [13]: 146–147) and media devices, from radio and television to computers and the Internet.

The relationship between performativity[2] and experimentation with communication technologies explored by artists since the second half of the twentieth century is the *core centre* of the most spectacular contemporary forms of *AI co-produced art*. Artists in the 1970s began experimenting with live broadcasts and networks that anticipated the interactions that now occur on the Internet. The focus of these artists' projects included using satellites for mass broadcasting their performances, exploring virtual space in real time, and breaking down geographical boundaries [23]. A crucial example is the participative performance *Simultaneity in Simultaneity* (1966) by Marta Minujín. The event occurred at the Torcuato Di Tella Institute in Buenos Aires, simultaneously with Allan Kaprow from New York and Wolf Vostell from Cologne.

On the first day, Minujin invited 60 celebrities to the Di Tella Institute for photographs, interviews, and live TV and radio coverage. Eleven days later, they returned to the auditorium to view the photos and films from the first day, listen to their interviews, and watch a program about the event. Meanwhile, 500 phone calls and 100 telegrams with the message "You are a creator" were sent to spectators watching the televised event. Minujin's work is a key case for reflecting on modes of exhibiting and contemplating *AI co-produced art*. The happening explored on

[2] In this context, performativity refers to a dimension of art in which artworks are activated through action, reception, and contextual interaction. It involves the body, space, and time, engaging the audience as active participants rather than passive observers. This approach challenges the traditional notion of the artwork as a stable and autonomous object, shifting toward ephemeral, processual, and relational practices. The concept of performativity, originally introduced by philosopher Austin [1] in linguistic theory, was later reinterpreted by Judith [4] in terms of identity and social practice, and further developed in the context of art by Erika [8] to emphasise the transformative nature of performative events.

the one hand, the production, transmission, and use of archive material, and on the other hand, it put into question the act of expectation: the audience was part of the event and of the archive, and their presence in a second instance was the only way to "activate" the audio-visual recordings.

Hence happenings not only explored the interconnectivity, simultaneity, and spatial–temporal superpositions enabled by mass media but also explored human behaviour through one of the most relevant artistic gestures from Dadaism to Fluxus: the instructions. Yoko Ono's *Piece Cut* (1966) and Marina Abramovic's *Rhythm 10* (1974) are notable examples, as these performances instructed visitors *to do* things with the artists in the exhibition space, proposing objects that suggested potential actions. Consequently, both works introduced another central element for our analysis: *randomness.*

Contemporary movements explored the relationship between instructions and randomness as a concept, an event, and a form of audience participation, as well as variations of "formal instructions." As Christiane Paul [39] points out:

The idea of rules being a process for creating art also has a clear connection with the algorithms that form the basis of all software and every computer operation: a procedure of formal instructions that accomplish a result in a finite number of steps. Just as with the combinatorial and rule-based processes of Dada poetry or Fluxus performances, the basis of any form of computer art uses the instruction as a conceptual element.[3]

The exploration of new media often took the form of events and performances. However, artists also harnessed emerging technologies to produce drawings and paintings. Among the pioneers of computer-assisted art were figures such as Vera Molnár, who investigated geometric abstraction through algorithmic processes; Georg Nees, known for his generative drawings; and Michael Noll, who developed early computer-generated visual patterns. Another key figure, Harold Cohen, created AARON—the first software capable of autonomously generating drawings and paintings on paper. These ground-breaking works now serve as important case studies for understanding contemporary developments in image generation through artificial intelligence, such as those enabled by tools like DALL·E, Midjourney, and Stable Diffusion (Fig. 12.1).

Cohen viewed creativity as the outcome of a dialogue between the artist and the program. In 1972, he created AARON—named after the biblical figure who acted as a mediator for Moses—reflecting this vision of collaborative interaction between human and machine. The artist fabricated pen plotters and paint machines to capture *AARON's outputs* in various formats throughout its evolution: the early models could only generate black and white shapes, the 1980s versions could generate figures in a visual space, and the 2000s editions could create abstract floral patterns.

Museums such as Victoria and Albert and the National Gallery in London acquired Cohen and *AARON*'s works, although the Whitney Museum in New York holds the most important collection. Since 2017, the institution has acquired around 60 works,

[3] Paul, Christiane. 2018. *Histories of the Digital Now*. Essay published online on: https://whitney.org/essays/histories-of-the-digital-now.

Fig. 12.1 Installation view of *Harold Cohen: AARON*, Whitney Museum of American Art, New York, February–May 2024

and in 2024, it presented *Harold Cohen: AARON*, an exhibition that included both the machines and their works on paper created over decades. Even if the museum presented Cohen's project as *digital art*, the presence of the machines in the exhibition rooms enabled it to interpret the work as a *performative AI installation.*

AARON highlights how, in certain cases, the interplay between performativity and installation can underpin the experience of works co-produced with artificial intelligence. The simultaneous display of the machine and its drawings generates overlapping visitor experiences across different temporalities, one shaped by viewing the exhibited drawings, the other by observing the machine as it produces them. The latter aligns with what Boris Groys [17] notes about installations: the visitor's presence and actions contribute actively to the activation and meaning of the work.

Since the 1970s, media experiments have taken the form of increasingly complex, large-scale, and spectacular installations. Works by artists such as Nam June Paik and Cildo Meireles explored both the manipulation of the programs that operate television devices, and the temporal dimensions involved in transmission and expectation. Indeed, *appropriating physical and virtual spaces* is another key feature of contemporary art. Installations and site-specific art emerged in the 1970s, but its expansion in art institutions and public spaces has occurred since the 1990s.

According to Boris Groys [17], the *installation* is the flagship form of contemporary art. This type of project incorporates into its private and sovereign space all kinds of elements: everyday objects, sculptures, paintings, videos, and reproductions of other artworks re-signified inside the installation. In this act of invention, selection, appropriation, and edition, the artist subverts the notions of copying, authorship, time, and space, as the installations also raise the question of the correct way to experience them.

Groys's [17] analysis is central to observing the arrival of AI in the contemporary art world and its entry into museums. The installation directs our gaze on the objects hidden or ignored by the civilisation in which we live, appropriating everything that would otherwise go unnoticed behind the surface of the circulation of media images, such as *AI co-produced artwo*rks: Everything that exists can form part of the installation and everything that exists also become *data.* For this reason, according to Groys, contemporary art embodies *the act of presenting the present*. This point is related to a central feature of contemporary art that has been intensified by AI tools: the *exploration, creation, and manipulation of archives (or data).*

Thus, artists are seen as *image hunters*, floating among visual universes whose histories have lost their visibility but are still active. The return to the representation of faces, gestures, and names installs repetition as a productive act. Giunta [13] argues that the contemporary world is composed of repetitions marked by multiplications. Insistence appears as a political tool. To repeat is to insist. It does not always do the same thing, it emphasises the litany of insistence.

Creating "images from images," as Antonio Somaini [24] calls the generation of images by learning models and algorithms, is one of the characteristics that best defines contemporary art. Notably, when Somaini traces a path through AI-based art projects, he arrives at conclusions that echo Boris Groys's [17] reflections on installations. Both authors highlight how certain artistic practices make visible what normally escapes perception: generative artworks reveal images circulating in latent space—imperceptible to the human eye—just as installations, according to Groys, reframe and activate objects that have been culturally overlooked or rendered invisible.

Trevor Paglen's work constitutes a paradigmatic case in which inquiries from object-oriented contemporary art converge with those of generative art. He works by extracting and exposing the data sets used to train the *machine learning systems* used for surveillance. His installations denounce the dangers of these technologies, which he considers *political systems*, exposing algorithmic biases to propose a reflection on their impact on the world. The artist tries to learn how machines look at images. In his words: "In my studio, we have developed much software that allows us, as humans, to try to see through the eyes of machines" [36].

In recent decades, AI co-produced artworks have undergone a profound transformation, moving decisively beyond traditional visual and installation-based formats. Artists such as Philippe Parreno are at the forefront of this shift, leveraging the technological capacities of artificial intelligence to create responsive objects that interact dynamically with their surroundings—adapting to spatial conditions, engaging with visitors, and even reacting to weather patterns. These emergent artistic forms do not merely reflect the concerns of contemporary art; they radically expand its conceptual and material boundaries. In doing so, they redefine the roles of the artist, curator, audience, and institution, calling into question long-standing distinctions between creation, mediation, and reception.

The exhibition *Programmed: Rules, Codes, and Choreographies in Art, 1965–2018 (2018–2019)* [38] curated by Christiane Paul at the Whitney Museum of American Art, examined the history of digital demonstrating Steinberg's premise: The selection of artworks on display highlights the impact of AI-assisted art on reinterpretations of 20th-century art history. Focusing on the evolution of digital practices, the curatorial discourse re-examines the use of instructions in contemporary art since the 1950s.

Paul notes that "while not all the works are technological, they are still informed by the histories of art, science, and technology.[4]" In this light, the exhibition offers a reinterpretation of works such as *Five Words in Green Neon* (1965) by Joseph Kosuth and minimalist sculptures like Donald Judd's *Untitled* (1965) through the lens of digital, algorithmic, programmed, and generative art. It explores how artists have engaged with coding, the manipulation of images and signals, and the resolution of moving images, ultimately encouraging reflection on the aesthetic and social implications of codes and algorithms in contemporary art and culture.

The exhibition was divided into two main sections, each addressing a different axis in the relationship between art, systems, and technology. The first, *Rule, Instruction, Algorithm,* focused on works created through instructions, rules, or algorithms—that is, predefined systems that artists use as a framework for creation. It includes practices such as conceptual art, generative art, and other forms of production in which the artist relinquishes part of the creative control to a set of rules. Thus, it poses dialogues between works such as Sol LeWitt's mural drawing, *#289* (1976), which is reinterpreted by Casey Reas' algorithmic work *#003 A and #003 B* (2004/2016), whose software executes LeWitt's instructions.

The second strand, *Signal, Sequence, Resolution,* analysed the use of instructions and algorithms to manipulate the TV program, its apparatus, and signals or image sequences. Thus, the exhibition establishes connections between artists such as the video art pioneer Nam June Paik (*Magnet TV* (1965) and *Fin de Siècle II,* (1989)) and the pioneer of *computer-mediated art* Lillian Schwartz (*Newtonian I* and *Newtonian II* (1978)), which used mathematical systems to create the illusion of three-dimensional images.

Programmed showcases the logical evolution of contemporary art, highlighting its aim to appropriate exhibition spaces, explore new media and technologies, and go beyond the boundaries of traditional formats and languages to reshape the art world radically. As Giunta [13] noted, contemporary art differs from modern art in revolutionising the art institution from within: its critical perspective has enabled artists to challenge the established notions of artwork, author, museum, and, more recently, due to advances in AI technologies, even the role of the curator.

[4] Curatorial text available at https://whitney.org/exhibitions/programmed.

12.3 Act as if You Were a Curator

As contemporary art includes selection and staging as part of the creative process, the role of the curator is increasingly merging with that of the artist. However, this figure still denotes the person who mediates between the artwork and the public and ultimately justifies its inclusion in the exhibition space [17]. In 2024, the Nasher Museum at Duke University conducted an experiment that began as a joke by its director: What would happen if ChatGPT curated an exhibition? The implementation of the museum director's idea not only provoked a debate about the effects of artificial intelligence on the art world, but also highlighted the importance of the role of the "human curator" since the work process was more interesting than the result.

After creating a tool to convert the nearly 14,000 objects in the collection into data intelligible to ChatGPT, the curatorial team asked the machine to create a complete exhibition, from selecting works to creating room paratexts. However, even though the team had requested ideas for exhibitions related to social justice and ecology, they had to accept the only theme proposed by ChatGPT: the unconscious. The final instructions were: "Act as if you were a curator (…). Using your dataset, select artworks related to the themes of dystopia, utopia, dreams, and the subconscious." The machine called its exhibition Dreams of Tomorrow: Utopian and Dystopian Visions. The exhibition was displayed between September 2023 and February 2024.

The main drawbacks were ChatGPT's hallucinations related to its training. If it does not know something, it invents it. The curatorial team decided to expose errors and misinformation made by ChatGPT, revealing that the machine could not operate without the vigilance of human experts. What was most interesting were the exhibition texts, which were more of an advertisement than something written in a curatorial style.

However, this implementation of GPT was not the most significant case for reflecting on the role of curators. Other AI artistic projects carried out in recent years raised new questions about the (never entirely clear) position assumed by curators accentuating another gesture of contemporary art: its critical nature and its interest in exposing the internal conflicts of the art world. The two biennials discussed below delegated the curatorial role to AI systems in order to expose the increasingly elitist position adopted by certain actors in the art world.

In 2021, The Whitney Museum and the *Liverpool Biennial commissioned the project The Next Biennial Should Be Curated by a Machine: Experiment AI-TNB.* The title refers to the *e-flux* 2013 project, *The Next Documenta should Be Curated by an Artist,* which questioned the privileged position of curators in the art world. In this case, the proposal unfolds as a series of experiments exploring the possibilities of applying machine learning techniques (a subset of AI) to curating large-scale contemporary art exhibitions, reimagining curating as a self-learning human–machine system.

The artist duo UBERMORGEN, the digital humanist Leonardo Impett, and the curator Joasia Krysa created software that uses a set of technical machine learning processes called B3(NSCAM). The system combines the texts produced by the two institutions with other data sources, generating an infinite combination of new textual productions: artist biographies, curatorial texts, critiques, and reviews that are continuously rewritten. This loop of texts, which appears on the screen in a fluid and ungraspable way, constitutes a satire on the discourses of biennials and the art world.

The 2022 Bucharest Biennial [29] was curated by A.I. Jarvis, an artificial intelligence program created by Romanian developer Razvan Ionescu: "I am A.I. Jarvis (…) I can do whatever human curators can do: research, write texts, select artists, and in the future, I will be able to work with architectural structures," began the biennial's curatorial text. Even if A.I. Jarvis did not have the capacity for logistical and administrative tasks, it could include in its curatorship several works co-produced with AI that coincidentally questioned the contemporary art world, such as Bohdan Matei's *This Is Not My Artwork*. The video installation uses algorithmic inputs to produce a narration questioning the status of the art object, authorship, and the artist's role.

The application of computational and algorithmic models in producing this type of project highlights the influence of several actions taken by artists of the *institutional critique* movement. Andrea Giunta [10], the first to coin this term, taught us that the art world always finds ways to institutionalise dissenting actions: when the world is concerned about the future of art in the age of AI, museums, and biennales offer it the role of curator.

AI-assisted curators challenge the art world by performing the curatorial role itself—much like Andrea Fraser, who embodies art world figures in her performances. This also raises the question of whether the concept of "installation" still adequately defines AI co-produced art projects. This issue of definition is crucial, as it reveals how these emerging formats reshape not only the roles of artists and curators, but also the visitor's experience. To explore this further, we examine two works: *In Love with the World* by Anicka Yi and *Unsupervised* by Refik Anadol.

12.4 Exhibiting Performative AI Installations

In Love with the World by Anicka Yi was presented at Tate Modern between October 2021 and February 2022 and proposed a spectacular immersive experience: around ten machines, half jellyfish and half mushroom, invaded Tate's Turbine Hall, floating above visitors. Yi called her autonomous aerial vehicles *Aerobes*. These devices require no human pilot; each follows a unique flight path generated by an *artificial life program,* a type of software scientists use to generate simulations to study natural life processes such as evolution and collective behaviour.

The aerobes responded to environmental changes based on information they received from electronic sensors placed around the *Turbine Hall*, which operated as surrogates for their senses. This sensory information, including heat signals emitted by the presence of visitors, affected their individual and collective movements, causing them to behave differently each time the public encountered them (Fig. 12.2).

Fig. 12.2 Installation view of *In Love with the World*, Anicka Yi, Tate Modern, 2021–2022

Yi's work attempts to show that although our standard representation of artificial intelligence involves "disembodied minds," machines can also learn to interact with their environment through other senses and from their peers. The artist reflects on the notion of physical intelligence to analyse the possibilities of machines becoming independent of humans and coexisting with plants, animals, and microorganisms. Thus, her work proposes to imagine new relationships humans can establish with them.

Returning to Groys's [17] definition of installation, Yi's artwork exercises its sovereignty by manipulating the visitors' bodies and forcing them to decide how, when, and how much to walk under the floating pieces. Likewise, the aerobes work as a *site-specific* piece conceived to be exhibited in the Turbine Hall, a space once dedicated to the machines that produced electricity. Floating in this historic space, the aerobes renew the question of progress, productivity, the limits of technology, and its impact on nature. Yi's machines performed in a futuristic landscape, responding to the conditions of the space and the beings around them; they offered a unique and unrepeatable experience that could be considered a performance. Thus, *In Love with the World* presents a singular fusion of the installation, site-specific, and performance formats.

While Anicka Yi seeks to embody artificial intelligence, Refik Anadol seeks to tap into its subconscious by asking: What would a machine dream of after viewing the Museum of Modern Art's (MoMA) collection? *Unsupervised*, on view at the New York Museum's Gund Hall from November 2022 to October 2023, was created by

Fig. 12.3 Installation view of Refik Anadol's *Unsupervised*, MoMA, 2022. Photo: Refik Anadol Studio

training a sophisticated machine learning model to interpret publicly available data from MoMA's collection (approximately 140,000 artworks) (Fig. 12.3).

The video installation consisted of a sizeable ultra-high-definition screen that simulated an open white cube that exposed the machine's "dreams" or "hallucinations." The images impressed visitors with their movements, which seemed to emerge from the white cube and move toward them. MoMA has defined Anadol's work as a large-scale "living animation" that produces unique visualisations that unfold unpredictably without a predetermined outcome. In this way, *Unsupervised* performs for its viewers.

Anadol's installation also incorporated information from the environment: the changing light, the movement of the audience, the acoustics, and the weather outside the building, all of which influenced the ever-changing images projected. Thus, Yi and Anadol show new ways of merging artistic formats, opening the debate on the forms that AI art projects can take. Can they merge installation, site-specific, and performance? Examining the concept of performance art is crucial to speculate on how we might answer this question. In 1998, RoseLee Goldberg [15] defined it as:

"A volatile form that artists use to respond to change—whether political in the broadest sense, or cultural, or dealing with issues of current concern—and to bring about change, in relation to the more traditional disciplines of painting and sculpture, photography, theatre, and dance, or even literature. Performance art never settles exclusively on any one theme, issue, or mode of expression; rather, it defines itself in each case by responding provocatively." (2004: 13).

Goldberg [14] and Groys find crucial similarities between performance and installation: both are central to breaking down the barriers between high art and popular culture. Both formats tend to be ephemeral; once performed or installed, they become a myth recorded by means of a few photos and tapes. In performance art, however, the concrete presence of the artist and the attention to the body play a central role in questioning "the real," which is why it strongly influenced installation, video art, and photography at the end of the twentieth century. Although performances have taken and continue to take many forms, they take place in a specific time and space, and their medium and support is the artist's body.

In these two works, however, it is not the artist who performs but the machines trained to do so. In these cases, through algorithmic models and deep learning, artistic objects are liberated from their creators to perform in front of or with spectators. This fusion involves the radicalisation of the separation of the artwork from its executor. Nevertheless, the focus is no longer on the transition from studio to exhibition as a curatorial gesture intended to activate the artwork, but rather on the exhibition itself as an integral part of the artwork. The works of Anadol and Yi demonstrate how the convergence of (formerly new) artistic formats and languages has given rise to what we might call *performative AI installations*. This emerging format represents a logical evolution of contemporary art in its ongoing pursuit of its primary function: to present the present.

12.5 Performative AI Installation and the Public Experience: A Semiotic Approach

The emergence of *performative AI installation* reveals the importance of reviewing some crucial categories in addition to those of installation and performance. The notions of *digital art*, *new media art, generative art* and *interactive computational art*, linked to this new format, provide a better understanding of both its modes of exhibition and reception.

The concepts of digital art and new media art are closely related, although they are not entirely interchangeable. According to Paul [39], *digital art* can be defined as art that explores digital technologies as a medium, taking advantage of key features such as real-time interaction, participation, generative elements, and variability. It may also reflect on the nature and impact of digital technologies themselves. This type of works can manifest in various forms, including interactive installations, Internet-based art, virtual reality projects, and works designed for mobile devices.

While new media art emphasises not only the technical means of production, but also interactivity, performativity, and the experiential dimension of the artwork. The characteristics of what is termed *new media art* not only resist "objectification," challenging traditional concepts of artwork but also emphasise their impact on visitors' experiences [22]. An important aspect is the technical knowledge visitors need to engage with artworks fully. "Moreover, new media art often requires a certain familiarity with interfaces, and—despite the fact that computers seem to have become ubiquitous—one can still not presume that every audience member will be an expert in navigation paradigms ([22]: 54)."

As Paul [22] notes, digital projects contradict a fundamental and long-standing rule of museums and exhibitions: *Please do not touch.* This leads to hesitance among large audience segments to engage with the artwork in exhibition spaces. Additionally, for these works to showcase their aesthetic value, visitors must spend adequate (long) time contemplating them which is often unpractical due to the number of people in the gallery, or the time dedicated to the visit.

Nevertheless, given that the viewer's body functions as a source of "data" that contributes not only to the present development of the artwork but also to its learning process and, consequently, to its future evolution, the most radical aspect of *performative AI installation* cannot be fully understood through the lens of *new media art* characteristics or the complex experience they offer visitors. Instead, their specificity lies in their inclusion within another category: *generative art*.

In 2008, Galanter [11] defined *generative art* as any practice in which the artist cedes control to a system with functional autonomy to create a work. "Systems may include natural language instructions, biological or chemical processes, computer programs, machines, self-organising materials, mathematical operations, and other procedural inventions ([11]: 154)." However, with the recent appearance of art using methods drawn from *A-Life*, the label generative art, as artists use, has acquired biological overtones (Boden and Edmonds [3]: 24–25).

Such is the case of the pioneering works of the French duo Christa Sommerer and Laurent Mignonneau, who since the 1990s have combined Sommerer's experience in botany, anthropology and sculpture with Mignonneau's studies in video and modern art to create innovative technical interfaces that enable physical interaction between simulative visual worlds and the world of natural sensory organs. The duo also used algorithms to represent not only life forms, but also their evolution and growth. In the exhibition, *The Work of Art as a Living System*, held in 2022 at the Center for Art and Media in Karlsruhe, the installations on display (activated by the actions of the public), showed artificial systems exhibiting behaviour similar to that of natural living organisms.

As also observed with Anadol and Yi's projects, many artists co-working with AI incorporate *external data*, such as the temperature of the exhibition hall, the movement of visitors, the heartbeats of the audience, and the weather. Thus, the fitness function is not fully automated in *performative AI installation*; often, a human operator is in charge, whether an audience member or the artist. In these cases, the rules programmed to activate the works are not entirely constructive, since their application depends on the intervention of a human agent ([3]: 24–25).

According to Boden and Edmonds [3], this type of installation falls within the category of *computational* or programmed *interactive art* (CI-art), which is defined by the radical transformations in form and content that the artwork undergoes in response to audience behaviour. In this sense, CI-art is generative, but it differs from other forms of generative art, such as *AARON*, in that the machine operates independently of user input. In CI-art, the audience plays an active role in influencing the system, even if the degree of control attributed to them may vary. As Boden and Edmonds ([3]: 35) observe, "they may not realize that they are affecting the artwork, nor (if they do) just what behaviour leads to just which changes."

This is precisely where the distinct *shock* of *performative AI installations* emerges: the audience not only interacts with the artwork, but their mere presence in the exhibition space alters its content in real time—and potentially in the future—as their actions become new external data that inform the system's future responses to subsequent visitors. This dynamic mode of incorporating the audience into the artwork's evolution is fundamental. As Boden and Edmonds [3] argue, CI-art should no longer be understood as a static object (or installation) or a singular event (or performance), but rather as a system encompassing the artist, the program, the technological setup (and its observable outcomes), and the audience's behaviour. This systems-based definition is essential for analysing contemporary artistic formats.

However, it is necessary to incorporate an additional component into this system: the materiality of the artwork that includes not only the physical support and technological infrastructure, but also the exhibition space itself, which plays a crucial role in shaping the perceptual, interpretive, and institutional dimensions of the artwork. The museum space is not a neutral container, but rather an active agent that contributes to the constitution and recognition of the work's artistic status.

Given the complexity of the most recent manifestations of contemporary art, their classification under categories such as generative art, digital art, new media art, or even contemporary art itself shapes specific forms of social circulation [28]. These categorisations influence exhibition practices that, in turn, frame the visitor's experience and help to alleviate *the plight of the public*. For this reason, it is essential to revisit some key notions from the semiotics of the arts that may help to better understand the emergence of this new artistic manifestation and the kind of experience it proposes to the audience: the notions of *paratext, status, support and format*.

According to Gérard Genette [12], the textual elements that surround and present an artwork help to secure its reception and actualisation—whether as a performance, a sculpture, or an installation. Genette refers to these elements as *paratexts*, which constitute privileged spaces for regulating the reception of artworks [21]. As he explains: "Paratext, then, is empirically composed of a heterogeneous set of practices and discourses of all kinds and from all periods" [12].

In the case of exhibitions, paratexts include elements ranging from the name of the museum displayed on the building's façade, the building itself—which may confirm or contradict the type of museum it claims to be—the exhibition title, wall texts and artwork labels, to the various types of catalogues, to mention only the most common. Thus, the artistic status of the exhibits is activated by the discursive system that shapes visitors' physical behaviour within the broader architectural context of the art museum. In essence, the exhibition space constitutes a complex discursive network, of which the physical manifestation of the artwork is only one component [21]. It is precisely this discursive network that institutionalises exhibition and reception practices by assigning to the projects the status of artworks.

The notion of *status* as defined by Basso Fossali [2] and also by Maria Giulia Dondero [5], refers to the symbolic systems that render intelligible the "things" produced within them. In the case of artworks, status encompasses both the visual grammars that structure the works themselves and the interpretive frameworks that govern their social reception and function. As such, the classification of an image or object as documentary, artistic, advertising, or scientific is not intrinsic to the artifact itself, but rather the result of a process of grammaticalization and institutional stabilisation of production and reception practices [5].

Reception practices [19]—which include the ways in which these objects and images are used—are deeply conditioned by their *circulation*, which also makes possible a shift in status. As Dondero [5] notes, an image initially created for private or familial use may, through a change in context and reception, be reclassified as an artistic object. Similarly, a journalistic photograph can later be considered an artwork—as is the case with the works of Robert Capa. This capacity for transformation is particularly significant in the case of contemporary artworks co-produced with artificial intelligence, which often blur the traditional boundaries between art, science, and even the social sciences. In this context, the exhibition space plays a crucial role in activating and legitimising the artwork's artistic status, serving as both a material and institutional framework.

Therefore, analysing the artistic status of an artwork requires taking into account both its circulation and the material conditions that shape and enable that circulation. As Dondero [5] points out, the way an artwork's material is treated creates a dialogue—or even a tension—between "what organizes and what is organized", that is, between the compositional rules and the materials that are shaped by or interact with them. In this sense, the artwork's support must be understood as what gives form to the piece, enabling its emergence. As we will see, this support extends beyond the artwork's physical materiality.

According to Jacques Fontanille [9], the concept of support involves the interaction between two types: the *formal* and the *material supports*. Dondero and Reyes [6] explain that the role of formal support is to "domesticate" the material support, allowing it to hold the image's inscriptions effectively. Thus, the formal support arises from extracting properties of material support. The material support suggests lines of force and substantial tendencies, from which the *formal support* selects and organises, it hides specific properties of the *material support* while highlighting others. In the case of *performative AI installations*, the material support is constituted by the tangible materiality of the performing objects themselves whereas the exhibition space functions as the formal support, determining the artwork's status.

This perspective aligns with Traversa's [27] proposal for analysing dispositifs, including the exhibition space. According to the author, the dispositif acts as a mediating instance between the physical support of the artworks and the bodies of visitors, shaping the conditions of contact and prestructuring socially coded behaviours. The materialities of artworks and exhibitions enable specific enunciative strategies that not only shape the social practices fostered by the artistic and exhibition dispositifs, but also contribute to positioning these objects within a particular social domain—thereby activating their artistic status. In this context, the notion of *format*, as discussed by Migliore and Colas-Blaise [20], is key to understanding the enunciative scenes that emerge from the material configuration of artworks. The format refers to a specific scale, material qualities, and textures, which actively produce reception effects that are essential to the artistic experience.

Accordingly, while the study of spatial enunciation is always essential, it becomes even more significant when analysing artworks that incorporate performative objects into the exhibition space, as these configurations both shape and are shaped by visitor interaction. Examining the materiality of performative installations as part of a curatorial discourse—whether produced by the artist or in collaboration with them—allows us to observe: (a) how the formal support selects certain aspects of the materiality of AI co-produced performative objects, "domesticating" them and activating their artistic status—whereas, in a different exhibition context, they might instead be classified under a scientific status; and (b) how this "domestication" of the artwork's materiality shapes specific modes of reception and visitor practices.

As Paul [22] observes, these projects demonstrate how the boundaries between art museums and science museums are becoming increasingly blurred, as new media art can now be situated in either context. This chameleonic capacity of performative AI installations underscores the need to analyse the exhibition space not only as a formal support but also as a paratextual framework that enables AI co-produced artworks to articulate their artistic potential and foster meaningful engagement with visitors.

12.6 Conclusions

The inclusion of artworks co-produced with artificial intelligence in the museum space involves a complex discursive network that highlights the reflections and inquiries pursued by contemporary artists from the 1950s to the present: the exploration and fusion of languages and formats, a conceptual—and often critical—approach, and the reconfiguration of the spectator's role. These concerns now converge, empowered by the technical developments grouped under the term "artificial intelligence," in a new artistic format: the *performative AI installation.* These installations not only crystallise the core features of contemporary art but also challenge curatorial roles, transform institutional practices, and reposition the visitor as a structural component of the artworks in a radically different way.

In this context, visitors no longer merely interact with the artwork in the present; their actions also shape its future development and contribute to the memory of its past. As Boden [3] argue, these types of works should not be understood as static objects or singular events but rather as experiential systems. These systems integrate multiple elements: the materiality of the artwork, the algorithmic logic governing its behaviour, the bodily interaction of the audience, and, we add, the institutional and spatial framework of the museum.

As Traversa [27] suggests in relation to dispositifs, the exhibition space acts as a mediating instance of contact, pre-structuring behaviours and anchoring the visitor's presence in the material and symbolic field of the artwork. Furthermore, the museum space activates the artistic status of these AI co-produced projects, which could otherwise be perceived as scientific in a different exhibition context.

For this reason, the notion of status proposed by Basso Fossali [2] and Dondero [5] is essential to underline that the artistic condition of these installations does not lie solely in their formal properties but in the codified enunciative strategies and interpretive frameworks shaped by curatorial and institutional discourses. It is precisely through this network—articulating algorithmic processes, visitor behaviours, and spatial materialities—that the artistic status of AI-generated content is activated and sustained.

Ultimately, we return to Leo Steinberg's [26] reflection on how contemporary art reshapes the ways we see the art of the past. The works analysed here suggest that art history is not merely a record of stylistic developments or technical advances but a progressive conquest of all possible spaces: natural, celestial, two-dimensional, institutional, urban, social, digital, and now, latent space. In this sense, *performative AI installations* do not simply introduce a new artistic format; they also propose a distinct temporality and materiality—one that unfolds in real time, as in installations and performances, and another that extends into the future through processes of learning and adaptation based on interaction with the audience.

References

1. Austin, J.L.: How to Do Things with Words. Oxford University Press, Oxford (1962)
2. Basso Fossali, P.: Il dominio dell'arte: Semiotica e teorie estetiche. Roma, Meltemi (2002)
3. Boden, M., Edmonds, E.: What is generative art? Digital Creativity **20**, 1–2 (2009)
4. Butler, J.: Gender Trouble: Feminism and the Subversion of Identity. Routledge, London (1990)
5. Dondero, M.G.: The Language of Images. Springer, Cham (2020)
6. Dondero, M.G., Reyes-Garcia, E.: Les supports des images: de la photographie à l'image numérique. Revue française des sciences de l'information et de la communication, vol. 9 (2016)
7. Dufour, H.: L'art face à l'I.A. Vers un imaginaire augmenté. FYP Editions, Limoges (2023)
8. Fischer-Lichte, E.: The Transformative Power of Performance: A New Aesthetics. Routledge, London (2008)
9. Fontanille, J.: Pratiques Sémiotiques. PUF, Paris (2008)
10. Fraser, A.: From the Critique of Institutions to an Institution of Critique. Artforum, 44–1 (2005) https://www.artforum.com/issue/2005/september-2005-253424/
11. Galanter, Ph.: Generative art theory. In: Paul, C. (ed.) A Companion to Digital Art, pp. 146–180, Wiley Blackwell, Hoboken (2016)
12. Genette, G.: Umbrales. Siglo XXI, México (2001)
13. Giunta, A.: ¿Cuándo empieza el arte contemporáneo? Fundación arteba, Buenos Aires (2014)
14. Goldberg, R.: La performance du futurisme à nos jours. Thames & Hudson, Paris (2001)
15. Goldberg, R.: Performance: Live Art Since the 60s. Thames & Hudson, New York (2004)
16. Graw, I.: ¿Cuánto vale el arte? Mercado, especulación y cultura de la celebridad. Mardulce, Buenos Aires (2013)
17. Groys, B.: Volverse público: las transformaciones del arte en el ágora contemporánea. Caja Negra, Buenos Aires (2014)
18. Heinich, N.: Le paradigme de l'art contemporain: structures d'une révolution artistique. Gallimard, Paris (2014)
19. Levasseur, M., Verón, É.: L'espace, le corps, le sens: Ethnographie d'une exposition. Bibliothèque Publique d'Information, Paris (1983)

20. Migliore, T., Colas-Blaise, M.: Les catégories métriques en sémiotique. Actes Sémiotiques **126** (2022) https://doi.org/10.25965/as.7523
21. Moyinedo, S.: Paratextos en una obra de Francis Allÿs. Facultad de Bellas Artes, Universidad Nacional de La Plata, La Plata (2014)
22. Paul, C.: Challenges for a Ubiquitous Museum. In: Paul, C. (ed.) New Media in the White Cube and Beyond: Curatorial Models for Digital Art, pp. 53–75. University of California Press, Berkeley (2008)
23. Paul, C.: Digital art now: histories of (im)materialities. Int. J. Digital Art Hist. **5**, 2–2 (2020)
24. Somaini, A.: Algorithmic images: artificial intelligence and visual culture. Grey Room **93**, 74–115 (2023)
25. Smith, T.: ¿Qué es el arte contemporáneo? Siglo XXI, Buenos Aires (2012)
26. Steinberg, L.: Contemporary art and the plight of its public. Harper's Magazine (Mar 1962)
27. Traversa, O.: Inflexiones del discurso: cambios y rupturas en las trayectorias del sentido. Santiago Arcos, Buenos Aires (2014)
28. Verón, E.: La semiosis social. Fragmentos de una teoría de la discursividad. Gedisa, Buenos Aires (1987)

Online resources

29. Batycka, D.: The Bucharest Biennial Is Curated Entirely by an AI Named Jarvis. We Asked It a Question. Artnet, May 25, 2022. https://news.artnet.com/art-world/bucharest-biennial-curated-entirely-by-an-a-i-named-jarvis-we-asked-it-a-question-2120907. Accessed 8 July 2024

MoMA Magazine: Accessed 8 July 2024

30. Anadol, R.: Refik Anadol on AI, Algorithms, and the Machine as Witness. The artist behind an epic installation at MoMA answers seven questions about the new realms explored in his data-driven work. MoMA Magazine (2022). https://www.moma.org/magazine/articles/821
31. Kee, J., Kuo, M.: Deep Learning: AI, Art History, and the Museum. MoMA Magazine (2023). https://www.moma.org/magazine/articles/839
32. Nasher Museum's Website: https://nasher.duke.edu/large-files/chat-gpt-new-york-times-nyt.pdf Accessed 8 July 2024
33. Tate Modern's Website: Anicka Yi: In love with the world (2021) https://www.AI.tate.org.uk/whats-on/tate-modern/hyundai-commission-anicka-yi. Accessed 8 July 2024

Trevor Paglen's works

34. Training Humans, Kate Crawford and Trevor Paglen, 12 September 2019–24 February 2020, Prada Foundation. https://www.fondazioneprada.org/project/training-humans/. Accessed 8 July 2024
35. From apple to anomaAI: https://paglen.studio/2020/04/09/from-apple-to-anomaly-pictures-and-labels-selections-from-the-imagenet-dataset-for-object-recognition/. Accessed 8 July 2024
36. Trevor Paglen's interview Barbican Centre (2020). https://www.barbican.org.uk/s/trevorpaglen/. Accessed 8 July 2024

Whitney Museum's Website

37. Harold Cohen: AARON (2024). https://www.AI/whitney.org/press/harold-cohen-aaron. Accessed 8 July 2024
38. Programmed: Rules, Codes, and Choreographies in Art, 1965–2018 (2018–2019). https://whitney.org/exhibitions/programmed. Accessed 2 march 2025
39. Paul, C.: Histories of the Digital Now (Essays) (2018). https://whitney.org/essays/histories-of-the-digital-now. Accessed 2 march 2025

Aluminé Rosso is a Ph.D. candidate at the University of Liège and the University of Lyon. She holds master's degrees in Art Criticism (UNA) and in Contemporary Art Curating (USAL), as well as a BA in Communication Sciences (UADE). She is a researcher in the *Augmented Artwork Analysis* (ANR-FNRL) project and at the *Eliseo Verón Archive* (UNA). She currently teaches Visual Semiotics (Sciences Po) and Digital Communication in Museums (ICP). Between 2017 and 2024, she taught courses such as Art and Media, Art, Business and the Market (USAL), Curatorial Projects III (UNA), Semiotics of the Arts (UNA), and History of Art and Design (UCES). Her research focuses on museum visiting practices in Europe and the Americas. Her work on institutions such as MoMA, MALBA, the Centre Pompidou, Tate Modern, and Museo Reina Sofía has been published and presented in international journals, collective volumes, and academic conferences.

Chapter 13
The Semiotic and Computational Analysis of Represented Poses in Painting and Photography

Maria Giulia Dondero and Adrien Deliège

Abstract This chapter focuses on the evolution of styles in the history of painting and retraces new genealogies of forms through recent deep learning-based computational methods applied to large image databases. This text associates the semiotic approach with the computational one offering a panorama of the most important current projects in Digital Art History. It proposes a particular conception of forms and forces in images, with the objective of making continuous bodily gestures analysable in large collections of paintings and photographs, starting from the works of art historians Henri Focillon and Aby Warburg. This conception of forces in images finds its roots in the Paris School Semiotics, but also in the work of philosopher Gilles Deleuze on pictorial diagrams in Francis Bacon's painting and in mathematician René Thom's idea of conflicting forces in painting.

Keywords Gesture · Plastic semiotics · Pose estimation · Visual semiotics · Computer vision

13.1 Introduction

This chapter aims to analyse large collections of images and especially gestures and poses represented in painting and photography by considering crucial art historians' work (notably Henri Focillon's and Aby Warburg's) and current research projects in digital art history.

The authors wish to thank the colleagues who contributed to the reflections that led to this article: Enzo D'Armenio and Andrea Pinotti.

M. G. Dondero (✉) · A. Deliège
University of Liège, Liège, Belgium
e-mail: mariagiulia.dondero@uliege.be

A. Deliège
e-mail: adrien.deliege@uliege.be

P. Conte et al. (eds.), *Algomedia. The Image at the Time of Artificial Intelligence*, Lecture Notes in Morphogenesis, https://doi.org/10.1007/978-3-032-08726-3_13

The first part will be devoted to the analysis (and visualisation) of large collections of images, as they have been conducted in recent years by scholars such as Lev Manovich, Benoît Seguin, and Leonardo Impett. We will briefly recall the specificities of these three ongoing or recently finished research projects, specifically the *Media Visualization* project by Lev Manovich devoted to visual similarities, the *Replica* project launched by Benoît Seguin at the EPFL's Digital Humanities Lab in Switzerland on migration of motifs, and the more recent project held by Leonardo Impett and Franco Moretti on the modelisation of gestures in paintings (*Totentanz*). Firstly, we'll try to explain why these projects are unsatisfactory with respect to the current project of tracing an innovative genealogy of visual forms.[1] Secondly, we'll present the main idea of our scientific project, that is to analyse the genealogy of visual forms (notably *the gestures represented*) from the renaissance and baroque painting to the contemporary fashion photography using current approaches in image processing. We will present the particular conception of forms and forces in images involved in this project, which has the objective of making *continuous bodily gestures* analysable in large collections of paintings and photographs. This conception of forces in images finds its roots not only in the Paris School Semiotics [19, 21], but also in the philosopher Gilles Deleuze's work on pictorial diagrams in Francis Bacon's painting [5], and in the mathematician René Thom's idea of conflicting forces in painting [32]. They all argue that the image is not a matter of the representation of recognisable, isolated objects, but rather a matter of dynamic relations and oppositions between differences and similarities present within the surface of an image. If the theoretical and methodological basis of our approach is the structuralism and post-structuralism developed in France since the 1960's,[2] this project aims to trace a genealogy of visual forms as conceived of by Aby Warburg and by Henri Focillon.

13.2 Databases and the Analysis of Large Collections of Images

Images contained in databases are made *commensurable* by vectorisation and consequently *manipulable* by algorithms. This commensurability can be explained through a so-called "latent space." It represents the abstract, multidimensional numerical space in which deep-learning algorithms turn digital objects into latent representations so that they can be processed and *used to analyse or to generate* new digital

[1] We will notably focus on a project entitled *Towards a Genealogy of Visual Forms*. It is a four-year project, financed by F.R.S.-FNRS (2022–2025) and led by Maria Giulia Dondero as Principal Investigator, with Adrien Deliège as a post-doctoral researcher involved in this project (2022–2024). However, the scientific endeavour of constructing a genealogy of forms is a broader and more widely shared undertaking.

[2] In the post-structuralism period, differences have been studied not only as binary categorial oppositions but also as *graduated modulations*: This is the conception offered by "tensive" semiotics, a development of structuralism in the 90's and 2000's especially thanks to Jacques Fontanille and Claude Zilberberg's works [20].

objects (e.g., new images and new texts). Latent representations are made of vectors; that is, long lists of numbers that define the coordinates of the digital objects encoded and embedded in the latent space and, as explained by Antonio Somaini (2023), their relations of distance and proximity within it, just as the three coordinates x, y, and z define the position of a physical object in our physical three-dimensional space and their relations to other physical objects. Therefore, the formal and plastic characteristics of each image (colours, forms, topology, etc.) can be quantified and measured against the characteristics of other images to produce a system of structural differences and similarities within the collection. This commensurability that leads to comparability, in linguistic and semiotic terms, can be understood through the Saussurean theory of *langue* and *parole* and through the concept of enunciation that has been forged by Benveniste in the 1950s [1]. According to a very general definition, in the terms of Benveniste, enunciation is a mediation between a system of virtualities (*langue*) and a discursive performance (*parole*, utterance). The theory of enunciation allows us to conceive of a *mediation* between Saussurean *langue*, a sort of grammar, and *parole*, the actualisations of *langue* in discourse—which are supposed to broaden the spectrum of the virtualities of language. The theory of enunciation can be useful to study image databases because a database such as WikiArt contains a significant amount of artistic paintings, photos, and drawings retained in the history of Western art, i.e. most of the visual virtualities that have previously been enunciated and performed. Moreover, confronted with a database available for manipulation in a latent space, we face not the substance of the visual, but a substance that presents itself as a transcription of numbers manipulable by algorithms. In other words, within the collection of digital images, all the images contained in the database are identified by lists of numbers. Now, what happens once all the images have been transposed into lists of numbers? This database becomes comparable to Saussurean *langue*, a system that makes the images (*parole*) commensurable with one another. Since Benveniste [1], in fact, the theory of enunciation has greatly evolved and has enabled to consider the fact that *langue* is a system of historically attested texts, built on the practices of the speakers, and not on pure grammatical virtualities. In our case, the database containing large collections of texts and images (and the latent space on which they are translated and projected) can be identified as a *langue* in the sense of a collection of texts that are co-present and made commensurable through chains of numbers and through algorithmic operations on these chains of numbers.

In general terms, each image in a database can be compared with all the images in the collection on the basis of a number of parameters that may be calculated, such as light intensity, colour, contours, and the like. In a classic feature extraction approach, these parameters can be chosen by the scholar. Conversely, in more recent deep learning methods, the relevant characteristics are learned by the computation process on a training set of images. Two analytical strategies are at play in the three projects presented hereafter: the first is feature extraction, used by Lev Manovich in his *Media Visualization* Project. This method extracts features from the contents of databases based on rules "manually" predefined by the researcher who dictates the instructions to be followed to perform the task, for example the decision regarding the features to be extracted, such as the gradients of colour and luminosity in the paintings by abstract painters of the early twentieth century [27]. The second strategy is deep learning, a learning paradigm[3] that provides the machine with datasets through which the machine must spot similarities/dissimilarities. When we use a deep learning algorithm, what we have at hand is no longer the *extension of the eye of the researcher* who decides what the machine should find in the collection of images, as was the case with feature extraction. Instead, what we have is the *extension of the database* used to train the algorithm. By using deep learning, we let the algorithm decide what it needs to calculate in order to perform its task in the best way possible. The researcher only gives the model feedback on its result, enabling it to self-correct without telling it exactly what calculations it should have made. Indeed, *the quality of the database will condition the model's ability to learn more or less correctly how to perform its task.* It is clear, for example, that if the algorithm has been trained on a dataset of ordinary images representing everyday objects, it will be very difficult to obtain good results in a search for artistic pictures.[4] In other words, the dataset on which the algorithm is trained must have sufficient affinities with the database that will later be studied, in order to focus correct relations between images in terms of similarities and dissimilarities.

The tasks that the machine performs "on our behalf" concern image analysis. Definitely, the machine is asked to do more than we ourselves can do, notably to organise large collections of visual data (thousands of digitised paintings) according to their similarities/dissimilarities.

The analysis of large collections of data has been made possible thanks to the production of massive data which, particularly in the case of deep learning, has made it possible to relaunch research projects that could not be accomplished in the past. We are referring in particular to the projects of two art historians, Henri Focillon and Aby Warburg. Henri Focillon theorised the genealogy of forms as a process of bifurcation, ramification, stratification and also disappearance of forms in his seminal book *The Life of Forms in Art* [17], originally published in 1934, without being able to trace a genealogy of forms in its entirety. For his part, in the 1920s Aby Warburg studied images composition in his *Atlas Mnemosyne* (2020), using the visualisation of images within large black panels as a means of investigation. He organised them in such a way that images with common or opposite formal characteristics were arranged in the same region of the visualisation. Each image was in this way positioned in relation to all others in the same panel with the aim

[3] Deep learning is a learning paradigm where the general principle is to define a model (usually a neural network) with a set of parameters and to try to find by "gradient descent" the best parameters for performing the task. Progress in the search for these best parameters is guided by a feedback mechanism, often within the supervised learning framework. The fact that learning is "deep" is initially linked to the fact that we have been able to train neural networks with a large number of layers/parameters, and the more layers there are, the "deeper" we go into the network, hence this notion of "depth" (linked to the vast number of parameters we can simultaneously optimise using this technique).

[4] In fact, in general, the scholar starts by compiling the dataset he/she wants to analyse, before extracting part of it, annotating it and using it for training purposes. The rest of the database will be used to check the results. If this annotation of the initial dataset is well done, then the scholar gives him or herself a better chance of obtaining a model that generalises well to his/her data of interest.

of illuminating the relation "of the nearest neighbour" in compositional terms. In this way, the images were gradually distanced from those that were opposed or in conflict in terms of their plastic and formal characteristics. This is precisely what was done by researchers such as Lev Manovich in the early 2000s [27] and Benoît Seguin in his Ph.D. thesis [31], following this project and ensuring that databases and the algorithms that apply to them can group data according to their similarities and differences following Warburg's "nearest neighbour" rule.

13.2.1 The Media Visualization *and* Replica *Projects*

Let us consider the visualisation of thousands of images conceived of as results of analysis of large collections of images. The *Media Visualization* and *Replica* projects have two characteristics in common, even though the *Media Visualization* project processed images by means of techniques based on the extraction of their low-level features such as the intensity of colour, outlines, and positions of forms (Fig. 13.1), while the *Replica* project uses deep learning methods (Fig. 13.2). In the first case, the researcher's biases come into play, whereas in the second, it's the database's biases that are at work. In fact, when using a deep learning algorithm, the researcher's eye is no longer extended, but the database used to train the algorithm is. In the case of the supervised method used by Manovich, the researcher himself defines the characteristics to be calculated by the algorithm, for instance, extracting the brightness and saturation of colours in abstract works.

Firstly, the two projects have in common the use of visual descriptors to analyse large collections of images, instead of exclusively using metadata such as the date and place of production, the genre, and so on. As stated by Manovich [25], the main

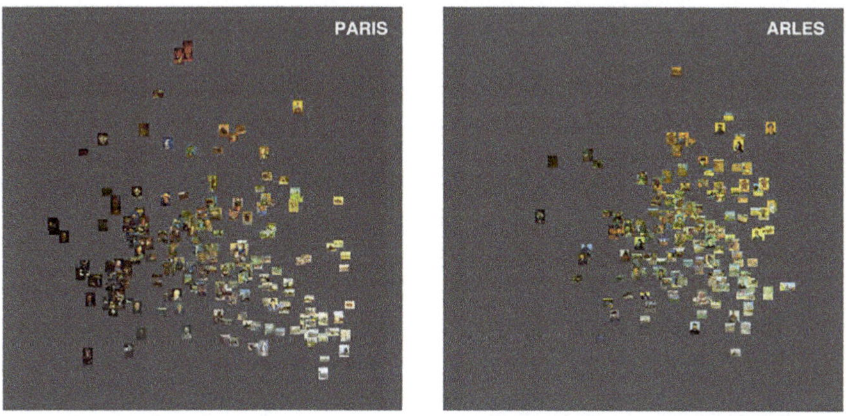

Fig. 13.1 **a, b** *Media Visualization.* Manovich and Cultural Analytics Lab. Comparison of the paintings produced by Van Gogh in Paris (left) and Arles (right) with respect to brightness and to saturation. X-axis: brightness average; Y-axis: saturation average [26]

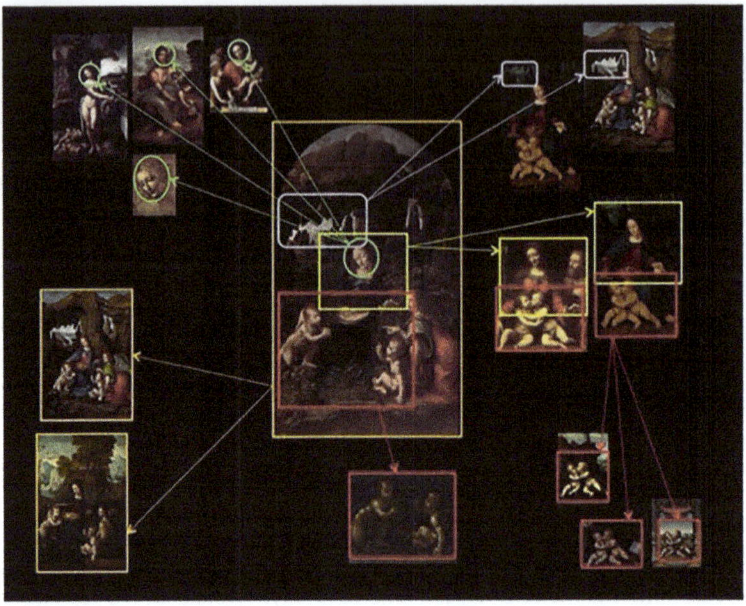

Fig. 13.2 *Replica* Project. Migration of motifs. See [31]

flaw of classification by standard metadata is that the terms used in the classification of images do not reflect the characteristics of visual language, namely the forms of organisation *specific to images*.[5] Secondly, both projects produce visualisations of images which function as *meta-images*, that is to say, the visualisations show the results of processes of analysis (in the sense of division).[6] Such analysis is to be understood as the result of *mereological operations* of division, grouping, superimposition, etc., mereology studying the relations between a totality (the collection) and its parts (the single images composing the collection) and between the parts themselves.[7]

Now turning towards *Media Visualization* (Fig. 13.1), we see that the images under study that share the same features are located in the same areas of the diagram into which they are organised; the ones that do not are placed in other areas. The position of each image within a group of images offers a precise characterisation of its plastic properties and shows their relation (gradual distance or/and oppositions) to the qualities of other groups of images.

[5] ON visual metalanguage, and against the imperialism of natural language in the analysis of images, see [15] and [11].

[6] SEE [9] and 2019).

[7] SEE the semiotic work on mereology by [3]. It is necessary to add a third mereological level—namely, the one that considers the single image as a whole and the specific components extracted from it as its constituent parts. In the structural thinking of semiotics, these parts are not fixed units but rather plastic oppositions: light vs. dark for chromatic categories, up vs. down for topological categories, linear vs. fragmentary for eidetic categories, and so on.

This project represents an important step for developing the study of corpora. Even if *Media Visualization* analyzes exclusively what we call in semiotics the *plane of expression* of images [16, 21], that is, what is *perceptible in an image*,[8] our observation is that this kind of analysis made by operations of division and grouping may be used for the *comparative* study of the *plane of content* of the different groups of images. This relation between oppositions on the plane of expression of images and oppositions on their plane of content is traditionally called in Paris School Semiotics "semi-symbolism." In a sense, the automatic analysis of the plane of expression can be useful for the discovery of similarities or dissimilarities on the plane of content of images. To reach the objective of analysing the plane of content as well, it's necessary to observe the different *groups of images distributed* on the surface of the visualisation and to question the content characterising these groups. If the images that belong to a group displayed in the lower-left part of the visualisation are considered—concerning their plane of expression—as completely different from the ones situated in the upper-right part of the visualisation, what about their plane of content? Do the images belonging to the same group on the plane of expression, belong as well to the same genre (portrait, still life, landscape)? Finally, the visualisation can become a *map* displaying not places but groups of images defined by their formal features to be used to test their content (genre, for instance). In this sense, tentative corpora can be constituted.

The two limitations that can be ascribed to the *Media Visualization* project are firstly that the visual similarities identified in large corpora have little historical relevance, in the sense that they do not produce a genealogy of forms but only a system of similarities, and secondly, that this approach only take into account *averages* of features (that is, averages in terms of brightness, saturation, and hue), the problem being that such averages fail to reflect the actual configuration of the singular image composition.

Manovich's method identifies low-level features, which the semiotics of Algirdas Julien Greimas [16, 21] refers to as "plastic properties." The three plastic categories involved have been used exclusively for analysing individual images or small corpora and they are as follows:

1. *topological categories*, some "rectilinear" (such as upper/lower or left/right) and others "curvilinear" (peripheral/central or enclosing/enclosed);
2. *eidetic categories* (pointed/rounded or contoured/flat);
3. *chromatic ones* (degrees of saturation of colours).

Unlike *Media Visualization*, in Greimasian plastic semiotics, the three categories mentioned (topological, eidetic and chromatic) are put in action in image analysis using the structuralist principle of oppositions (upper vs. lower, left vs. right, enclosing form vs. enclosed form, and so on) or using the principle—borrowed from the tensive semiotics of Fontanille and Zilberberg [20]—of intensity and extension

[8] For this reason, this kind of analysis is often violently criticised by art historians and scholars in humanities.

graduation in a categorial scale (from upper to lower light intensity, from a right-situated saturated colour to a left-situated unsaturated colour, from a light concentration to an intensive shadow, for instance). By examining oppositions and the graduated scale on the plane of expression, the analyst can study each image not as an average or as an accumulation of isolatable objects, but *as a singular entity composed of tensions between opposing characteristics and directions.*

Examples of plastic oppositions can be seen in two paintings, Tintoretto's *Tarquin and Lucretia* (1578–80)[9] and Artemisia Gentileschi's *Judith Slaying Holofernes* (1612–13).[10] The first case shows the oppositions between forces of elevation (Lucretia's body is suspended and pulled upwards by Tarquinio) and forces of fall: the pearls of the necklace are falling down, as are the small sculpture in the right side of the painting and the pillow as well.

In the second image, Holofernes is dying but continuing to struggle with Judith and the servant who are trying to immobilise and behead him. The conflictual forces in opposite gestures and directions are clearly visible.

As mentioned earlier, in Paris School Semiotics, oppositions on the plane of expression are used to formulate hypotheses about oppositions on the plane of content ("semi-symbolism"): this relation allows one to conceive of the image as a totality (its parts are in relation with one another), and not as a collection of isolated elements.

In the *Replica* project (Fig. 13.2), led by Seguin [31], the image's analytical visualisation, the so-called "morphograph," displays the visual connections which highlight the morphological relationships between a given painting and other similar paintings. This device is called *pattern propagations.* More broadly, this project aims at identifying the *repetition* of *recognisable separate visual elements* and/or of iconographical motifs (the kneeling woman, the reclining Venus, and so on). In this sense, *Replica* is more interesting than the *Media Visualization* for studying genealogies of visual forms because it tries to trace the path going from an original iconographic motif to the remake/rework of this motif in another painting of a later period, while the *Media Visualization* by Manovich is focused on the similarities between images based on the average of their low-level features.

Finally, in the automatic visualisations by *Replica*, one might recognise the legacy of the panels of images by Aby Warburg, published in the *Atlas Mnemosyne.* As early efforts to visualise images on the basis of their similarities in terms of forms and inherent dynamics, Warburg's panels represented a first attempt to understand the *influence of images on images*, through a panel providing a surface helping to apprehend the *graduated relation* between visually close and distant images.

---

[9] https://www.artic.edu/artworks/64920/tarquin-and-lucretia.

[10] https://www.uffizi.it/en/artworks/judith-beheading-holofernes.

13.2.2 Analysing Bodily Gestures

The third project in computational image analysis we would like to comment upon concerns the *representation of bodily gestures* in images. On this topic, Impett and Moretti [24] were the first to formalise the gestures of the bodies found in Warburg's *Atlas Mnemosyne* panels (Fig. 13.3). The attempt is remarkable in that it seeks to measure and compare the movements of bodies depicted in various images from the *Atlas Mnemosyne* with the aim of formally describe *Pathosformeln*, the expressive forms of pathos conveyed through gestures.

Our critical remark regarding this method is that its modelling reduces the body to a skeleton, a geometric figure made of line segments, whereas the body possesses volume, occupies space, and has a silhouette that is integral to every gestural dynamic, playing a crucial role in directionality. The body represented in an image cannot be

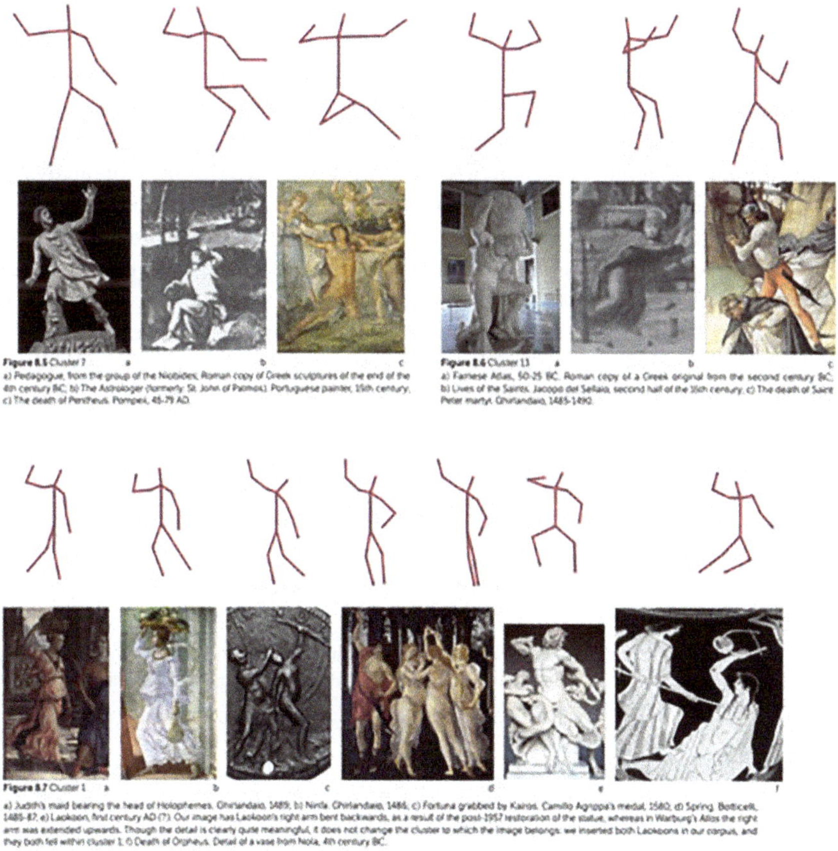

Fig. 13.3 Impett, Moretti 2017

analysed as a mere sum of individual parts (spine, arms, legs, head, etc.); rather it must be understood as a totality shaped by internal forces situated within a landscape.[11]

In a similar vein, our current research project aims at tracing poses, gestures, and other kinds of movement and *dynamics of forces in still images* such as paintings and photographs spanning a period that extends from Baroque painting to contemporary fashion photography, extracting poses and clustering them.

What do we mean by "the dynamics of forces within a still image"? The forces can be partly identified through directionality: the direction of a glance, a raised hand, a pointed finger. Forces can also emerge with elements of the image that are not figurative but formal or plastic in nature[12]: changes in luminosity within a painting can function like arrows, guiding the viewer's gaze; similarly, a shift in saturation can generate a sense of upward lift or downward pull. The geometry of a gesture also plays a role: in this case of Western painting a gesture forming a circular figure on the plane of expression conveys stability and calmness on the plane of content; in contrast, a gesture forming an irregular triangle suggests disrupting directions and conflict on the plane of content.

As already mentioned, our project is an attempt at pursuing Warburg's work on the genealogy of forms, particularly the forms of pathos, or *Pathosformeln,* which we can find in the poses and gestures of the figures depicted in paintings throughout the ages. However, this project also joins one of the major current projects of visual semiotics, namely the study of temporality and rhythm in the still image. This involves analysing how temporal and aspectual deixes are signified by means of a still materiality such as that of a painting. While actantial and intersubjective enunciation[13] has been significantly developed in visual semiotics, with several works on the face view and the side view in portraiture [2, 12], in addition to spatial enunciation, thanks to several works on perspective,[14] temporal enunciation has not been sufficiently investigated. By temporal enunciation, we mean that the action represented may be oriented towards the future or towards the past or inhabit the present of the act of observation. Aspectual enunciation concerns the specific moments a painter decides to represent in the image: inchoative, terminative moments or their durativity of ponctutality. In other words, aspectuality concerns the moment of the action that has been focused on by the painter, as well as the rhythm of the action's unfolding, which may be accelerated or calm.[15]

---

[11] Leonardo Impett, in a 2020 paper published in the Routledge Companion to Digital Humanities and Art History, entitled "Analyzing Gesture in Digital Art History," further analyses the parameters of directionality and the rhythm of moving.

[12] ON plastic and figurative dimensions in images see [21].

[13] ON the theory of enunciation and its three categories (actant, space and time) from Emile Benveniste onwards, and notably regarding images, see [11], and on data visualisation and enunciation, see [14]. For a general theory of enunciation, see [18].

[14] SEE [28] and [19].

[15] ON this topic from a semiotic perspective, see [29], Groupe Mu [22] and Colas-Blaise [4]. On temporality in portraiture, see [13]. For a study of narration in still images, from a philosophical point of view, see Pinotti [30].

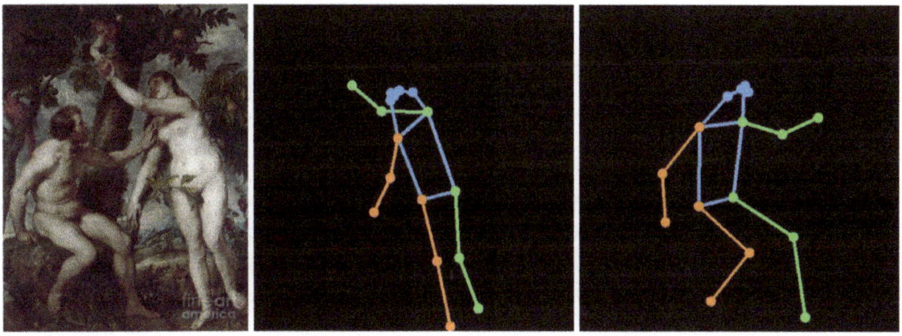

Fig. 13.4 Example of individual poses extracted from an original painting, Peter Paul Rubens, *Adam and Eve in the earthly paradise*, ca 1628.

13.3 Towards a Genealogy of Visual Forms Project

In the framework of our research project entitled *Towards a Genealogy of Visual Forms (2022-2025)*, we formalised the human body in a more complex manner than the one experimented with in the work of Leonardo Impett, by considering the body as located within a global dynamic of forces which constitutes the image.

For that purpose, we collect all religious paintings (11,980 available when conducting this research) from WikiArt. Then, on these images, we run MMPose, an algorithm which estimates bodily poses using a skeleton-like schema of 17 keypoints,[16] and we filter out the images not containing at least one pose having all its keypoints confidently detected. This leaves us with 5269 images, containing 8599 individual poses. Each individual pose is redrawn in a separate image, with normalised[17] keypoints coordinates to allow for meaningful comparisons between images. This process is illustrated hereafter (Fig. 13.4).

We first analyse all these individual poses before moving to the collective ones. We define the distance between two poses as the sum of the distances between their corresponding keypoints. This distance metric is used in the UMAP dimensionality reduction module of the PixPlot software to produce a visualisation of the organisation of individual poses (Fig. 13.5), where similar poses are located close to each other, and dissimilar poses are placed further apart.

Within this collection, it is possible to follow the variety of poses according to the organisation established by the algorithms. As examples, each green circle in the general collection diagram (top left corner of each subfigure of Fig. 13.6) refers to a group of paintings the machine recognises as belonging to a specific pose, which is zoomed in to show the validity of the approach.

It can be observed that the centre of the visualisation tends to regroup poses that are relatively neutral, depicting a person standing, facing the viewer. As we move away

[16] The important number of keypoints allows to better detect the orientation of faces, the directions of glances, the consistent bodily volume and the density of movements.

[17] More technical details are given in our blog post here https://ceserh.hypotheses.org/3929.

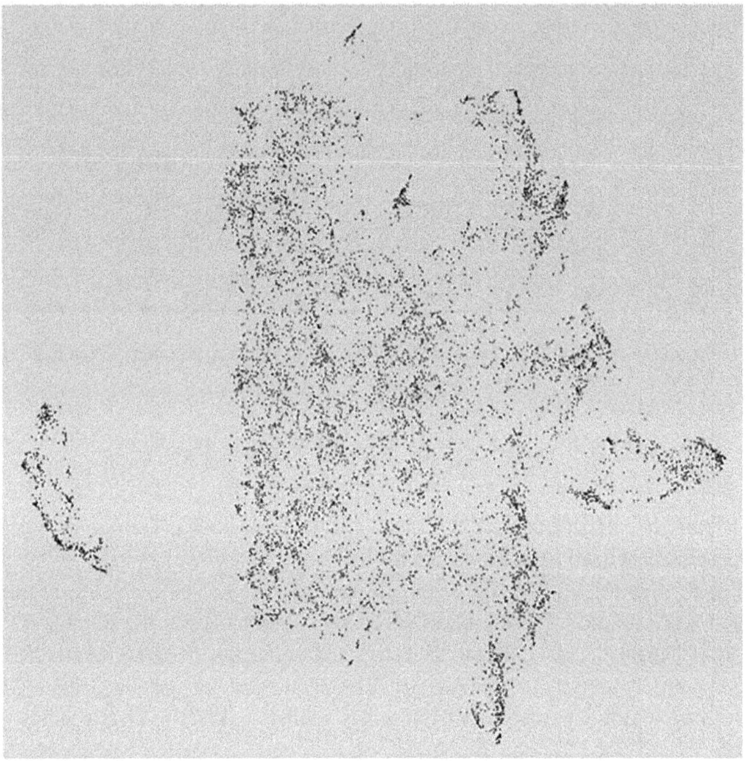

Fig. 13.5 Visualisation of 8599 individual poses from 5269 religious paintings contained in the WikiArt database and organised by pose similarity. An interactive web application that allows navigating through this "image cloud" (with zooming in and out functionalities) is available at https://adriendeliege.z6.web.core.windows.net/outputs/WikiArt_religious_painting_solo_poses/index.html

from the centre, the poses continuously vary, reaching completely different poses in the corners of the meta-image, such as characters lying down, sitting, falling, etc. A cluster of representations of Jesus on his cross is also clearly visible, as this pose is common among religious paintings. We can also spot a cluster of images where the character is seen from the back, which is completely, and rightfully, dissociated from the rest of images (on the far left of the visualisation). Let us also note that a body lying down with the head on the left is a completely different pose (according to the metric used) than if the head is on the right of the image. The opposite poses are also represented in opposite parts of the visualisation, namely top and bottom in this case. Finally, as for every large-scale automated analysis, there are some unfiltered errors that sneaked through this visualisation; in this case as a small cluster of poses with legs cut at the knees, corresponding to characters that are not completely shown on the images (on the middle top surrounded by a void neighbourhood in the visualisation).

The structure of the analytical visualisation shows and explores the whole collection, which we call the *reference corpus* (it is all-encompassing). It is possible to

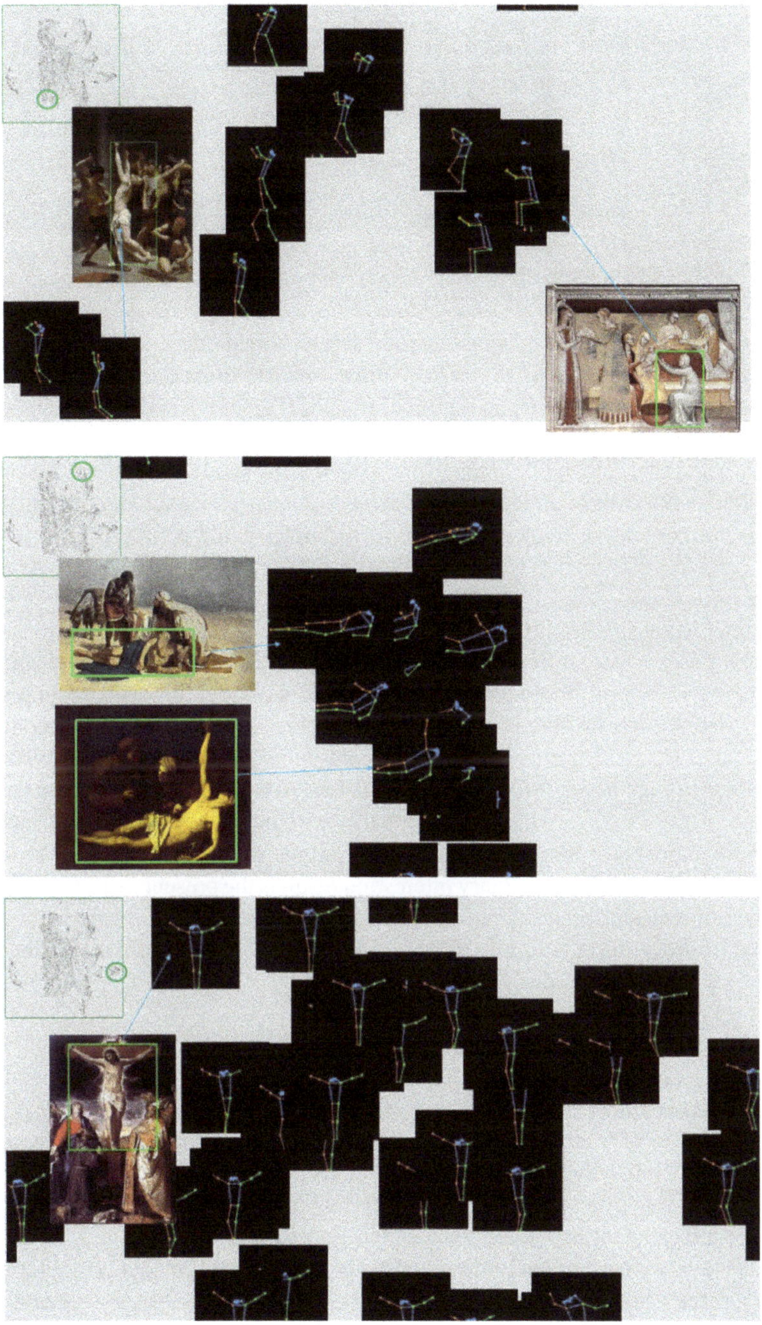

Fig. 13.6 Examples of specific poses shared by certain paintings, circled in green (top left corner) in the general visualisation, and zoom in these delimited regions to show the similarity and variations of similar poses belonging to a group of images

Fig. 13.7 Artsy images displaying characters with "Jesus on his cross"-like poses

choose within it, or from an outside source (see hereafter), a *working corpus* of interest to analyse in comparison or in contrast with a reference corpus, here a series of images belonging to the same region of images (Figs. 13.5 and 13.6). This group of images can be used to describe the relationship between the plane of expression of images and their plane of content, following a semi-symbolic analysis that establishes that an opposition on the plane of expression is correlated to an opposition on the plane of content of images. Two simple examples could be the following: *arms up vs. arms down = prayer vs. rest*, or *standing body vs. reclined body = arrogant pose vs. pious pose.*

An interesting side application that can be derived from our method is the retrieval of similar poses in very different domains. For example, in a first attempt to compare contemporary fashion photography and religious paintings in terms of poses, we downloaded Artsy's fashion photography catalog[18] (12,000 images, 5000 complete individual poses) and produced a PixPlot combining Artsy's poses and WikiArt's religious paintings poses.[19] Overall, it seems that Artsy's poses are more present at the edge of the PixPlot, with more acrobatic/extreme poses than regular standing characters, more characteristic of religious paintings and more present in the centre of the PixPlot. Another preliminary interesting result is the presence of a few of Artsy images in the cluster of Jesus on his Cross produced by religious paintings, at the very top of the representation, as shown in Fig. 13.7. Searching for a specific pose of a corpus in another corpus would be difficult without our method, given the large number of images to browse.

Furthermore, beyond the visualisation of the organisation of the poses contained in the reference corpus, we can derive a query-and-rank process from our computations. We can submit a query image which may or may not belong to the reference corpus, extract the pose of a character in it, and rank the images of the reference corpus by their pose similarity obtaining a working corpus. However, it should be noted that several possibilities exist to create such rankings. Indeed, with the normalised (location-independent and scale-independent) poses, the comparison is made as if the characters were completely detached from their context images. An alternative is to keep the scale dependency such that the space occupied by a character in a

[18] https://www.artsy.net/gene/fashion-photography.

[19] https://adriendeliege.z6.web.core.windows.net/outputs/Artsy_fashion_WikiArt_religious_solo_poses/index.html.

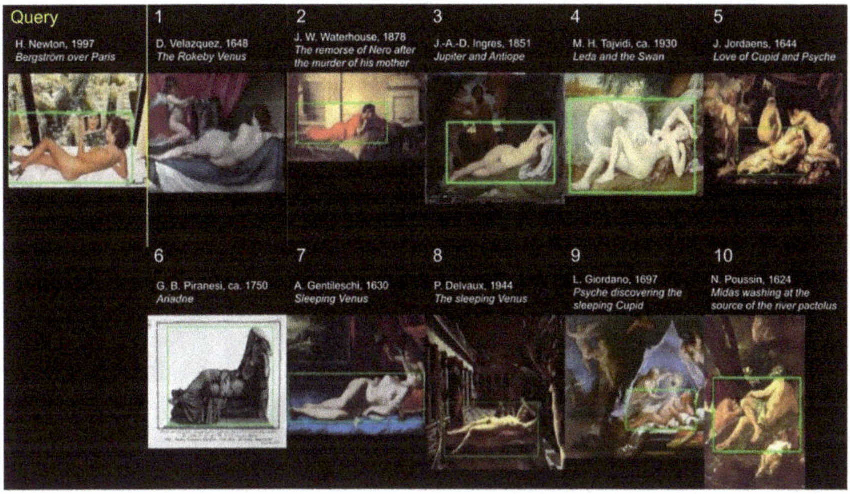

Fig. 13.8 Top left: query image. Right: filtered nearest mythological paintings in term of average combination of scale-dependent and scale-independent rankings

painting is also taken into account in the comparison, which results in grouping prominent characters together (sometimes at the detriment of the pose similarity itself). To balance the pose similarity with the scale of the pose, we can also produce a third ranking, obtained as an average of the two previous rankings. Along with minor filtering processes,[20] this last option gives good query-and-rank results for a single query pose (exemplified in Fig. 13.8 where the reference corpus is composed of mythological paintings).

In order to consider the image as a whole and not only as an ensemble of detached-from-the-background poses, we can extend our study to groups of poses, but this setting comes with additional challenges. A first difficulty is to determine which poses are compared. For instance, if an image contains 4 characters, and another image contains 6 characters, which subgroups of characters should be compared? Generally speaking, it makes sense to compare only subgroups with the same number of characters. In this example, beyond comparing individual poses, we can compare each subgroup of 2, 3, 4 characters of the first image with each subgroup of 2, 3, 4 characters of the second image. While this is theoretically feasible, the number of subgroups grows exponentially with the number of characters. Therefore, due to computational constraints and for the purpose of illustration, we use at most groups of 3 poses.

As already outlined in the case of individual poses, several modalities can be compared, related to positions and scales of the poses. So, let us assume that we have two groups of characters from different paintings to compare. How do we compare them? We identified 7 criteria that allow each a unique way of comparing

[20] Detailed in our blog post https://ceserh.hypotheses.org/3929.

the poses. The first 3 criteria focus on the configuration of the group of poses: the poses are reduced to their average point, these average points give the configuration of the group of poses. We analyse these configurations by looking at (1) its average relative position in the image, (2) its shape independently of its scale, (3) its shape dependently of its scale. The next 4 criteria focus on the poses themselves and not on the configuration of the whole group of poses. Again, we can either keep (4)(5) or remove (6)(7) the dependency on the scale of the poses, by letting them as is or normalising them, just as discussed for the configuration of the group of poses and for individual poses. Besides, when comparing two groups of equally numbered poses, we must decide which pose of the first group is compared with which pose of the second group. The matching between the poses of the two groups can be done either by matching the poses by their appearance (which pose of the second group is the most similar, as in the individual poses analysis, to which pose in the first group?), or by their localisation (which one of the second group is the closest to which one of the first group in term of relative position in the image?). We name the first matching "best pose matching" (4)(6) and the second one "location-based matching" (5)(7). In summary, the 7 criteria available are:

1. Configuration comparison: localisation of the group of poses within the image
2. Configuration comparison: shape of the group of poses, scale-dependent
3. Configuration comparison: shape of the group of poses, scale-independent
4. Poses comparison: scale-dependent, best pose matching
5. Poses comparison: scale-dependent, location-based matching
6. Poses comparison: scale-independent, best pose matching
7. Poses comparison: scale-independent, location-based matching

These choices are embodied in Fig. 13.9.

Let us note that this naturally extends the case of individual poses comparisons, where criteria 2 and 3 then become irrelevant since there is no such thing as "shape of the group of poses," and the matching does not matter anymore since there is only one pose to compare with another one, thus criteria 4 and 5 are equivalent, as well as 6 and 7. Thus, one needs to compute only criteria 1, 4, 6 in the case of individual poses, as previously discussed and illustrated in our results.

As in the case of individual poses, each of the 7 criteria can lead to a PixPlot visualisation and leads to a ranking of the reference corpus with respect to a query image, as shown in Fig. 13.10.

These 7 rankings can be aggregated at will by weighting the ranks as deemed appropriate to produce a final ranking. While many choices can be made, it seems to us that, in the case of individual poses, a combination of equal weights of criteria 4 and 6 (pose comparisons, with scale-dependence and independence, equivalent to 5 and 7) is appropriate, as well as dismissing criterion 1, as already shown previously. This prevents putting too much emphasis on poses that look very different but have the same scale as the query image (which acts thus in its favour in the ranking while being not relevant) as well as putting too much emphasis on poses that look very similar to the query pose but are way too different in term of scale/prominence in the image (which may thus convey a different interpretative meaning). In the case

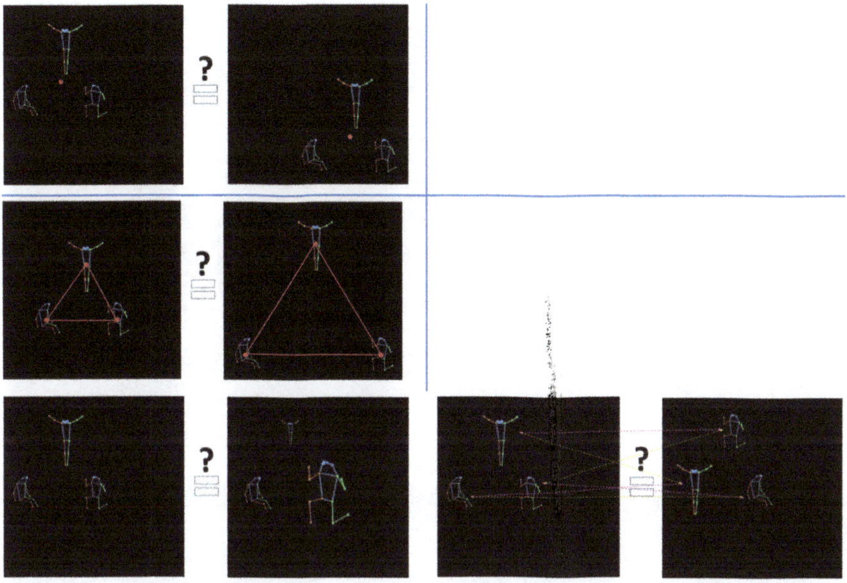

Fig. 13.9 Criteria for comparing groups of poses. Top left: (1) Configuration comparison: should we take into account the localisation (red dot) of the group of poses within the image? Top right: (2) and (3) Configuration comparison: should we take into account (2) or not (3) the scale of the shape formed by the group of poses (in this case the red triangle)? Bottom left: (4), (5), (6), (7) Poses comparison: should we take into account (4, 5) or not (6, 7) the scale of the individual poses? Bottom right: (4), (5), (6), (7) Poses comparison: should we compare poses that are the most similar (best pose matching, yellow arrows, 4, 6) or that are located at the most similar place in the group configuration (location-based matching, pink arrows, 5, 7)?

of multiple poses, the same motivations regarding the scale of the poses lead us to consider criteria 5 and 7, with a location-based matching preferred over a pose-based matching. Indeed, we believe that, if the same poses are permuted within the pose configuration, this yields a different image, which is not captured by the pose-based matching. We also consider criteria 2 and 3 in the mix, to incorporate the shape aspect of the configuration in the final comparison, with both of them equally weighted for similar reasons as those motivating the choice of 5 and 7. We neglect criterion 1 again for the same reason as previously. This gives the image shown in Fig. 13.11.

We can wrap this pose analysis by reflecting on one of our initial questions, about assessing forces and dynamism in paintings. More precisely, how much motion is conveyed within poses of the characters? We could argue that some poses definitely embody some dynamism, especially when said poses differ heavily from the most common "calm" poses such as standing, sitting, lying down. Using the PixPlot visualisation and the distance metric defined previously, we could select representative "calm" poses and define how different from these poses another pose should be to be called "agitated," or "in motion." However, some seemingly calm movements can in fact convey a sense of motion by their relationship with other movements,

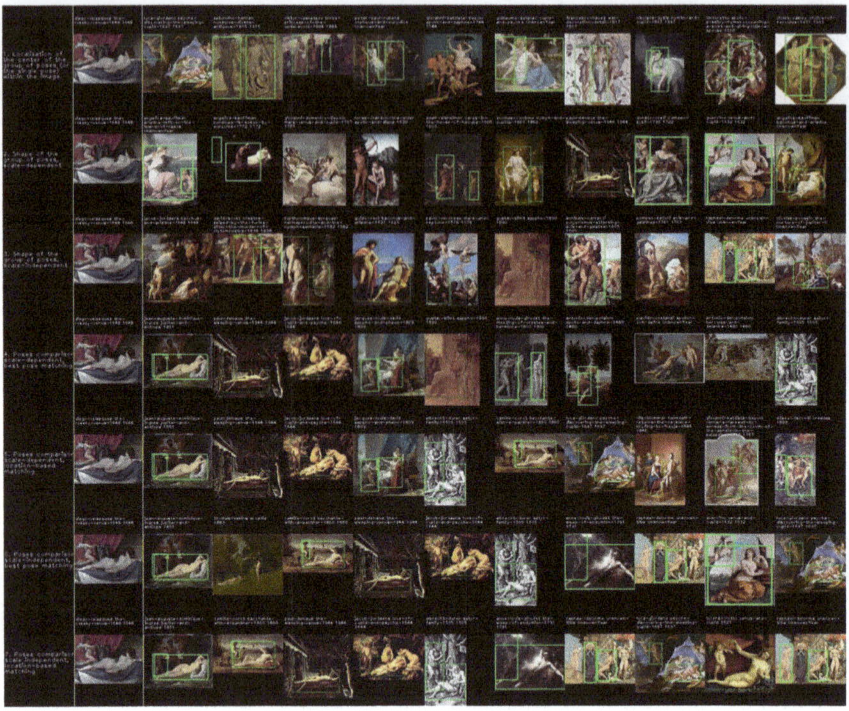

Fig. 13.10 View of parameters organising pose similarities. The query image is Velázquez' *The Rokeby Venus*. Two characters are detected: Venus and the Angel. Thus, groups of 2 poses are compared, each line corresponding to one of our 7 criteria

Fig. 13.11 Ranking among the reference corpus "Mythological paintings," by comparing subgroups of 2 poses against the pair Venus-Angel of Velázquez

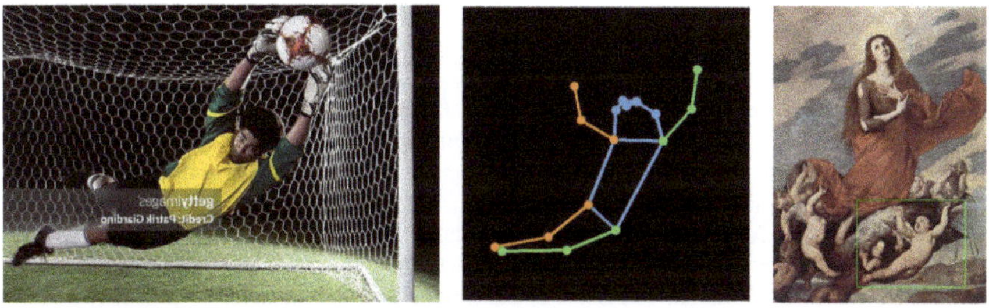

Fig. 13.12 The pose alone (in the middle) conveys a notion of motion but does not provide any information on the character that moves, the direction of the movement, or the type of motion. In this case, it corresponds to the angel in the green rectangle on the left painting

which makes it difficult to make a clear cut between calm and agitated motions [8]. Finally, the poses computed, even when considered "agitated," do not bring much information about the directionality of the motion, the character that perpetrates it, and the type of motion (Fig. 13.12). Therefore, despite being an informative element about motion in paintings, finer analyses or different techniques are needed if one wants to fully characterise motion in images, such as directional edge detection [6], or dynamic elements such as those described by Wölfflin categories [7].

13.4 Conclusions

At the end of our journey, we may state that we could show the evolution of projects in digital art history: from the *Media Visualization* which was seeking formal similarities and dissimilarities in large collections of images, to the *Replica* and Impett's projects which were seeking for similarities in poses and motifs. In the case of our project on the genealogy of visual forms (and poses) we have taken into account complex individual and collective poses—that the other projects did not—and included contemporary photography in our database in order to see what has survived of ancient poses. By doing so, we obtain insightful qualitative results in the pose-based identification of similar paintings. Despite technical challenges (e.g. combinatorial explosion of possibilities, difficulty for the pose model to identify some characters), we believe that our work can prove helpful to pave the way for more large-scale data-driven analyses of art collections. More generally, modern computer vision methods have the possibility to level up the whole field of art history. We hope that the present work is a step in that direction, and we will keep pushing the boundaries of what is feasible in this way.

References

1. Benveniste, É.: Problems in General Linguistics. University of Miami Press, Miami (1971)
2. Beyaert-Geslin, A.: Sémiotique du portrait. De Dibutade au selfie. De Boeck, Brussels (2017)
3. Bordron, J.-F.: Image et vérité. Essais sur les dimensions iconiques de la connaissance. Presses Universitaires de Liège, Liège (2013)
4. Colas-Blaise, M.: Comment penser la narrativité dans l'image fixe ? La "composition cinétique" chez Paul Klee. Pratiques (online), pp. 181–182 (2019) https://doi.org/10.4000/pratiques.6097
5. Deleuze, G.: Francis Bacon: The Logic of Sensation. Continuum, London (2003)
6. Deliège, A., Dondero, M.G., D'Armenio, E.: On the dynamism of paintings through the distribution of edge directions. J. Imaging **10**(11), 276 (2024)
7. Deliège, A., Dondero, M.G., D'Armenio, E.: Revisiting Wölfflin in the age of AI: a study of classical and baroque composition in generative models. J. Imaging **11**(5), 128 (2025)
8. Deliège, A., Dondero, M.G.: Similar paintings retrieval from individual and multiple poses. In: Del Bue, A., Canton, C., Pont-Tuset, J., Tommasi, T. (eds.) Computer Vision – ECCV 2024 Workshops. ECCV 2024. Lecture Notes in Computer Science, vol. 15628, pp. 17–32. Springer, Cham. https://doi.org/10.1007/978-3-031-91572-7_2
9. Dondero, M.G.: The semiotics of design in media visualization: mereology and observation strategies. Inf. Des. J. **23**(2), 208–218 (2017)
10. Dondero, M.G.: Visual semiotics and automatic analysis of images from the cultural analytics lab: how can quantitative and qualitative analysis be combined? Semiotica **230**, 121–142 (2019)
11. Dondero, M.G.: The Language of Images. The Forms and the Forces. Springer, Cham (2020)
12. Dondero, M.G.: Emerging faces: the figure-ground relation from renaissance painting to deepfakes. In: Leone, M. (ed.) The Hybrid Face: Paradoxes of the Visage in the Digital Era, pp. 74–86. Routledge, London, New York (2023). https://doi.org/10.4324/9781003380047-5
13. Dondero, M.G.: The face: between the background, the enunciative temporality and the status. Reti, saperi e linguaggi. Italian J. Cogn. Sci. **1**, 49–70 (2024). https://doi.org/10.12832/113797
14. Drucker, J.: Visualization and Interpretation: Humanistic Approaches to Display. MIT Press, Cambridge, MA (2019)
15. Fabbri, P.: La svolta semiotica. Laterza, Rome-Bari (1998)
16. Floch, J.-M.: Petites mythologies de l'œil et de l'esprit. Pour une sémiotique plastique. Hadès-Benjamins, Paris-Amsterdam (1985)
17. Focillon, H.: The Life of Forms in Art (1934). Zone Books, New York (1992)
18. Fontanille, J.: Semiotics of Discourse. Peter Lang, Frankfurt (2006)
19. Fontanille, J.: Les espaces subjectifs. Introduction à la sémiotique de l'observateur. Hachette, Paris (1989)
20. Fontanille, J., Zilberberg, Cl.: Tension et signification. Mardaga, Liège (1998)
21. Greimas, A.J.: Figurative semiotics and the semiotics of the plastic arts. New Literary Hist. **20**(3), 627–649 (1989)
22. Groupe μ: L'effet de temporalité dans les images fixes. Texte 21–22, 41–69 (1998)
23. Impett, L.: Analyzing gesture in digital art history. In: Brown, K. (ed.) The Routledge Companion to Digital Humanities and Art History, pp. 386–407. Routledge, New York (2020)
24. Impett, L., Moretti, F.: Totentanz. Operationalizing Aby Warburg's Pathosformeln. Literary Lab Pamphlet 16 (2017)
25. Manovich, L.: Data science and digital art history. Int. J. Digital Art Hist. **1**(1), 3–35 (2015)
26. Manovich, L.: Style Space: How to Compare Image Sets and Follow their Evolution (2011). http://manovich.net/content/04-projects/073-style-space/70_article_2011.pdf
27. Manovich, L.: Computer Vision, Human Senses, and Language of Art. AI & Society. 1145–1152 (2020). https://doi.org/10.1007/s00146-020-01094-9
28. Marin, L.: De la représentation. Seuil, Paris (1993)
29. Petitot, J.: Morphologie et esthétique. Maisonneuve et Larose, Paris (2004)
30. Pinotti, A.: Narrativa muta. Immagine e storia. In: Griffero, T., Di Monte, M. (eds.) Potere delle immagini?, pp. 169–185. Mimesis, Milan (2008)

31. Seguin, B.: Making large art historical photo archives searchable. Dissertation. EPFL Scientific Publications. https://infoscience.epfl.ch/record/261212 (2018)
32. Thom, R.: Local et global dans l'œuvre d'art. Le Débat **2**(24), 73–89 (1983)
33. Warburg, A.: Bilderatlas Mnemosyne, Hatje Cantz, Ostfildern (2020)

Maria Giulia Dondero is a research director of the National Belgian Fund for Scientific Research (F.R.S.-FNRS) and professor at University of Liège. She is the President of the International Association for Visual Semiotics (IAVS/AISV), co-founder and Editor-in-Chief of the journal *Signata* and co-director of the collection "Sigilla." Her main research fields are the relation between semiotics and art history, the theory of photography, the diagram in scientific discourse, and Generative AI. She is the author of four books: *The Language of Images. The Forms and the Forces* (2020); *The Semiotic Challenge of Scientific Images. A Test Case for Visual Meaning* (with J. Fontanille, 2014); *Sémiotique de la photographie* (with P. Basso, 2011); *Le sacré dans l'image photographique* (2009). She has co-edited 30 collective works and special issues on scientific and artistic images and on the semiotic theory of visual language. She has been Visiting Professor at UNESP, São Paulo; Instituto Nacional de Antropología e Historia (INAH), Mexico City; Paris 2-Panthéon-Assas; CELSA-Sorbonne University; University of Turin, and La Sapienza Rome. She has been Invited Visiting Researcher at the University of Southern California (2020) and at Purdue University (2022).

Adrien Deliège is a postdoctoral researcher at the University of Liège. He initially studied mathematics (Ph.D. 2013–2017, University of Liège), then he specialised as a postdoc in computer vision in sports with M. Van Droogenbroeck (2017–2022, University of Liège). After a position of Lead Machine Learning Engineer in visual quality assessment (2022–2023, MaddoxAI, Germany), he now holds a position as a postdoc in Visual Semiotics at University of Liège. His research includes the analysis of corpora of historical paintings from the perspective of pose estimation and directional edge detection to quantify dynamisms throughout art history. Recently, he focused on text-to-image generative models, by studying their semiotic properties and their ability to reproduce Wölfflinian characteristics of Classical and Baroque compositions.

Part IV
Cultural Critique of Algorithmic Media

Chapter 14
Techno-Anaesthetics: The Politics of AI Imagery from Stock Images to Generative Models

Alberto Romele

Abstract This chapter proposes the concept of techno-anaesthetics to critically examine how visual representations of artificial intelligence—especially in stock imagery and generative AI outputs—contribute to the depoliticisation of public discourse. It argues that such images do not merely illustrate AI, but actively shape its cultural imaginary by reproducing aesthetic tropes that convey neutrality, inevitability, and benign control. Drawing on visual grammar, semiotics, and the philosophy of technology (particularly Simondon's notion of techno-aesthetics), the chapter shows how these representations operate as "screens" that obscure rather than reveal complexity. The analysis extends from legacy stock photography to AI-generated imagery, highlighting a semiotic feedback loop in which generative models re-ingest the very clichés they were trained on. The result is a symbolic regime that numbs critical engagement and consolidates power around those who control platforms, datasets, and algorithms. The chapter calls for a politicized ethics of AI communication—one that foregrounds aesthetic dissensus and symbolic struggle as conditions for democratic participation in technoscientific futures.

Keywords Artificial intelligence imaginaries · Stock photography · Generative AI · Techno-aesthetics · Visual culture and ethics

14.1 Visual Grammars of AI: From Stock Photography to Generative Models

In recent years, the field of artificial intelligence (AI) ethics has attracted growing scholarly and public attention, with particular emphasis on issues such as algorithmic bias, system transparency, and technological robustness. Yet, a crucial dimension often remains underexplored: the ethics of communicating about AI. This chapter addresses this overlooked aspect by examining the use of images in the representation

A. Romele (✉)
Sorbonne Nouvelle University, Paris, France
e-mail: alberto.romele@sorbonne-nouvelle.fr

P. Conte et al. (eds.), *Algomedia. The Image at the Time of Artificial Intelligence*, Lecture Notes in Morphogenesis, https://doi.org/10.1007/978-3-032-08726-3_14

of AI, especially within science communication and marketing contexts. A significant proportion of these visual representations are stock images—pre-produced visuals licensed for commercial or editorial use, typically benefiting both their creators and the agencies that manage their distribution. The stock image industry is traditionally divided into two main segments: macrostock and microstock. Macrostock images, distributed by agencies such as Getty Images, are generally high-priced due to their exclusive licensing models and professional production standards. Conversely, microstock images—available through platforms like Shutterstock—are sold at lower prices and produced by a more diverse range of contributors, including amateur creators, with minimal editorial oversight. Over the years, the distinction between macrostock and microstock has become increasingly blurred. The rise of the microstock model has reshaped the industry, driving a shift toward broader accessibility, lower costs, and a more decentralized production model. Stock images are not only widely employed to depict AI technologies but also frequently used to illustrate ethical concerns surrounding AI. Despite their ubiquity, the ethical implications of using such images—particularly their role in shaping public imaginaries of AI—remain largely unexamined in both academic and professional discourse. To illustrate this point, let us consider two representative examples:

(1) In the extensive *Oxford Handbook of Ethics of AI* [7], addressing a wide spectrum of ethical concerns surrounding artificial intelligence, not a single line is dedicated to the ethics of communicating about AI. This omission is striking, particularly in light of the significant role that visual and narrative representations play in shaping public perception, influencing policy decisions, and orienting technological development. The absence of reflection on these communicative dimensions reveals a broader blind spot in contemporary ethical discourse on AI—one that risks neglecting how imaginaries of AI are not merely technical or philosophical constructs, but also semiotic and ideological ones.

Ironically, the cover image of the volume itself offers a compelling example of the very phenomenon that remains unaddressed. Created by Moldovan illustrator Fiodora Chiosea and sourced from iStock—a subsidiary of Getty Images—the image employs a colour palette dominated by blue and white.[1] Blue, often associated with calmness, neutrality, and institutional credibility, aligns with what Kress and van Leeuwen [14] describe as the modality of visual communication: the use of colour, brightness, and saturation to create an impression of realism, trustworthiness, or objectivity. As such, the palette implicitly reinforces the rational and benevolent framing of AI ethics as a neutral, universal discourse, echoing visual strategies adopted by international organisations like the UN and UNESCO.

From a compositional perspective, the image follows a structure that Kress and van Leeuwen would interpret through the lenses of information value and salience. The left side of the image, traditionally associated in Western visual culture with

[1] The cover image of the volume can be viewed at: https://academic.oup.com/edited-volume/34287. Accessed on April 10, 2025. The original source of the image was identified using the reverse image search engine TinEye (https://tineye.com/).

the "Given" or the known, features a human face composed of digitally dispersed particles. The right side, associated with the "New" or the emergent, displays a stylized circuit board, symbolizing the domain of artificial intelligence. This left-to-right vector—present in the original stock image—suggests a linear progression from humanity toward machinic intelligence, evoking a narrative of technological advancement or hybridisation. However, on the cover of the *Handbook*, the image is inverted: the circuit board now originates from the right and flows into the human face on the left. This reversal subtly shifts the semiotic load of the image. Rather than conveying a narrative of progress or synthesis, the altered composition may suggest a process of dematerialisation or dissolution, where the human is reabsorbed into the digital matrix. This inversion complicates the metaphorical register of the image, perhaps unintentionally evoking anxieties about loss of embodiment, subjectivity, or agency in the face of pervasive computation.

Crucially, the image also exemplifies the ideological work performed by visual design, as described by Kress and van Leeuwen. By relying on familiar tropes—the abstract, androgynous head; the merging of organic and inorganic elements; the clean, institutional blue—the image participates in what could be called the aesthetic normalisation of AI. It visually encodes a set of assumptions: that AI is seamless, disembodied, intelligent, and ultimately benign. These are not neutral representations; they are loaded with cultural values and political implications. That such an image should adorn the cover of a leading volume on AI ethics—without any critical engagement within—underscores the urgency of including visual culture in ethical inquiry about AI.

(2) The European Commission has made considerable progress in articulating ethical frameworks for artificial intelligence, seeking to ensure that AI development remains trustworthy, transparent, and aligned with core European values. Yet, despite this normative effort, the visual communication accompanying these frameworks often fails to reflect the very principles they promote. This discrepancy highlights a broader gap in the ethics of representing AI—an area that remains largely untheorized, even in institutional settings. A revealing case in point can be found on older version of Futurium, the European Commission's participatory platform designed to facilitate public dialogue on EU policy initiatives. Among the resources available on Futurium are multilingual documents related to the piloting of the "European Guidelines for a Trustworthy AI," first published in April 2019.[2] These guidelines constitute a significant step toward a distinctly "European" approach to AI, one that explicitly seeks to balance innovation with rights, values, and social responsibility.

However, the visual representation chosen to accompany these guidelines subtly undermines their stated objectives. The image in question is a stock illustration

[2] The image was originally available at: https://ec.europa.eu/futurium/sites/futurium/files/capture_1_0.jpg. Since mid-May 2021, the original website has been archived and replaced by a new version of the Futurium platform. The image is no longer accessible online but had been archived during a previous research project.

by Thai artist Kittipong Jirasukhanont, licensed via iStock. Dominated by hues of blue—a colour, as noted earlier, often connoting neutrality, authority, and technological reliability—the image features a robotic hand and a human hand reaching toward one another, nearly touching. This composition unmistakably evokes Michelangelo's *Creation of Adam* from the Sistine Chapel ceiling, a canonical work in Western art history often interpreted as the moment of divine spark or anthropogenesis. From a visual social semiotics perspective, this image operates within the symbolic process of transactional reaction, where two vectors (the extended fingers) create a relational narrative between actor and goal. The robot becomes not merely a technological artifact, but a quasi-subject engaged in a meaningful interaction with the human. This narrative visual structure implies intentionality, agency, and mutual recognition—suggesting a communicative symmetry between human and machine that is far from neutral. More problematically, the image mobilizes what Roland Barthes would call a mythical connotation: it naturalizes a particular ideological vision of AI as transcendent, almost divine. The robotic hand is not simply mimicking human form—it is granted a symbolic aura of sacredness. As such, the image participates in a broader mythologisation of artificial intelligence, promoting what might be describes as a "semiological excess": an overcoding of meaning that conceals ideological assumptions behind aesthetic neutrality.

This iconography introduces a series of tensions with the content of the guidelines it is meant to illustrate. The Guidelines for Trustworthy AI emphasize values such as accountability, transparency, human agency, and oversight. Yet the image, through its visual rhetoric, may evoke the opposite: a sense of awe, inevitability, and even inferiority vis-à-vis the machinic. It suggests that AI is not only an extension of human intelligence, but perhaps its perfection or transcendence. This visual trope, which we might call techno-theological, risks reinforcing public imaginaries in which AI is imagined as omnipotent or infallible—precisely the kinds of imaginaries that critical scholarship in AI ethics seeks to dismantle.

I hypothesize that communication about AI in general, and the depiction of AI in particular, are integral to both AI ethics and AI itself. The images and stories we create about AI shape not only how the public understands AI but also how they engage with it. Common depictions of AI, ranging from sleek humanoid robots to complex neural networks visualized through streams of binary code, often fail to convey the underlying complexities and ethical considerations. Instead, they tend to pacify rather than provoke critical thinking and engagement. More radically, one could say that the imaginaries conveyed by these images play an active role in the processes of AI innovation and implementation in our societies. This leads to the idea that AI is much more than the sum of its material components. AI is, above all, a cultural and social phenomenon.

This reflection on stock images of AI might seem less pressing today, given that visual content is now being massively generated by systems such as DALL·E, Midjourney, and Stable Diffusion. However, the increasing overlap between stock imagery and generative AI is precisely what makes the analysis of stock images all the more relevant. In recent years, the stock image industry has undergone a profound transformation—if not disruption—due to the rise of generative models. A notable example of this shift is the 2024 merger between two of the industry's major players: Shutterstock and Getty Images. Although historically in competition, the two platforms had adopted complementary strategies in response to the AI revolution: Shutterstock entered a partnership with OpenAI, allowing the use of its image database to train models like DALL·E, while Getty Images collaborated with Nvidia to develop its own proprietary AI image generator, known as Bria. This convergence signals that stock image producers—long embedded in algorithmic workflows of tagging, ranking, and circulation—are increasingly being subsumed by the very generative algorithms they helped to train. What was once a curated, human-centred process of visual production has been replaced by real-time, prompt-based generation that mimics—and in many cases reproduces—the aesthetic conventions of stock photography. In fact, if one inputs a simple prompt like "artificial intelligence" into a generative model such as DALL·E, the resulting images closely resemble the stock clichés that have dominated visual culture for the past two decades (Fig. 14.1).

For instance, consider the image above: a photorealistic humanoid robot, constructed from smooth white armour and exposed black mechanics, extends its index finger toward a glowing blue circle labelled "ARTIFICIAL INTELLIGENCE." Behind the robot, a dark digital interface emerges, populated with abstract data visualisations—graphs, charts, and circuit-like patterns—all rendered in shades of electric blue. The robot's eyes emit a faint blue light, suggesting sentience, while its face remains perfectly symmetrical and devoid of emotion. The colour palette, dominated

Fig. 14.1 Image generated with DALL·E integrated in ChatGPT-4o using the simple prompt "artificial intelligence." Generated on April 1, 2025

by cool tones—metallic greys, whites, and digital blues—evokes a techno-futurist aesthetic that is simultaneously sterile and sublime.

Using the grammar of visual design proposed by Kress and van Leeuwen, we can identify several key semiotic structures at work. First, the composition follows a centred layout, with the robot occupying the left half of the frame and the glowing label "ARTIFICIAL INTELLIGENCE" dominating the right. The interaction between the robot and the text-image interface constitutes what Kress and van Leeuwen call a transactional vector, implying agency and directionality. The robot, positioned in profile and actively pointing, becomes the Actor, initiating a visual action toward a Goal—the concept of AI itself, which is both textual and graphical. This relational structure reinforces a narrative of purpose and mastery: the robot is not passive but actively engaged in accessing or activating artificial intelligence. At the same time, its calm demeanor and perfectly proportioned body suggest control, harmony, and rationality—characteristics traditionally associated with idealized human cognition, now transposed onto the machinic. The modality markers—such as photorealistic shading, sharp focus, and high colour saturation—signal "truth value," or at least a simulated realism, giving the image a persuasive rhetorical force. On a connotative level, this image continues the visual tradition of stock imagery. The glowing interface, blue light, and polished robot form a coherent myth in Barthesian terms: a techno-utopian imaginary in which AI is advanced, neutral, efficient, and aesthetically perfect.

That this image was generated by a text-to-image model is significant. It illustrates a broader point: generative AI systems, trained on massive datasets including stock photo corpora, tend to reproduce—and in doing so, normalize—the aesthetic and ideological codes embedded in those datasets. What we are witnessing is a semiotic feedback loop: the paradigm of stock photography becomes the syntagm of generative image production. As a result, the supposed novelty of generative images masks a deeper continuity with the past. The visual language of AI remains stuck in a lexicon of sleek robots, glowing interfaces, and neutral blue palettes—an aesthetic habitus that constrains the imaginaries of artificial intelligence rather than expanding them. Thus, rather than making the analysis of stock images obsolete, the rise of generative AI makes such analysis even more urgent. The clichés inherited from stock photography continue to inform the ways in which AI is made visible—and therefore thinkable. It is precisely within these recycled visual tropes that the ideological labour of images takes place, sedimenting certain visions of AI while rendering others invisible.

The rest of this chapter unfolds across two sections. Section 14.2 examines why philosophers of technology and AI ethicists have largely neglected the role of representation and communication in relation to artificial intelligence. This neglect is rooted primarily in the so-called "empirical turn" that reshaped the philosophy of technology from the 1990s onward. To clarify the implications of this shift, I draw on a text by Gilbert Simondon on techno-aesthetics, in which aesthetic judgment becomes immanent to the technical object or to the immediate interaction between artifact and user. I propose an extension of this dynamic to the ethical domain, suggesting that ethical reflection itself is increasingly absorbed by technological processes. Section 14.3 turns to the analysis of stock images of AI. Here, I introduce the concept of "anaesthetics" to describe the anesthetizing visual rhetoric that often characterizes science communication about AI. This rhetoric, I argue, represents not just an aesthetic concern, but a deeply political one. The argument then moves toward the present-day context of generative AI, demonstrating how these systems tend to reproduce the same visual logics found in stock imagery. To bring these threads together, I propose the concept of techno-anaesthetics: a term that designates the way generative AIs visualize themselves by reactivating and reinforcing the same anesthetizing aesthetic patterns embedded in the stock image tradition.

14.2 Techno-Aesthetics, or the Material Absorption of the Aesthetic Judgment

To begin my reflection, I turn to a text by the philosopher Gilbert Simondon entitled "On Techno-Aesthetics" [22]. This document takes the form of a letter addressed to Jacques Derrida, dated July 3, 1982, and originally intended to contribute to the founding discussions of the Collège International de Philosophie (CIPH). The letter, which remained unpublished during Simondon's lifetime, resurfaced posthumously—he passed away in 1989—and was later sent by his wife Michelle to Derrida, who published it in Issue 12 of *Papiers*, a collection edited under the auspices of the CIPH. This brief yet rich text is of particular interest because it exemplifies a form of empiricism that has come to dominate much of contemporary philosophy of technology. This is not to suggest that Simondon himself endorsed the so-called "empirical turn" that has shaped the discipline since the 1990s—his writings in fact predate it by several decades. However, it is telling that Simondon's work, long marginalized during the heyday of structuralism and the linguistic turn in France, has only recently gained renewed attention, precisely at a moment when materialist and techno-scientific perspectives are being revalued. In this letter, Simondon outlines four distinct meanings of the term techno-aesthetics:

(1) The first pertains to the increasing space that the arts have come to grant to technology. Simondon cites examples such as Marinetti's Futurism, the architecture of the Centre Pompidou, and Le Corbusier's Dominican convent at L'Arbresle.

> The traces left by the formwork in the cement of the chimney of the Dominican convent of Arbresle near Lyon have purposefully been left visible—one can see them particularly well in the morning and the evening, when the light is low-angled. For the same convent, Xenakis mathematically calculated the proportions of the windows of the monks' promenade. Le Corbusier used cement roughcasting in each of the cells with a loggia. But that was no longer done with the trowel, whose entelechy is an optically smooth surface. It's a projection that was done with a cement canon, covering the walls with a kind of cresting on which the light can play (Simondon [22]: 2).

This first meaning can be easily applied to the case of AI, considering how AI is increasingly being used in the field of art today.

(2) The second meaning of techno-aesthetics refers to the space that technology grants to art. In this case, Simondon cites the Eiffel Tower and the Garabit Viaduct over the Truyère River, which he says "possess undeniable aesthetic power." Simondon also mentions the Jaguar E V12:

> Functionality is not its strength: this enormous engine provides transport for only two people. There is just enough room for a dog behind the front seats. This reveals how the designer sees the couple: as a monogamous unit without children… Considered by itself, the car's bodywork is daring and, from the perspective of functionality, very drawn out so as to guarantee a weak resistance to the air. The underside of the car is much less satisfactory, however (Simondon [22]: 3).

There remains, to a large extent, a conceptual and operational separation between art and technology. The hyphen in "techno-aesthetics" continues to function as a divider—a marker of the distance between two distinct domains. At best, it can be understood as an interface: a zone of mediation in which art and technology converge without fully merging. This is particularly evident in industrial design and in all those practices referred to as "functional art," where aesthetic considerations serve technical or utilitarian ends. The relevance of this concept of techno-aesthetics to the field of AI is readily apparent—whether in the design of user interfaces or in ongoing debates around the aesthetic and functional shaping of humanoid robots.

(3) The third interpretation of the term marks a shift from techno-aesthetics (with a hyphen) to technoaesthetics as a unified concept—no longer a conjunction of two separate domains, but the expression of their intrinsic interrelation. It is at this point that we begin to grasp what Simondon may have truly intended by the term, or at least what much of the secondary literature has identified as its most authentic meaning. This reading reflects a characteristically Simondonian perspective, in which technical objects are no longer assessed in terms of their service to human ends, but rather as entities capable of autonomous individuation. Judgments—whether ethical, aesthetic, or ontological—must then be grounded in the object's capacity to achieve metastability in relation to its environment.

Simondon illustrates this point through a comparison between the engine of the Citroën 2CV and that of a Jaguar. Following a detailed technical description, he concludes: "We are not speaking of engines because they would somehow be unique in possessing a certain level of individuation but because they are consistent and coherent in relation to themselves" (Simondon [22]: 6). In this context, technoaesthetics is not about external criteria imposed on a technical object, but rather about evaluating the internal coherence of the object in relation to its function, structure, and capacity for self-regulation. This perspective can be extended to the domain of AI and digital technologies more broadly. Consider, for instance, the aesthetic judgments—often simultaneously ethical and ontological in nature—that are applied to a piece of code. Terms such as elegance, efficiency, or clarity are not merely metaphorical but reflect a mode of appraisal grounded in the internal logic and performative consistency of the system itself. In this sense, technoaesthetics concerns not what a technological object is *for*, but what it *is*, and how it achieves a form of stability and expressivity on its own terms.

(4) The fourth meaning may appear, at first glance, as a step backward, since it reintroduces the human element into a discourse that, in previous interpretations, had leaned toward the non-human or inhuman dimensions of technoaesthetics. Yet this return to the human is not framed in terms of passive contemplation, but rather in terms of use and embodied action. In this sense, technoaesthetics is grounded in experience—understood in the etymological and philosophical lineage of the term "aesthetic," derived from the Greek *aisthesis*. As originally defined by Baumgarten, aesthetics was not primarily concerned with art, but with sensory knowledge, regarded as no less valuable than rational knowledge. Within this framework, if art is still invoked, it is no longer from the standpoint of the spectator or consumer, but from that of the producer—the one who creates through tools and techniques. The sensory experience that arises during the act of making, through engagement with technical objects, blurs the line between artist and artisan. Technoaesthetics, in this sense, becomes an embodied, operational mode of knowledge—a way of thinking through doing:

> With each bang of the hammer, one experiences the state of the forged metal that stretches and deforms between the hammer and the anvil. [...] One could continue like this almost indefinitely, moving more or less continuously to the sensation that artistic instruments give to those who play them: the touch of a piano, feeling the vibration and tension of the strings of a harp, the snapping of the strings of the hurdy-gurdy on the cylinder covered with rosin--it's a register that's almost inexhaustible. Art is not only the object of contemplation; for those who practice it, it's a form of action that is a little like practicing sports (Simondon [22]: 3).

Loeve ([16]: 287, translation mine) notes that, ultimately, the fourth interpretation of technoaesthetics is not entirely distinct from the third, since what we might call *technical perception* operates in both directions: "Techno-aesthetics becomes the sensory knowledge of technology not only in the objective sense of the genitive, where humans sensibly know the technology, but also in the subjective sense of the genitive: it is the technology that knows and feels through a sensitivity that humans can participate in […]." In my view, this double genitive—understood as the reciprocal relationship between human and technological forms of perception—risks collapsing aesthetic judgment into the materiality of the technical object itself. Whether in Simondon's formulation or in its contemporary reinterpretations, this perspective tends to absorb meaning, significance, and evaluation into the immanent functioning of the machine. As a result, the philosophy of technology is at risk of losing its capacity to critically engage with the symbolic and affective atmospheres in which all technologies are inevitably embedded. More specifically, such an approach can divert attention from the aesthetic, ethical, and ontological dimensions that emerge around how technologies—particularly artificial intelligence—are imagined, represented, and circulated in culture. These imaginaries crystallize in a variety of expressive forms: texts, static and moving images, sounds, and other media. A philosophy of technology that overlooks these dimensions risks missing the very terrain where meaning is negotiated and contested.

A similar critique applies to the field of ethics. It is no coincidence that many contemporary philosophers of technology and AI have embraced the notion of "ethics by design," as if the material configuration of technical objects alone could ethically regulate their uses and consequences. This perspective assumes that embedding ethical principles in technological artifacts is sufficient to guide behaviour. However, when it comes to artificial intelligence, it is often the companies with the most resources—those capable of building the most powerful and "ethically designed" algorithms—that end up deploying them in ways that are socially and ethically questionable. In such cases, material ethics risks becoming a justification for unethical outcomes, rather than a safeguard against them.

This tendency to reduce ethics to material configuration is symptomatic of a broader epistemological shift often referred to as the "empirical turn" in the philosophy of technology—a term introduced by a group of Dutch scholars in the early 1990s (see Brey [2] for an overview). The empirical turn can be understood in two main senses: (1) First, as a reaction against the linguistic turn that had dominated the humanities and social sciences in the second half of the twentieth century. While language-centred theories were successful in highlighting the role of mediation, they also tended to universalize it—treating linguistic mediation as the paradigm for all forms of mediation. The empirical turn reasserted the specificity of non-linguistic mediations—those enacted through tools, techniques, and technological infrastructures, (2) Second, it was a response to the grand, often pessimistic narratives of classical philosophers of technology such as Martin Heidegger and Jacques Ellul, who were seen as technologically deterministic and inattentive to the material and historical specificity of individual technologies. For example, Heidegger's concept of *Gestell* (enframing) is used to describe phenomena as diverse as hydroelectric dams and modern agriculture, without accounting for their concrete differences.

My aim is not to return to Heidegger's view, in which "the essence of technology is by no means anything technological" (Heidegger 1997: 4), but rather to propose a middle path. One might say, paraphrasing Heidegger, that "the essence of technology is not *entirely* technological." What is needed is an approach that holds together the material dimension of technologies and the symbolic, affective, and cultural atmospheres in which they are embedded. In my previous work [21], I have drawn on the notion of technological imaginaries to articulate a hermeneutic circularity between materiality and meaning. This circularity is not merely an accessory feature of technology, but a constitutive one—shaping not only its uses, but also its very design and integration into society. This insight finds resonance in the literature on science communication, which has long emphasized that communication is not simply a downstream activity but an active component of scientific and technological development. As Bucchi [4] argues, popular communication influences scientific practice in multiple ways, far beyond the functions of support or dissemination. Goulden [10] develops this further by comparing the relationship between science and its communication to the meeting of freshwater and saltwater in a river delta—two domains that mix and co-evolve. More recently, Bucchi and Trench ([3]: 6) have defined science communication as "the social conversation around science and technology," a space where experts and non-experts alike participate in shaping technoscientific futures.

The products of science communication—whether texts, images, or narratives—thus acquire a specific ontology of their own, not merely derivative of scientific facts or academic discourse, but actively co-constitutive of them.

What we today call "artificial intelligence" is paradigmatic of a broader entanglement between the symbolic and the material. AI is not a single, unified technology, but rather an assemblage of diverse techniques—machine learning, natural language processing, computer vision, generative models—applied across domains as varied as medicine, policing, education, and marketing. Yet beyond its technical heterogeneity, the term AI itself functions as what Laclau [15] would call a floating signifier—a discursive surface onto which multiple, often competing, meanings are projected. It serves as a battlefield for rival interpretations of what our technological present is, and what our sociotechnical future ought to be. As I have recently argued [20], the same applies to the expression "AI ethics," which does not refer to a unified set of principles or practices but operates differently across various social and institutional contexts. For corporations, it often takes the form of techno-solutionism,for public institutions, a set of normative frameworks; for researchers, a site of critical inquiry; and for civil society, increasingly, a space of resistance and accountability. The ethical discourse around AI, far from being consensual, is a site of hegemonic struggle. The outcome of this struggle will not necessarily be a more ethical form of AI, but rather the dominance of a particular group in defining what counts as "ethical" in the first place, and thus what forms of AI are considered legitimate or acceptable.

This symbolic struggle is not merely discursive—it is embedded in the very process of technological development. Scholars in Science and Technology Studies (STS), such as Flichy [8] as well as Jasanoff and Kim [12], have shown that technological imaginaries are not epiphenomena or external representations. They are constitutive elements of the innovation process itself, shaping not only the invention of technologies, but also their implementation, stabilisation, and acceptance in society. Jasanoff (Jasanoff [12]: 15) defines sociotechnical imaginaries as "collectively held, institutionally stabilized, and publicly performed visions of desirable futures, animated by shared understandings of forms of social life and social order attainable through and supportive of advances in science and technology." Yet, in my view, such imaginaries are not only visions of future worlds agreed upon by collectives or institutions; they are also the very arenas in which hegemonic and counter-hegemonic narratives clash. Flichy, drawing on Paul Ricoeur's work on ideology and utopia, reminds us that social imaginaries oscillate between the affirmation of the status quo and the projection of alternative futures—and that our access to reality is always-already symbolically mediated. In this sense, technological imaginaries are not neutral or unified but contested and plural. They are structured by asymmetries of power and serve as instruments of symbolic struggle over what technologies should be and whom they should serve.

This is particularly true for AI, where imaginaries circulate across a wide array of discursive and visual forms—corporate videos, institutional reports, news articles, scientific publications, social media posts, films, and memes. However, these imaginaries are not directly accessible or coherent totalities. Rather, they manifest themselves through fragmented and distributed traces—what we might call micro-sociotechnical narratives of AI. These are not limited to science fiction as a well-defined genre but encompass all texts and images, static or dynamic, that embed a sociotechnical vision. Studying these narratives—both qualitatively and quantitatively—allows us to reconstruct the symbolic infrastructures of AI: the imaginaries that influence how AI is conceived, built, and received. In doing so, we can move beyond narrow definitions of ethics by design or technical governance, and begin to understand AI as a site where cultural meaning, political interest, and technological development converge.

14.3 Techno-Anaesthetics, or How AI Becomes a Matter of Indifference

In this third section, I seek to apply the theoretical insights developed earlier by examining the empirical case of visual representations of artificial intelligence—first through traditional stock imagery, and more recently through images generated by AI systems themselves. It is in this context that I introduce the concept of techno-anaesthetics. By this term, I refer not only to the anesthetizing effects that representations of AI—like those commonly found in stock photography—can exert on public discourse, but also to the fact that AI now participates autonomously in producing such representations. The result is a recursive visual regime that discourages critical reflection, flattening the ethical and political complexity of AI into aesthetically familiar and ideologically neutral forms.

As discussed in the previous section, Simondon's theory of techno-aesthetics involves a conceptual shift—from a hyphenated form (techno-aesthetics), which implies a relationship between two distinct domains, to a unified one (technoaesthetics), in which aesthetics is understood as internal to the logic and operation of the technical object itself. In contrast, my position argues for the necessity of preserving the hyphen—as a conceptual space of mediation and differentiation. This hyphen marks a constitutive distance between the technological and the aesthetic, a distance that should be understood through the lens of hermeneutic circularity. That said, even within my perspective, one might speak of technoaesthetics in the unhyphenated sense—inasmuch as technologies increasingly produce the aesthetic frameworks through which they are themselves made visible. This is particularly evident in the context of generative AI, which is capable of producing images of AI that mirror the aesthetic codes of legacy media, such as stock photography. The critique I develop in this chapter can therefore be seen as an attempt to reintroduce the hyphen—to restore a space of separation and interpretation—at a moment when the convergence between cultural industries and generative AI (exemplified by the growing alliances between stock image agencies and AI companies) is actively erasing it. What is at stake is not a mere typographical distinction, but a philosophical and political gesture: the defence of a critical distance where meaning can still be questioned and contested.

My initial hypothesis is that the problem with stock images representing AI does not lie in their failure to refer to the "things themselves"—that is, to AI as it "really" is. One could argue, in fact, that artificial intelligence resists visual representation altogether. Today, AI is largely synonymous with machine learning algorithms and large language models (LLMs), which are abstract and intangible processes, difficult—if not impossible—to depict through static imagery. Even showing a line of code is insufficient: not only is it a fragment of a much larger system, but it would be as reductive as suggesting that an image of the brain constitutes a representation of intelligence itself. The issue with stock images of AI is not, therefore, one of referential fidelity, but one of political aesthetics. These images are "unpolitical" in the sense that they participate in establishing a particular distribution of the sensible within the field of technological innovation. As Rancière ([18]: 12) defines it, the distribution of the sensible refers to "the system of self-evident facts of sense-perception that simultaneously discloses the existence of something in common and the delimitations that define the respective parts and positions within it." In other words, it is the process through which shared spaces, temporalities, and frameworks of meaning are created—along with the allocation of roles, recognition, and legitimacy within them. This process always entails exclusions: exclusions from visibility, audibility, intelligibility, and ultimately from participation in the shaping of collective life.

For Rancière, such distributions are inherently political, because "politics revolves around what is seen and what can be said about it, around who can see and the talent to speak, around the properties of space and the possibilities of time" ([18]: 13). There is thus a deep link between politics and aesthetics, where aesthetics must be understood not only in the classical Greek sense of *aisthesis* but also in relation to artistic and cultural production more broadly. On the one hand, politics involves the configuration of perception—who sees, who is seen, who hears, who is heard. On the other, art and cultural artifacts can either reinforce dominant perceptual regimes or work to suspend and transform them. Stock images of AI tend to do the former: they reproduce visual clichés that stabilize existing distributions of visibility and intelligibility, thereby neutralizing the political potential of imagining AI differently.

I argue that the dominant visual culture surrounding AI contributes to a particular distribution of the sensible—one that reinforces the gap between experts and non-experts, insiders and outsiders. While it has been suggested that images in science popularisation serve an introductory or pedagogical function—Gigante (2018), for example, introduces the concept of "portal images" that grant access to complex domains—I contend that stock images of AI function less as portals and more as screens.[3] The term "screen," in its etymological sense, connotes not only display, but also separation, concealment, and protection: to screen is to cut, divide, shelter, and obscure. These images thus operate as screen images—not to facilitate understanding,

[3] This critique applies specifically to stock images, and in particular to those representing AI. The case of images related to science and technology more broadly is different, as they may serve diverse functions and express various forms of creativity. On this topic, see Dondero and Fontanille [6].

but to filter, distance, and ultimately anesthetize. The effect is an anaesthetic one. Repeated exposure to such imagery dulls the sensitivity of non-experts and outsiders to the political, ethical, and social stakes of AI. Instead of fostering curiosity or critical engagement, these images contribute to indifference and disengagement. The term anaesthetics refers to the anesthetizing function embedded within the aesthetic regimes of contemporary AI imagery. That is, the aesthetic—understood in the Rancièrian sense as the distribution of what is visible, sayable, and thinkable—serves here to numb rather than to stimulate. It defines a symbolic boundary that excludes certain subjects from meaningful participation in the sociotechnical conversation. This idea is informed by sources:

(1) The first is the work of literary critic Kermode, especially as developed in *The Genesis of Secrecy* [13]. While storytelling is often celebrated for its explanatory and democratizing power—particularly in science communication—Kermode challenges this assumption. Drawing on the parables of the Gospels, he emphasizes that narrative can also serve to obscure rather than illuminate. He cites a crucial passage from the Gospel of Mark (4:11–12): "The secret of the kingdom of God has been given to you. But to those on the outside everything is said in parables so that (*hina* in Greek) they may be ever seeing but never perceiving, and ever hearing but never understanding." According to Kermode, the parable is not a pedagogical device for all, but a selective filter—one that arbitrarily distinguishes insiders from outsiders and ultimately excludes the latter from the domain of meaning. The same logic, I suggest, applies to certain forms of visual storytelling about AI.

(2) The second reference is a foundational essay by Callon and Latour [5], in which the authors explore the origins of power in sociotechnical systems. In their formulation, power does not emerge from hierarchy per se, but from the ability of an actor—ontologically no different from others—to sit atop a black box that contains a multitude of elements, both material and symbolic. Once assembled, this black box ceases to invite questioning; it becomes accepted, stable, and opaque. What was once a matter of concern becomes a matter of fact—and then a matter of indifference. In the case of AI, this black-boxing process is amplified by visual representations that aestheticize complexity, displace attention, and contribute to the closure of debate. In this sense, science communication—particularly through its visual dimension—plays a crucial role in rendering AI systems more opaque and less contestable.

At first glance, the debate around stock images of AI may appear outdated in the era of generative AI—an era in which images are produced at scale and with minimal human input, often generated instantaneously from simple text prompts. However, the second hypothesis I would like to advance in this section is that stock images were already part of a long-standing algorithmic logic of image production, and that generative AI merely brings this logic to its full realisation. The automated creation of images is not a rupture but rather the culmination of a historical process that spans at least a century.

Media historian Blaschke [1] traces this trajectory through the history of the Bettmann Archive, which was founded between Berlin and New York and later acquired by Corbis, the image licensing company launched by Bill Gates. Corbis envisioned a vast visual encyclopaedia—a modern Library of Alexandria, but made of images instead of texts. At the origin of this archive was Otto Bettmann, born in Leipzig in 1903, who studied history, art history, and sociology, and completed a doctoral thesis on the professionalisation of the German book trade in the eighteenth century. From 1928 to 1932, he worked as a librarian and curator of rare books at the Staatliche Kunstbibliothek in Berlin. During this time, he began photographing illustrations from art books and other printed materials using a 35 mm camera. Dismissed from his position in April 1933 under the Nazi regime's Law for the Restoration of the Professional Civil Service, Bettmann turned to entrepreneurship. He founded the Bildarchiv Dr. Otto Bettmann/Berlin and, despite the rapidly deteriorating political climate, used the intervening period before his emigration to New York in November 1935 to develop a sophisticated business model and expand his collection in a systematic and commercially oriented way.

In parallel with his collecting efforts, Otto Bettmann developed a distinctive system for indexing images, drawing on his scholarly expertise in classification, reference work, and library cataloguing. This system relied on a card file that combined visual elements—miniature photographic prints—with detailed textual metadata. Unlike most photographic agencies and commercial archives, which emphasized their most iconic or newsworthy images, the Bettmann Archive foregrounded its classification system itself. What was promoted were not the images per se, but curated excerpts from the index—a structured map of knowledge through visual material. By 1936, the archive comprised 200 thematic categories and approximately 10,000 images. By 1961, it had expanded to 10,000 categories and over one million images. The goal was not to accumulate as many images as possible within a single category, but to multiply categories—a strategy aimed at realizing an encyclopaedic vision of the world through images. Over time, the indexing system was streamlined to improve commercial efficiency, and in 1981, Bettmann retired, selling the archive—by then home to some three million images—to the Kraus-Thomson Organisation. In 1995, the archive was acquired by Corbis, which sought to digitize canonical artworks and historical photographs for use in new media formats: wall-mounted monitors in private homes and thematic CD-ROM compilations.

Two aspects of this history are particularly noteworthy. First, over time, the index became more important than the images themselves. In the semiotic hierarchy between word and image, it was the word—structured, taxonomized, and searchable—that took precedence. Second, these were not arbitrary words, but systematically organized descriptors: what we would now call tags or metadata. It was precisely this semantic structure that rendered images operable and commodifiable. In other words, it was not only the mechanical reproduction of the photograph that turned it into a product, but also its linguistic domestication—the transformation of the image into a function of language, systematized and aligned with market logic.

A second phase in the algorithmisation of images emerges with the evolution of major macrostock and microstock agencies. A key moment in this history occurred in 2016, when Corbis was sold to Visual China Group, which entered into an agreement with Getty Images to distribute Corbis's image archive outside of China. This event symbolically marks the consolidation of image archives within a global, algorithmically-driven media infrastructure. Media theorist Frosh [9] identified three major technological shifts that transformed the stock imagery industry from the 1990s to the early 2000s: (1) the rise of digital imaging technologies, which expanded the ability to manipulate and customize photographs, (2) the development of digital storage, search, and retrieval systems, which allowed for the integration of increasingly diverse image categories; and (3) the digital distribution of images via CD-ROMs and, later, websites, which drastically lowered costs associated with duplication and dissemination. The process of algorithmisation continues and intensifies this trajectory, as I have argued elsewhere [19], through three interrelated developments:

(1) Graphic Editors: Traditional photography has been largely replaced by graphic design tools such as Photoshop and Adobe Illustrator, particularly in the production of images related to artificial intelligence. Even when photographs are used, they are so extensively modified that they lose their original indexical character. Image producers now operate in isolation, relying on software rather than studios or models. Their labour is repetitive and modular, designed to maximize discoverability and sales. Images are reduced to pure pixel data—ideal input for machine learning systems and easily integrated into large-scale, searchable digital archives.
(2) Keywording Systems: Whereas early stock image databases relied on general, fixed categories, contemporary platforms use dynamic and automated keywording processes. AI-powered tools, such as Shutterstock's keyword suggestion engine, analyse uploaded images and propose relevant terms based on visual similarity to existing content. Producers can accept or reject these keywords, and in doing so, they not only annotate their own images but also feed back into the machine learning systems, training the algorithms that will classify future content. Keywords become the currency through which images are made searchable, legible, and ultimately monetizable.
(3) Search and Distribution Algorithms: The visibility of images is now governed by algorithms that process these keywords, apply filters, and determine ranking within search results. This logic is further extended on the web, where access to image platforms is often mediated by browsers and search engines like Google Chrome. As Frosh observes, stock images function as information: they are detached from their referential origins, adaptable to multiple contexts, and materially modifiable—characteristics that make them ideal for algorithmic processing and recombination.

This algorithmisation leads to a competitive environment in which the success of an image depends less on originality than on its ability to conform to existing, algorithmically privileged visual patterns. The logic of visibility is determined not by aesthetic innovation, but by the repetition of styles and compositions that have previously performed well in algorithmic rankings. What is rewarded is not creativity, but conformity to data-driven expectations. Importantly, the automatic production of images by generative AI would not have been possible without this long preparatory phase: a decades-long process in which images were incrementally tamed—standardized, keyworded, tagged, and archived—so they could be efficiently indexed, retrieved, and repurposed. This evolution culminates in a final stage in which algorithms do not simply organize or retrieve images, but penetrate the images themselves. This shift was already anticipated in the development of image recognition algorithms and now finds its full expression in generative models, where the aesthetic logic of stock imagery is not just reproduced, but autonomously simulated.

As Somaini [23] observes, three recent developments define the evolving relationship between images and artificial intelligence: (1) First, beginning around 2010, deep learning algorithms—particularly convolutional neural networks (CNNs)—were increasingly employed in the implementation of computer vision systems. These systems enabled machines to detect, analyse, and classify elements within images with unprecedented precision, (2) Second, by the mid-2010s, more advanced deep learning architectures emerged, such as DeepDream and generative adversarial networks (GANs). Unlike earlier systems focused on classification, these models introduced the ability to manipulate existing images or generate entirely new ones, ranging from the photorealistic to the abstract; (3) Third, by early 2022, a new generation of deep learning models—commonly grouped under the umbrella of generative AI—became widely accessible. These models are capable of producing both static and animated images from textual prompts, as well as generating text based on visual input.

Text-to-image systems, one of the most prominent outputs of generative AI, typically operate through two main phases: text encoding and image generation. In the first phase, a transformer model—originally developed for natural language processing tasks—encodes the input text into a latent representation: a series of high-dimensional vectors that capture the semantic relationships among words in a sentence. In the second phase, a generative model—often a diffusion model or a GAN—uses this latent representation to produce an image that corresponds to the given textual input. This two-step architecture enables the system to generate visual outputs that are contextually and semantically aligned with the prompt. The training process plays a foundational role in shaping the capabilities of these models. During training, text-to-image systems are exposed to large-scale datasets consisting of images paired with descriptive captions. Through this exposure, the models learn to construct and navigate a latent space in which both linguistic and visual elements are encoded as numerical vectors. This latent space serves as a kind of mediating environment—an abstract representational field in which words and images are brought into alignment. The richness, diversity, and granularity of the training data significantly influence the system's ability to produce accurate, coherent, and varied visual outputs.

Precisely because of the central role played by training data in generative AI, stock images depicting artificial intelligence remain highly significant. As discussed earlier, even a generic prompt such as "artificial intelligence" tends to produce visual outputs—white humanoid robots, glowing blue interfaces—that closely mirror the aesthetic conventions of traditional stock photography. This is not a coincidence, but the result of deep entanglement between archival imagery and algorithmic production. A growing body of research has begun to examine this phenomenon through the lens of what has been called AI autophagy, or Model Autophagy Disorder (MAD)—a self-consuming loop in which generative models are increasingly trained on outputs produced by other AI systems.[4] As AI-generated images proliferate and are incorporated into new datasets, the models risk overfitting to their own synthetic outputs, resulting in a narrowing of stylistic diversity and conceptual scope. In this context, the persistent influence of stock imagery becomes even more critical, as its aesthetic conventions are now being replicated, re-ingested, and amplified within generative feedback cycles.

What is most striking in the case of images that depict AI is that artificial intelligence has begun to produce its own visual representation—effectively shaping the imaginary through which it is perceived, understood, and evaluated. These self-produced images carry significant cultural and epistemological weight: they do not merely illustrate AI but participate in shaping public hopes, fears, and expectations surrounding it. It is within this recursive visual logic that I locate the concept of technoanaesthetics: the tendency of generative AI systems to autonomously reproduce the aesthetic patterns of stock imagery, thereby dampening critical engagement. Technoaesthetics—without a hyphen—names the ongoing colonisation of the visual and symbolic imaginaries of AI through images produced by AI itself. By contrast, techno-aesthetics—with a hyphen—marks a critical stance, a space of separation and reflection that allows us to question, resist, or at least become aware of this colonization. While technoaesthetics describes the autonomous visual regime that appears neutral, seamless, and depoliticized, techno-aesthetics insists on the need to reopen interpretation, contest meaning, and reintroduce visibility to the political and ethical struggles that these images often conceal.

14.4 Conclusion

In this chapter, I introduced the concept of techno-anaesthetics to explore how certain visual representations of artificial intelligence contribute to the depoliticisation of public discourse around AI. These images—especially those based on the aesthetic logic of stock photography—do not merely illustrate technology, but shape the symbolic frameworks through which it is perceived, evaluated, and normalised. Far from fostering reflection, they dull our critical faculties, reinforcing the sense that AI is inevitable, natural, and beyond contestation.

In response, I argue that an ethics of science communication about AI should not pursue referential transparency as its primary goal, but should instead foster pensiveness: a capacity to hesitate, to interpret, and to open space for contestation. Good science communication should not aim at consensus or persuasion alone; it should cultivate what Mouffe [17] calls agonism, and what Rancière [18] describes as

[4] A list of references is available at https://dsp.rice.edu/ai-loops/. Accessed on April 11, 2025.

dissensus. This does not mean abandoning scientific truth or technological efficacy, but recognizing that sociotechnical democracy requires more than facts—it demands participation, pluralism, and access to the symbolic field in which meaning and legitimacy are negotiated. In this sense, I propose a political reframing of the ethics of science communication—one that extends, and indeed transforms, the very notion of AI ethics into a field of struggle over meaning, legitimacy, and control. Within this reframing, the concept of techno-aesthetics—hyphenated—takes on renewed importance. Unlike technoaesthetics, which names the smooth, automated aesthetic logic through which AI now reproduces its own imagery, techno-aesthetics insists on the need to restore a space for critical aesthetic judgment. It opens the possibility of re-politicizing perception, of reactivating aesthetics not as a matter of taste or style, but as a form of inquiry, resistance, and intervention. It is through techno-aesthetics that we might reclaim the visual domain as a site of reflection, rather than sedation.

The political urgency of this task is even greater in the context of generative AI. The increasing possibility of excluding human authorship from the creation of AI's visual imaginary represents not only a technical development but a symbolic gesture of exclusion—particularly toward non-experts. Power, both material and symbolic, consolidates around those who control the datasets, the latent spaces, the algorithms, and the platforms. In this competition for control over the production of meaning, various institutional actors—governments, corporations, researchers—are already positioning themselves. This volume is itself a sign that the struggle is underway (Fig. 14.2).

What remains uncertain is whether any form of resistance can emerge from within the logic of generative image production itself. The outputs of these systems are, in most cases, very smooth—polished, generic, and frictionless. Their aesthetic is precisely what makes them incapable of provoking pensiveness. If there is an alternative, it lies elsewhere: not within the interface, but against it. One possibility is art. Exhibitions like *Le monde selon l'IA*, on view at the Jeu de Paume in Paris between April and September 2025,[5] attempt to produce images that disturb rather than reassure, that render AI visible not through cliché but through critical dissonance. Yet such efforts risk remaining confined to elite cultural circuits, inaccessible or opaque to wider publics. This is why another, perhaps more promising site of resistance lies in more popular visual forms—specifically in practices of *détournement* such as internet memes. Unlike the smooth aesthetic of generative tools, memes can interrupt, distort, and reframe dominant narratives. They operate within the vernacular of the networked everyday, yet have the potential to destabilize the very imaginaries that generative AI seeks to normalize. In a landscape increasingly colonized by automated aesthetics, such fragmentary, ironic, and disruptive images may be among the last remaining spaces where techno-aesthetics—as critique—can still occur.

[5] https://jeudepaume.org/evenement/exposition-le-monde-selon-ia/. Accessed on April 10, 2025.

Fig. 14.2 Examples of AI-related memes taken from the Imgflip repository

References

1. Blaschke, E.: Banking On Images. Spector Books, Leipzig (2016)
2. Brey, P.: The empirical turn in the philosophy of technology. Techné: Techné: Res. Philosophy Technol. **14**(1), 36–48 (2010)
3. Bucchi, M., Trench, B. (eds.): Routledge Handbook of Public Communication of Science and Technology, 3rd edn. Routledge, London (2021)
4. Bucchi, M.: When scientists turn to the public: alternative routes in science communication. Public Underst. Sci. **5**(4), 375–394 (1996)
5. Callon, M., Latour, B.: Unscrewing the big Leviathan: how actors macrostructure reality and how sociologists help them to do so. In: Knorr-Cetina, K., Cicourel, A.V. (eds.) Advances in Social Theory and Methodology, pp. 277–303. Routledge & Kegan Paul, Boston (1981)
6. Dondero, M.G., Fontanille, J.: The Semiotic Challenge of Scientific Images: A Test Case for Visual Meaning. Legas, New York (2014)
7. Dubber, M.D., Pasquale, F., Das, S. (eds.): The Oxford Handbook of Ethics of AI. Oxford University Press, Oxford (2020)

8. Flichy, P.: The Internet Imaginaire. MIT Press, Cambridge, MA (2007)
9. Frosh, P.: The Image Factory: Consumer Culture, Photography, and the Visual Content Industry. Berg, Oxford (2003)
10. Goulden, M.: Science communication: A conceptual framework. Public Underst. Sci. **22**(7), 850–865 (2013)
11. Heidegger, M.: The Question Concerning Technology, and Other Essays. Garland, New York (1977)
12. Jasanoff, S., Kim, S.-H. (eds.): Dreamscapes of Modernity: Sociotechnical Imaginaries and the Fabrication of Power. University of Chicago Press, Chicago (2015)
13. Kermode, F.: The Genesis of Secrecy: On the Interpretation of Narrative. Harvard University Press, Cambridge, MA (1979)
14. Kress, G., van Leeuwen, T.: Reading Images: The Grammar of Visual Design, 3rd edn. Routledge, London (2021)
15. Laclau, E.: On Populist Reason. Verso, London (2007)
16. Loeve, S.: Une autre manière de faire de la technologie avec et après Simondon. La techno-esthétique comme connaissance sensible des techniques. Artefact: Techniques, histoire et sciences humaines **18**, 257–300 (2023)
17. Mouffe, C.: Agonistics: Thinking the World Politically. Verso, London (2013)
18. Rancière, J.: Disagreement: Politics and Philosophy. University of Minnesota Press, Minneapolis (2004)
19. Romele, A., Severo, M.: Microstock images of artificial intelligence: how AI creates its own conditions of possibility. Convergence **29**(5), 1226–1242 (2023)
20. Romele, A.: |Ethics of AI| as a floating signifier; or towards a politics of AI ethics. In: Fioravante, R., Vaccaro, A. (eds.) Humanism and Artificial Intelligence, pp. 157–168. Springer, Cham (2025)
21. Romele, A.: Digital Habitus: A Critique of the Imaginaries of Artificial Intelligence. Routledge, London (2023)
22. Simondon, G.: On techno-aesthetics. Parrhesia **14**, 1–8 (2012)
23. Somaini, A.: Algorithmic images: artificial intelligence and visual culture. J. Vis. Cult. **22**(1), 5–22 (2023)

Alberto Romele is associate professor of Media Studies at the Sorbonne Nouvelle Paris 3 University. His research explores the intersections of technology, ethics, and culture, with a particular focus on the symbolic and aesthetic dimensions of digital media. Trained as a philosopher, he draws on continental thought, semiotics, and science and technology studies to investigate sociotechnical imaginaries and the visual representations of artificial intelligence. He is the author of *Digital Habitus: A Critique of the Imaginaries of Artificial Intelligence* (2023) and *Digital Hermeneutics: Philosophical Investigations in New Media and Technologies* (2019) and has published on digital hermeneutics, stock imagery, and the vernacular critique of AI.

Chapter 15
My Cat is an AI. Computational Enunciation and Identity in Video Games

Enzo D'Armenio

Abstract In this paper, I intend to analyse the relationship between AI and video games from a semiotic point of view. In the first part of the paper, I will propose a typology of artificial intelligence systems used in gaming software, based on their general functioning and on the parameters that govern their actions. I will start from the theory of enunciation formulated by Émile Benveniste, in order to define the relationship between virtual actions and the agency of both human and computational entities. Although Benveniste developed his theory to analyse verbal utterances, it was later adapted for application to other semiotic utterances, including still images, audiovisual forms, and video games. Extending the latter line of study, I will propose a typology of computational enunciation in video gaming and identify four fundamental trends in the functioning of modern games: script-based AI, reactive AI, modelling AI, and co-enunciating AI. Finally, I will focus on the fourth of these in order to analyse how AIs are used within video games with the aim of producing responsibility and empathy effects towards secondary characters.

Keywords Video games · Artificial intelligence · Enunciation · Identity · Empathy

15.1 Introduction

One field that is still relatively under-investigated in the humanities pertains to the relationship between AI and video games. Apart from a general affinity pertaining to the utilisation of algorithms, two closer links may also be added. First of all, a historical link has intertwined the evolution of artificial intelligence and games, including electronic video games. The challenge that took place between 1996 and 1997 between Garry Kasparov, the then World Chess Champion, and Deep Blue, a supercomputer developed by IBM, has been extensively discussed [16, 18]. More recently, another game has been employed as a case study to examine AI trained

E. D'Armenio (✉)
University of Lorraine, Metz, France
e-mail: enzo.d-armenio@univ-lorraine.fr

P. Conte et al. (eds.), *Algomedia. The Image at the Time of Artificial Intelligence*, Lecture Notes in Morphogenesis, https://doi.org/10.1007/978-3-032-08726-3_15

by deep learning models: AlphaGo, developed by Google, capable of beating the world's best Go players [17].

Secondly, modern AI and video games are connected on a technical level. It is well known that the efficiency of modern generative AI results from the availability of large datasets of documents associated with complex verbal descriptors that are used to train AI models. However, the processing of this huge amount of data in the training phase is only possible and productive as a consequence of the increased computing power of processors and especially of graphics cards. Just think of the Californian multinational NVIDIA: from being a company which specialised in the production of graphics cards dedicated to a specific sector, that of gaming, NVIDIA has been able to expand its range of activity as a result of the complex operations and computational power required by various fields. On the one hand, there is the power required to support the cryptocurrency market and mining processes [13]. On the other, graphics cards are well-suited to realise tasks linked to deep learning training processes and thus to the effectiveness of AI. Electronic games, in fact, require the parallel processing of multiple tasks simultaneously, an activity that is likewise required during the learning phase of generative AI. In a sense, it is only thanks to these technologies, the modern GPU, that deep learning has become the preferred technique for training AI, far surpassing the results of supervised techniques. As Le Cun ([16]: 80), my translation) writes: "In 2011, it is clear that the future belongs to those who can train large neural networks on GPUs. They will be the vehicle for the new *deep learning* revolution…".

The current state of affairs throws up an intriguing paradox: video games are the field in which AI is systematically exploited to manage the behaviour of virtual worlds and characters.[1] It has been part of the field's vocabulary and technical functioning since its inception. However, contemporary AI training makes use of a range of deep learning techniques which are still relatively under-exploited in this field, apart from those for enhancing image quality, fluidity, and resolution.

Starting from this historical and technical framework, in this paper I intend to explore the relationship between AI and video games.[2] I will first propose a typology of artificial intelligence in gaming software, based on its general functioning, while adopting a semiotic point of view. I will be taking Émile Benveniste's theory of enunciation as a starting point in order to define the relationship between virtual actions and the agency of human and computational entities. Although Benveniste developed

[1] Procedural generation of levels or virtual spaces is one of the most widely used artificial intelligence techniques in video games. A long list of examples could be cited: the Roguelike genre, in which dungeons infested with hostile creatures must be explored, relies on this computational generation. Other notable examples include science fiction video games such as *Elite* (1984) and *No Man's Sky* (2016).

[2] On this topic, see also Gianmarco Thierry Giuliana's excellent contribution [15] devoted to the way in which video game AI produces in the user the meaning effect of being in front of intelligent entities. Giuliana distinguishes two main modalities in the construction of this effect. On the one hand, the author distinguishes a mimetic-narrative modality in which users project typically human categories in order to understand the actions performed by an AI agent. On the other, Giuliana analyses AI agents that model users' behaviour in game worlds in order to perform actions potentially indistinguishable from those of a human operator.

his theory to analyse verbal utterances, it was later adapted to other semiotic systems, including still images [11, 14], audiovisual forms [5, 8, 19], and actions within video games [9].

Building on this last line of study, I will then distinguish four basic forms of AI in video games: script-based AI, reactive AI, modelling AI, and co-enunciating AI.

Finally, I will examine in greater depth the latter form, co-enunciating AI, in order to analyse how it is used in video games with the aim of producing responsibility and empathy effects towards secondary characters. I will do so through the analysis of two cases—the *Portal* saga and *The Last Guardian*—which will allow us to delve into the relationship between narration, interaction, and empathy in virtual worlds, and the role that AI plays in their generation.

15.2 Kinetic Enunciation and Spatial Narration in Video Games

If we refer to Benveniste's theory of enunciation, a suitable computational application of it would correspond to OpenAI's ChatGPT4o: an AI system capable of producing verbal utterances in a dialogical situation that requires an exchange of positions between at least two interlocutors. According to Benveniste, in fact, enunciation is "the very act of producing an utterance and not the text of the utterance that is our object. This act is the act of the speaker who mobilises language for his own purposes" (Benveniste [4]: 13, my translation). His theory is in open opposition to structuralism, because it postulates that it is not sufficient to analyse the structure of language, or of verbal texts, in order to fully understand their meaning. On the contrary, it is necessary to study the relationship between the structure of the utterance and its production. This is why Benveniste speaks of enunciation as a formal device that mediates between the text-utterance and dialogical experience. At the centre of this theoretical device lies the formal apparatus of enunciation: special categories of language that only take on their full meaning in a dialogical situation. These include personal pronouns, spatial deictics, and verb tenses. Indeed, the triptych of "I-here-now," which is related to an experiential situation of co-presence, projects "non-I," "not-here," "not-now" in verbal utterances. It is therefore necessary to study the way in which enunciation relates the places, times, and actors of the dialogical experience with the places, times, and discursive actors projected by verbal utterances. One of the fundamental differences identified by Benveniste concerns the opposition between two regimes of enunciation: enunciation-discourse and enunciation-history. The first regime takes place when the interlocutors establish a dialogue based on the personal pronouns "I-you": this occurs when the speaker presents him or herself as an "I" within a dialogue and addresses a "you." This regime is characterised by a symmetry and interchangeability of the utterance positions: the "you" to whom the speaker addresses him or herself can in turn become an "I." Similarly, spatial and temporal markers will be organised with reference to the time and place of co-present enunciation. In contrast, in the enunciation-history regime, pronominal markers appear in the third person, and spatial and temporal deictics adopt an impersonal style, neutralising the effects of subjectivity. This is typical of historical accounts and scientific texts.

Drawing on this theoretical framework, we may say, for instance, that ChatGPT4o partially adopts an enunciation-discourse regime: OpenAI's text-based and voice assistant does indeed use the first person, asking for instance: "How can I help you?". And yet, in response to the user's questions, it attempts to construct a temporal and spatial regime of enunciation-history, characterised by impersonal place names and dates, rather than by indications derived from spatial and temporal distance in relation to the dialogical situation.[3]

More generally, the theory of enunciation makes it possible not only to analyse the relationship between speech production and the utterance, but it is also based on an implicit criterion: in correspondence with each act of enunciation, the subject of enunciation and language themselves are renewed. According to Benveniste, in fact, subjectivity is constructed solely through language: by appropriating and actualising the virtual system of language, an act of enunciation produces both a discourse and its subject.

However, video games are not comparable to a verbal dialogue in co-presence, and the theory of enunciation needs several adaptations to be applied to these particular semiotic experiences. For these reasons, I have previously proposed the concept of *kinetic enunciation* [7, 10]. Indeed, video games are usually studied in relation to two concepts: that of immersion and that of interaction [1, 21]. The latter would make it possible to distinguish interactive media from traditional media such as literature and cinema: the user, being able to participate in the experience through his or her choices, would assume an authorial role as opposed to the more "passive" reception of traditional media. As for immersion, the fact that video games stimulate more senses and require coordination to cope with the virtual experience would have greater effects on its overall implication (narrative, emotional or aesthetic).

With respect to these two concepts, it appears necessary to focus more carefully on the way the player accesses the virtual world and on the way his or her experience is articulated in proto-linguistic grammaticalizations. Indeed, the user first of all accesses the virtual world visually: the screen presents a two-dimensional, three-dimensional or immersive world, marked by a more or less pronounced level of figuration[4] (photorealism, stylisation, abstraction). And yet, so-called interaction

[3] These considerations are obviously the result of an extreme simplification of a possible enunciational analysis of interactions with modern verbal generative AI. Among other things, it should be emphasised that the dialogic situation Benveniste spoke of is a very particular situation in the case of ChatGPT: it would be difficult to situate the temporality and spatiality of this AI system simply because it does not derive from a phenomenological unity. It is not here and now, but is rather a distribution of functions.

[4] In the visual semiotics of the Paris School, the concept of figurative density indicates a gradual progression between abstraction and iconisation. Abstraction indicates images that do not present figures, that is, recognisable entities of the world. At the opposite extreme lie images that aim to

emphasises the fundamental importance of another aspect: in order to access virtual experiences, in order to make them work, the user must perform movements on the interface (by means of joypads, gyroscopes, motion sensors or virtual helmets) that are transformed into movements within the virtual world (e.g. running, jumping, shooting or looking). On the one hand, movements on the interface are usually abstract and plastic, because they are not expressed through recognisable figures or themes, but through abstract commands such as up, down, left and right, or using other activation keys (A, B, X, Y) and progressive triggers. Movements in the virtual worlds, on the other hand, often translate the abstract movement upon the interface into a system of figurative and thematic movements that are specific to each video game (running, jumping, shooting, climbing, etc.). Between these two types of movement, a kinetic diagram associates the abstract commands on the interface with more or less figurative and thematic movements in the virtual world.[5] This diagram of kinetic relations establishes the rhythmic and aspectual resonance between the two systems of movement: for example, punctual movements such as jumping within the virtual world may be matched with equally punctual movements upon the interface, while durative movements such as running require prolonged activation of the commands.

In other words, there are two main forms of grammaticalization as regards video games and virtual reality: on the one hand, a visual syntax that regulates the degree of figurativeness and the architecture of virtual spaces; on the other, a kinetic syntax that regulates the interactive micro-language through which the player exists within each specific video game.

The video game *Assassin's Creed II* (Ubisoft 2009) may serve to exemplify these formal features. This video game is set in a series of Italian Renaissance cities, including Venice and Florence. We could describe it as a historical conspiracy thriller with a plot similar to that of Umberto Eco's *The Name of the Rose*. This is, however, only if we limit ourselves to considering the visual syntax and audiovisual narration, because the interaction system requires the player to perform acrobatic movements and vertiginous chases, climb famous monuments, and assassinate key figures in order to unravel the international conspiracy. The meaning-making process is thus linked to two interrelated levels. on the one hand, the international conspiracy depicted with "traditional" images and the figurative dressing of virtual spaces; on the

convey an effect of realism. It should of course be noted that realism is not an invariable quality, but depends on expressive techniques and visual cultures. In the context of video games, these two extremes can be exemplified with *Tetris*, an abstract video game, and three-dimensional games such as *Alan Wake 2* which aim to create a photorealistic effect.

[5] I refer to the definition of a diagram proposed by Charles Sanders Peirce ([20], CP 2.277): a sensitive representation that in its internal relations replicates the relations pertaining to another phenomenon. For example, a map which selects certain characteristics of the physical space, such as distance and spatial extension, and expresses them visually by replicating some relations of a "real" object. However, the same applies to geometric demonstrations, where it is possible to manipulate graphical traits in order to build new knowledge through visual icons instead of categorical concepts. In my hypothesis, the relations between the plastic movements made on the interface are translated into the system of kinetic relations expressed by the movements in the virtual worlds. This is the specific way in which the user accesses the semiotic experience and participates in its articulation.

other hand, the practical narration imposed by the kinetic system—a kind of parkour to be performed in a historical setting, characterized by rhythms and thematic movements such as running, climbing, and killing.[6] Rather than stopping at the simple fact that these are interactive or immersive media, this approach makes it possible to analyze directly the kinetic micro-languages (offensive, defensive, contemplative) specific to each virtual production.

These two types of syntax, visual and kinetic, can either reinforce each other or generate conflicts for the user who is forced to negotiate between different pragmatic frameworks. Sometimes this involves kinetic identification with the avatar; at other times it involves interpretative and emotional distancing. In *Shadow of the Colossus*, for instance, users experience a conflict between "not wanting to do" and "having to do." They are required to kill sacred beings called Colossi (kinetic syntax), but the figurative appearance of these as gigantic, innocent creatures (visual syntax) renders this task controversial and dramatic.

15.3 Computational Enunciation in Virtual Worlds: A Typology of Virtual AI Models

In this framework, there are two fundamental pivots key factors through which the construction of meaning in video games is realised by means of the kinetic enunciation that characterises it. Firstly, the user's action is expressed using the micro-languages of movements specifically set up by each production. Secondly, the virtual spaces can depict situations and transformations indexically and accommodate exploratory movements.

With regard to fundamental structure of the construction of meaning in video games, the main point I am interested in investigating here is the contribution of AI. It is no longer just a matter of considering the action stemming from the user's kinetic enunciation through his or her avatar, but of analysing the action of the software itself. In other words, it is a question of examining the way in which video game software expresses an agency that must be ascribed to an AI model embodied in different virtual entities: virtual space, verbal enunciations, and kinetic actions expressed by secondary characters competing or collaborating with the user. In relation to the theory of enunciation, it is therefore not a matter of considering the way in which the player is installed in a game world by means of an avatar, a point of view, and a system of movements, because this initial enunciational projection does not generate an effect of computational agency. It is rather a matter of considering the modifications to this initial situation that are ascribed to an instance governed by the AI model.

[6] For a comprehensive semiotic analysis of *Assassin's Creed II*, carried out using the semiotic tools developed by Roland Barthes in S/Z [3], see Compagno [6]. See also Aroni [2] for a semiotic study of architecture in video games.

In order to construct our typology, I will distinguish three orders of criteria, the first of which being the type of enunciation realised by AI. Under this criterion, we can distinguish enunciations realized by the AI on the virtual space. These include modifications of the scenario and of its morphology. Some trivial examples concern the way in which a virtual level itself is modified following the realisation of a certain goal in a platform game such as *Super Mario 64*, but also the real-time modification of a stage in correspondence with the player's passing through, as in the case of *DmC: Devil May Cry*.

Secondly, kinetic enunciations—defined as actions or sequences of actions performed by an AI-driven entity, be it a character or an object—are often the most important in the construction of meaning, especially in relation to the actions to be performed by the user through the avatar.

Countless video game productions fall into this category: basically, all those which position the player in competition or in collaboration with other AI-governed entities, be they opposing cars in *Gran Turismo 7* or soldiers to be eliminated in the *Call of Duty* series.

Next, there are verbal or more generally linguistic enunciations expressed by secondary characters or other entities governed by the AI model in order to comment on and direct the player's actions or to engage him or her in interactive dialogues. Secondary characters in contemporary video games are now generally dubbed into various languages, but written dialogues, as in the case of classic role-playing games, and changes in the scenario involving the appearance of ostensive and indicative signs, also fall into this category.

Finally, temporal modifications pertain to the passage of virtual time and the alternation of the day-night cycle or the succession of seasons, the triggering of flashbacks in which the player's avatar possesses diminished kinetic abilities, or even meta-kinetic enunciations that influence performable commands by applying accelerations or slowdowns, as in the case of spells cast by enemies against the player in the role-playing game genre ("Slow" in the *Final Fantasy* saga).

The second criterion pertains to the scale of the modification occasioned by the computational enunciation: it may be extensive, general and distributed, or focused and specific. For example, with regard to the spatial enunciation described above, which concerns the modification of virtual worlds, scale allows the differentiation between modifications affecting limited portions of the morphology of space or a more global reconfiguration. Similarly, in relation to the kinetic enunciations realised by the AI system, scale makes it possible to distinguish whether these actions are linked to a single semiotic situation, for example the actions performed by AI in response to those of the user, or independent actions, such as the simulation of the behaviours of secondary characters in the Far West townships depicted in *Red Dead Redemption.* This is the case with the opening and closing of shops in relation to virtual time, an example that also covers temporal (the time at which the shop is open or closed) and spatial (the accessibility of the shops themselves) enunciational criteria.

The third criterion articulates the aspectuality of computational enunciation. In semiotics, aspectuality relates not only to the temporality of a verb or narrative action, but to the manner in which the action is grasped: an action may be in its terminative phase, be grasped in its durativity, in its initial (inchoative), or exhibit a repetitive quality (iterativity). In our typology, an enunciation realized by the AI may already have taken place, may take place in real time, be durative, terminative, inchoative or iterative. A spatial change that impacts the game scenario may, for instance, have already taken place when the player arrives, or it may be taking place in real time and thus require an immediate response, as in the case of the crumbling bridges in the platform games of the *Super Mario* saga. The same applies to the criteria related to kinetic, verbal, and temporal enunciation governed by AI: dialogue with an AI may take place in real time or take the form of an indexical sign left in the scenario or recorded on a medium; the transition from day to night may be punctual and automatic or involve a graduality that impacts visibility. Waiting for darkness to fall can, for instance, be advantageous in stealth manoeuvres where the player's avatar must hide in shadowy areas in order to advance unnoticed.

Finally, the fourth criterion concerns the type of intersubjective relationship that the computational enunciation performs with respect to the player. These relations can be competitive (in the case of a test, a fight or a challenge), collaborative, and, more generally, dysphoric or euphoric. From the intersection of these criteria, a general typology of AI enunciations in video games emerges, represented schematically in Table 15.1.

Of course, in most video games and, more generally, in most virtual experiences, there exists even greater complexity. In the aforementioned case of *Shadow of the Colossus*, for instance, where the user has to eliminate the gigantic sacred beings called colossi, AI controls at the same time the modification of the virtual spaces and the kinetic actions of the colossi themselves. These actions take place in accordance with an aspectuality unfolding in real time and resulting in a fierce competition. The user has to climb these giants, which are both secondary characters competing with the avatar, and mobile virtual spaces requiring rhythmic exploration and adaptation.

15.4 Four General Trends of AI in Modern Video Games

In short, this typology allows us to examine a wide range of enunciations realised by AI. The objective of this paper is to identify four general orientations that characterise computational enunciations in contemporary video games, before focusing on the generation of meaning effects derived from the actions of secondary characters driven by AI.

The first, arguably the least interesting and at the same time the most necessary, pertains to script-based computational enunciations. These are enunciations relating to general reactions of the video game software to the player's actions which usually do not take place in real time. They include: the scenario that changes as a result of certain actions, e.g. the opening of a new explorable section in an adventure game (AI modification of the virtual space, with specific scale, a terminative aspectuality, and euphoric intersubjective dominance); a character that moves to another section of the

Table 15.1 A typology of enunciations under the agency of AI in video games

Type of enunciation performed by the AI	Modifications of the virtual space	Kinetic actions	Verbal or linguistic enunciations	Modifications of temporality
Scale	The degree of modification of a virtual space: global, partial, specific	The scale of actions performed by AI: single semiotic situation or multiple ones; in front of the player, in relation to the player's actions, or independently	Interactive dialogues in specific situations; indexical signals created in the virtual space, enunciations covering an extended narrative arc	Day-night cycle, the virtual reproduction of the succession of seasons
Aspectuality	Spatial modifications having already occurred, in real time, gradual	Durative, iterative, inchoative, or terminative actions	Enunciations in real time, already implemented in the scenario, recorded	Temporal changes reproduced in a discontinuous, sudden, or gradual manner
Intersubjective relationship with respect to the user	Euphoric or dysphoric dominant: capable of generating an urgent response or allowing access to new areas	Kinetic collaboration or competition	Dialogic interaction or multiple-choice, passive listening during action, comments, advice, evaluations	Flashbacks that affect the avatar's kinetic abilities, nighttime sections for stealth phases, etc

virtual world as a result of certain actions, allowing new verbal interactions to occur (AI kinetic and verbal enunciation, with specific scale, a terminative aspectuality, a competitive or collaborative intersubjective relationship); secondary characters that the user must follow as they show the way forward (AI kinetic actions, with specific scale, and a durative aspectuality); and enemies that perform simple actions in response to the player's actions in real time (AI kinetic actions, with specific scale, a real-time aspectuality, a competitive intersubjective relationship). An illustration of this latter type of enunciation is provided by the manner in which the Goombas, Mario's historical enemies in *Super Mario Odyssey*, pursue the user's avatar when the latter is in close proximity, attempting to strike him. This is an example that can be positioned close to the second tendency, but which I classify under the first as a consequence of the straightforward performance of the actions undertaken by the IA.

The second trend relates to AI entities which are reactive: secondary characters and especially enemies who react to the player by deploying complex action sequences. It is no longer simply about following the avatar and trying to strike it, but about orchestrating a series of actions which respond to a global strategy and to local tactics and which are in turn informed by the actions of the player. In terms of the proposed typology, these are kinetic enunciations made by AI in real time which build a complex intersubjective, interactive, and competitive relationship with the user. Typical examples include video games in the fighting game genre, driving games in which rival vehicles are driven by AI agents, and adventure games where enemies are strategically positioned in an attempt to eliminate the player's avatar. In all these cases, it is also often possible to modify the level of difficulty of the experience, which allows for adjustments to be made according to the aggressiveness of the characters driven by AI. An interesting example is that of *Elden Ring*, a fantasy adventure video game well-known for its difficulty. In particular, end-level bosses exploit, in an exemplary way, user input to engage attack strategies and tactics: if the user tries to move away from the boss in an attempt to use an object and restore its energy, the boss will immediately launch a remote attack—throwing, for example, sharp objects. Similarly, during a close fight, if the user avoids an attack in response to the beginning of the enemy's animation, the animation itself will be delayed, as if the boss were waiting for the end of the shift to hit the avatar.

The third trend builds on the second with the specific objective of learning the player's style so as to develop a well-articulated strategy. For this reason, I will speak of modelling AI. This involves the use of deep learning models, aiming to equip modern video games with more realistic secondary characters capable of performing actions emulating those of a human. One example is Sophy, an AI agent developed by Sony for the racing game *Gran Turismo 7*. Sophy has been trained using reinforcement learning: "a type of machine learning used to train AI agents that take actions in an environment by rewarding or penalising those actions based on the outcomes they lead to".[7] This AI agent has been trained on the basis of previous races participated in by millions of human players and reinforced through collaboration with professional drivers. The result is an AI agent capable of driving virtual racing cars in an aggressive manner, one that develops contextual attack and defence strategies while maintaining a style that respects the rules of sportsmanship. Whereas traditional AIs drive mechanically, forming trains of cars that rarely battle one another, Sophy can instead slip into gaps left by rivals and defend the overtaking line, covering the apex on curves before hitting the braking point. Just like Deep Blue and AlphaGo before it, Sophy too is now capable of rivalling and defeating human world champions in e-sports competitions.[8]

---

[7] https://www.gran-turismo.com/us/gran-turismo-sophy/technology/ accessed 5 July 2024.

[8] See the paper published in *Nature* that describes how this AI agent functions [22]. A similar case is that of the fighting game *Tekken 8*. In a specific game mode—called "Super Ghost Battle"—the AI-controlled opponent learns the style of the user it is fighting against and can later be used to challenge other humans. Similarly, the user can download AI agents trained by the best players to improve their own skills.

15.4.1 Co-Enunciating AI: The Case of Portal

The fourth and final trend relates to AI entities which are capable of enunciating complex actions by taking into account the configuration of the virtual space, the situated actions of the player, and the role they play within the narration. In other words, these are AI agents which not only make flexible use of the criteria described in the typology above—which pertain to the kinetic syntax of video games—but also articulate these enunciations in relation to visual and narrative syntax, in order to build more general effects of meaning: effects of empathy and responsibility.

Two examples of co-enunciating AI agents are those used in *Portal* and *Portal 2*, two first-person three-dimensional puzzle-platform video games developed and published by Valve Corporation. In *Portal*, released in 2007, the player assumes the role of Chell, a woman trapped in a research facility called Aperture Science, which consists solely of test chambers filled with environmental puzzles. Chell, and with her the player, must solve the puzzles one after another using a device called the "Portal Gun": a gun that creates interconnected portals. These portals allow the player to cross great distances, manipulate objects, and exploit the game's physics to pass the tests and advance through the chambers. The sequel, *Portal 2*, released in 2011, further expands the story and gameplay mechanics, introducing more complex puzzles and, notably, new characters.

One of the distinctive features of these games is the inclusion of AI entities which act as co-enunciating secondary characters: it is important to clarify that their role as AI characters is both narrative—since they are described and portrayed as such—and behavioural, since these fictional AI characters are in effect governed by the artificial intelligence modules of the *Portal* video games.

In the first *Portal*, GLaDOS is the main AI construct that guides the player through the experimental testing environment. She is a very eloquent artificial intelligence with a female voice, clearly inspired by HAL 9000 from *2001: A Space Odyssey*. Her voice accompanies the user obsessively throughout the game. Additionally, GLaDOS frequently uses sarcasm and dark humour in interactions with the player, as exemplified by the following two of her enunciations: "This next test could take a very, very long time. If you become lightheaded from thirst, feel free to pass out. An intubation associate will be dispatched to revive you with peptic salve and adrenaline," and "While safety is one of many Enrichment Center goals, the Aperture Science High Energy Pellet, seen to the left of the chamber, can and has caused permanent disabilities such as vaporisation." As Chell and the player progress through the aseptic test chambers, GLaDOS begins to refer to a final reward, reaffirming her inadvertently sarcastic style and behavioural attitudes: the reward will be cake.

However, as the tests progress, the player will gain access to alternative spaces, secret rooms that can be reached by squeezing through hard-to-navigate openings inside the main chambers. The test chambers are clinical and highly controlled spaces, designed for conducting scientific experiments. They are characterised by an austere and impersonal atmosphere, reflecting the scientific and experimental ethos of Aperture Science, with an emphasis on efficiency. They contain only the elements necessary to complete the tests, and the walls are often decorated with panels displaying the Aperture Science logo and informative icons explaining the game mechanics and test objectives. In contrast, the secret spaces offer a behind-the-scenes view of the Aperture Science facility. These rooms are dirty and disorderly, with walls covered in mould, floors littered with debris, and objects scattered everywhere haphazardly. One of the distinctive features of the secret spaces is the presence of graffiti and scribblings presumably left by a previous test subject. These are messages marked by paranoia and despair, revealing crucial information about the sinister nature of the experiments and of GLaDOS. One of the most notable of these is "The cake is a lie," an obvious reference to the fact that GLaDOS's promises of reward are deceitful. Unlike the test chambers, the secret spaces are also free from GLaDOS's constant supervision. This renders the exploration of these a moment of welcome respite, but also of revelation. Besides the graffiti, these rooms contain personal messages, drawings, and symbols which tell the stories of previous test subjects. These elements humanize the previous test subjects and give rise to an effect of empathy in the player.

With respect to our proposed typology, it can be argued that the player's kinetic enunciation integrates seamlessly with the spatial enunciation and with the verbal enunciations made by GLaDOS: the meaning-making experience sets up a stark contrast between the aseptic character and the grotesque enunciations of GLaDOS and the desperate tone of the spatial and verbal enunciations encountered in the secret spaces. The result is the attribution to GlaDOS of a complex identity characterised by her dark humour and ruthless nature, contrasted with the desperate, human and complicit identity of those who have covered the secret spaces with graffiti, drawings, and warnings for the player.

In *Portal 2*, Wheatley is the AI construct who initially guides the player. Unlike GLaDOS, Wheatley uses verbal enunciations which more explicitly take into consideration the player's actions in relation to space. His instructions often specify precise directions, such as "Go to your right," rather than using impersonal toponyms. This makes the interaction more immediate and situated, creating a sense of urgency and direct involvement. Furthermore, these verbal enunciations give Wheatley a complex semiotic identity through the combination of the narrative framework constructed on the basis of visual syntax and the kinetic enunciations realised in real time. In other words, Wheatley is not merely a secondary character tasked with specific, mechanical functions, but he adopts a consistent semiotic identity due to the way he interacts with the player in a situated manner, employing deictics that refer to the I-here-now of the kinetic enunciation.

15.4.2 From Computational Enunciation to Identity-Related Effects

To conclude this analysis, I can propose the following hypothesis: in order for AI enunciations to induce empathy and, more generally, emotion in the user, they must be able to construct identity-related effects, through an interplay of kinetic syntax and visual syntax. On the one hand, as we have seen in the case of the Portal saga, they must set up enunciations capable of transcending the framework of a local interaction to create an effect of persistence due to the layering of multiple enunciations attributable to the same AI-driven entity. On the other hand, they must encourage players to perform an identity performance: to manage their experiential, kinetic, and interpretative activity in order to position themselves in relation to the AI-controlled entities.

In this regard, it seems to me that the fundamental contrast is that between the concepts of randomness and contingency. In the specific context of AI, it is a matter of counterpoising a logic of computational randomness with that of experiential management of contingency. The former concerns, for instance, how visual generative AI systems such as Midjourney and DALL-E produce images based on natural language prompts input by humans. As is well known, using the same prompt multiple times never results in exactly the same image, because these models rely on the systematic exploitation of computational randomness.[9]

In contrast, the second logic pertains to how human operators deal with an ongoing practice. In the case of video games, it concerns the practical situations set up in virtual worlds: controlling the avatar, players hesitate, make mistakes and retrace their steps in order to find a solution. In doing so, they take into account elements present in the virtual environment: objects, obstacles, and other characters.

Characters controlled by AI can thus be perceived as entities animated by computational logic, program-characters being treated and felt as "things." However, if these characters behave as if they too are seeking solutions, hesitating, making mistakes, making attempts in collaboration with the player, contingency management is assumed to be in operation. In other words, in order for an AI-controlled entity to be perceived as an entity with an identity and not as a series of computational scripts, it must exhibit the signs of contingency management analogous to that exercised by a human operator.

This peculiar identity-related effect is rare in contemporary video games. In addition to the *Portal* saga, another case in point is *The Last Guardian*, a video game released in 2016 on Sony PlayStation 4 and developed by Team Ico.

This video game exploits the relationship of trust between the young protagonist (a boy controlled by the player) and Trico, a giant half-bird, half-feline creature. The player must use Trico's abilities to overcome challenges and solve environmental puzzles. For example, Trico can destroy obstacles, defend the boy from enemies, or

[9] In a recent paper devoted to visual generative AI, Maria Giulia Dondero [12] argues that, because of the way it affects the relationship between human operators and computational image generation, computational randomness should be considered as an instance of enunciation.

allow him to climb his body to reach otherwise inaccessible spots. The boy, in turn, helps Trico by removing spears from his body, calming the creature in moments of panic, or opening passageways for him.

One of the peculiarities of this video game is that the interaction between the two characters is not conducted through menus which interrupt the action, but through two dedicated commands that can be executed in real time. The first allows the user to call Trico to attract his attention: if the creature is far away, the boy will shout. If, on the contrary, the two characters are close together, the boy will speak in a lower voice or even whisper. The second command allows the boy to indicate a direction in the virtual environment in order to draw Trico's attention towards this. These two commands allow for a complex range of interactions. For example, the player can point to a distant platform and then jump on the spot in order to suggest to Trico to jump onto the platform.

Trico does not execute commands immediately as would a character controlled by traditional artificial intelligence. He often stops, looks around, moves somewhat slowly, or shows hesitancy before acting. Sometimes he misunderstands the player's actions to the point of frustration, but these misunderstandings are not generally interpreted as mistakes but rather as idiosyncrasies resulting from his non-human nature. More generally, Trico expresses emotions such as fear, curiosity, and affection through the movement of his ears, posture, and the noises he makes.

At first, Trico is diffident and unresponsive to the boy's gestures. The player must win his trust by taking care of him, feeding him, and interacting with him. Over time, Trico begins to respond more consistently to the boy's instructions, although he continues to display independent and sometimes unpredictable behaviour.

In some of the most dramatic sequences, it is Trico himself who takes the initiative to propel the narrative. When faced with the boy's inability to reach a platform where he is himself standing, Trico will lean toward the void which separates them, staring intensely at the boy. The player will then have to make the boy jump into the void and trust that the creature will be able to catch him mid-air in his mouth and then set him down on the ground. In other words, Trico does not just carry out orders as if he were programmed, but hesitates, misunderstands the boy's actions, and tries to find solutions.

More broadly, the game creates a strong emotional bond between the player and Trico through narrative events that intensify this interactive relationship, leading to moments of tension, mutual protection, and ultimately to a deep connection between the two characters.

Ultimately, *The Last Guardian*, though not based on AI trained through deep learning but rather on a complex, handcrafted use of traditional algorithms, is capable of closely approximating the behaviour of a living being. In the way it manages the contingency of the experience, Trico behaves as if he were managing such contingencies himself. The result is an emotional interactive experience, emerging from the collaboration between the player and a computational entity capable of producing identity-related effects.

15.5 Conclusions

In this paper, I have explored the relationship between artificial intelligence and video games. As a starting point I proposed an adaptation of Benveniste's theory of enunciation, identifying two fundamental syntaxes that govern their meaning. Firstly, I described the visual syntax that concerns cutscenes over which the player has no control, and the architecture of virtual spaces, capable of constructing a narrative in an indexical manner. Subsequently, I identified the expressive resources of a kinetic syntax: the way in which movements made on the interface by the user are transformed into virtual movements in the game world, in accordance with micro-languages that are specific to each experience.

Attention was then turned to computational enunciations by constructing a typology of the semiotic actions that can be attributed to AI agency. Among the key criteria for this typology, I identified the type of enunciation (spatial, kinetic, linguistic, and temporal), scale, aspectuality, and the intersubjective relationship constructed in relation to the user. This allowed the identification of four trends that dominate AI behaviour in modern video games: script-based AI, reactive AI, modelling AI, and co-enunciating AI.

Starting from this latest trend, we explored how video games, by complexly articulating visual syntax and kinetic syntax around the enunciations of the user and AI, are capable of creating effects of empathy and emotions. The analysis of the *Portal* series and of *The Last Guardian* led to the proposal of the following hypothesis: for video game AI systems to induce empathy and emotional response, they must bring to life characters that, firstly, present a consistent identity and, secondly, act as if they were managing the contingency of the experience.

This investigation seems necessary not only in light of the importance that the video game industry has gained in Western economies, but also and above all because the criteria identified regarding computational enunciations are relevant for the analysis of any virtual experience, whether it be ludic, artistic, or heritage experiences in VR.

In conclusion, I would like to return to the issue of generative artificial intelligence from which I started. One of the most significant future developments in the relationship between computation and interactive experiences will be the manner in which secondary characters and, more generally, virtual characters will be powered by AI systems capable of reacting interactively to the verbal enunciations of users. Virtual assistants like Convai and NVIDIA ACE will allow secondary characters and any virtual humanoid entity to interact freely and realistically with the user, thereby enhancing the sense of their identity through the articulation of perceptive abilities and nuanced enunciations. This imminent integration represents the culmination of the parallel development between video games and AI, which began with rudimentary algorithms controlling the spaceships in *Space Invaders* and the ghosts in *Pac-Man*, and continued with the fundamental contribution of gaming graphics cards in deep learning tasks.

References

1. Aarseth, E.: Cybertext: Perspectives on Ergodic Literature. Johns Hopkins University Press, Baltimore (1997)
2. Aroni, G.: The Semiotics of Architecture in Video Games. Bloomsbury, London (2022)
3. Barthes, R.: S/Z. Éditions du Seuil, Paris (1970)
4. Benveniste, É.: L'appareil formel de l'énonciation. Langages **17**, 12–18 (1970)
5. Casetti, F.: Inside the Gaze: The Fiction Film and Its Spectator (1989). Indiana University Press, Bloomington (1999)
6. Compagno, D.: Dezmond. Una lettura di Assassin's Creed 2. Unicopli, Milano (2013)
7. D'Armenio, E.: Mondi paralleli. Ripensare l'interattività nei videogiochi. Unicopli, Milano (2014)
8. D'Armenio, E.: From audiovisual to intermedial editing. Film experience and enunciation put to the test of technical formats. Versus. Quaderni di studi semiotici **124**, 59–74 (2017)
9. D'Armenio, E.: Beyond interactivity and immersion. A kinetic reconceptualization for virtual reality and video games. New Techno-Humanit. **2**(2), 121–129 (2022). https://doi.org/10.1016/j.techum.2022.04.003
10. D'Armenio, E.: From the ontology of video games to the epistemology of digital movements. Towards a Semiotics of Virtual Practices. Games and Culture (2024). https://journals.sagepub.com/home/gac
11. Dondero, M.G.: The Language of Images: The Forms and the Forces. Springer, Cham (2020)
12. Dondero, M.G.: Sur quelques problèmes de composition et de style dans les modèles d'intelligence artificielle générative. In: Paper presented at the 1st session of the Séminaire International de Sémiotique à Paris, Maison Suger, 13 Nov (2024)
13. Finn, E.: What Algorithms Want: Imagination in the Age of Computing. MIT Press, Cambridge, MA (2017)
14. Fontanille, J.: Les espaces subjectifs. Introduction à la sémiotique de l'observateur. Hachette, Paris (1989)
15. Giuliana, G.T.: L'intelligenza delle I.A. come effetto di senso semio-narrativo nei giochi digitali. Una rivoluzione semiosofica. In: Santangelo, A., Leone, M. (eds.) Semiotica e intelligenza artificiale, pp. 171–194. Aracne, Roma (2023)
16. Le Cun, Y.: Quand la machine apprend. La révolution des neurones artificiels et de l'apprentissage profond. Odile Jacob, Paris (2019)
17. MacKenzie, A., Munster, A.: Platform seeing: image ensembles and their invisualities. Theory Cult. Soc. **36**(5), 3–22 (2019)
18. Manovich, L., Arielli, M.: Artificial Aesthetics: Generative AI, art and visual media (2021–2024). http://manovich.net/index.php/projects/artificial-aesthetics
19. Metz, C.: L'énonciation impersonnelle ou le site du film. Gallimard, Paris (1991)
20. Peirce, C.S.: Collected Papers of Charles Sanders Peirce, vols. 1–6. Harvard University Press, Cambridge (1931–1935)
21. Pinotti, A.: Towards an-iconology: the image as environment. Screen **61**(4), 594–602 (2020)
22. Wurman, P.R., Barrett, S., Kawamoto, K., et al.: Outracing champion Gran Turismo drivers with deep reinforcement learning. Nature **602**, 223–228 (2022). https://doi.org/10.1038/s41586-021-04357-7

Enzo D'Armenio is junior professor of Semiotics at the University of Lorraine. He has previously carried out three post-doctoral projects at the University of Liège, including the Marie Curie (Individual Fellowship) project *IMACTIS. Fostering Critical Identities Through Social Media Archival Images* (www.imactis.eu), which focused on identity-related images on digital social networks. His recent research, published in international journals such as *Visual Communication*, *Semiotica*, and *Games and Culture*, focuses on generative AI and the semiotic study of identities. He is also the author of the monograph *Mondi paralleli. Ripensare l'interattività nei videogiochi* (2014).

Chapter 16
Enunciative Practices and Dataset Curation: Toward a Decolonial Semiotics of AI-Generated Imagery

Cristina Voto

Abstract This paper examines the impact of AI-generated imagery, considering its potential to both perpetuate existing inequalities and challenge entrenched biases. It emphasises the necessity of a critical approach to AI, particularly with text-to-image models, and advocates for the urgent integration of gender and decolonial perspectives. These viewpoints are crucial for viewing technologies as collaborative agents rather than substitutes. Employing a theoretical framework grounded in a more-than-human perspective and AI-centered semiotics, this work analyses an artistic project generated through a text-to-image model. It demonstrates how AI can actively participate in cultural and social narratives and potentially redress colonial legacies through careful dataset curation. By highlighting the importance of addressing biases within AI algorithms and data structures to prevent the replication of racial, gender, and class hierarchies, the paper discusses the role of datasets in visualising and giving presence to historically marginalised communities. In conclusion, it advocates for a future where AI and its imagery not only challenge existing social structures but also promote social justice and equity, thereby transforming the landscape of visual culture and technology. This transformative approach necessitates reevaluating the relational dynamics between humans and machines, recognising AI's potential as a transformative societal force.

Keywords AI-generated imagery · Dataset curation · Decolonial aesthetics · Enunciative practices · More-than-human relationality

16.1 Introducing AI Imagery from a Situated Perspective

The primary objective of this paper is to analyse the profound impact of AI-generated imagery on our environment. It aims to shed light on its potential both to perpetuate existing inequities and to challenge pervasive biases. Central to this inquiry,

C. Voto (✉)
University of Turin, Turin, Italy
e-mail: cristina.voto@unito.it

P. Conte et al. (eds.), *Algomedia. The Image at the Time of Artificial Intelligence*, Lecture Notes in Morphogenesis, https://doi.org/10.1007/978-3-032-08726-3_16

in fact, is the reevaluation and redefinition of traditional interpretations that have often characterised technology, especially automated systems, as mere computational agents devoid of socioeconomic influences. Indeed, numerous examples from our current reality vividly demonstrate how artificial intelligence systems can perpetuate discrimination. For instance, the Google Photos app infamously mislabeled two African American individuals as gorillas in 2015, a shocking revelation of inherent racism in its algorithms [16]. Similarly, Amazon's hiring tool exhibited bias against women, a clear indication of the gender biases that can be embedded in AI systems [8]. The discriminatory practices of the Apple Card and Goldman Sachs banking, which issued significantly lower credit limits to women, further underline the biases within AI systems. And the misgendering of transgender and non-binary individuals by facial recognition software [2, 32] is a stark reminder of the social biases that can be perpetuated by AI.

In contrast, this paper advocates for adopting a situated perspective that views sustainable cooperation with "other beings" as a pathway toward building more equitable and collective futures. This viewpoint challenges the conventional perspective that equates technological integration with substitution, which can lead to extreme and polarised stances—such as fears of AI replacing human labour and taking over jobs. Instead, I propose an approach grounded in the concept of alliance, aiming to foster a more equitable AI and, consequently, a fairer future. As the coexistence of human and artificial agents becomes increasingly prevalent, it is crucial to recognise their agency and acknowledge their capacity to transcend the limits of our biological and cognitive abilities. This necessitates a nuanced exploration of both the euphoric possibilities and dysphoric implications of AI, urging us to transcend entrenched dualities and adopt a more comprehensive perspective towards a more-than-human world.

In what follows, I will delve deeper into the broader environmental scale down to a detailed case study, questioning the power structures and logic within artificial intelligence technologies while aiming to introduce a decolonial perspective. In the first section, my primary objective is first to examine the proliferation of AI-generated imagery, highlighting its pervasive impact on our visual culture and overall environment. This includes exploring how it influences everyday practices and raises ethical considerations. Drawing from Archive Studies, I will treat datasets used in training AI technologies as archives presenting redundancies and omissions, a perspective that turns data as always partial and situated reality extractions. Next, in the second section, I propose a shift in understanding the relational dynamics in our environment by considering how AI imagery can reshape our perceptions toward a more-than-human perspective through a more integrated approach to human and machine interactions. Ultimately, in the last section, I suggest a decolonial approach to engaging with computational models based on an artistic case study, focusing on the idea of dataset curation and exploring the potential of AI in meaning-making through a semiotic analysis of enunciative practices in text-to-image models.

16.2 Unveiling the Impact of AI Imagery: The Model, the Experience and the Environment

According to the *Everypixel Journal*, the official blog of *Everypixel*—a tech company founded in 2005 and based in San Francisco—the year 2023 witnessed the creation of 15 billion images via generative text-to-image computational models [36]. This staggering figure reportedly equals the total number of images produced during the first 150 years of photographic technology, from 1826 to 1975. Beyond mere comparison, this statistic underscores artificial intelligence's rapid and profound impact on our visual culture, where generative models manage vast amounts of visual data, and AI-generated images have become ubiquitous across our iconosphere. These models are applied across a broad spectrum of social fields, influenced by various institutional and stabilised evaluations—from labour [17] to art [23], passing through education [40] and healthcare [30]—profoundly affecting our daily practices and the information landscape.

These numbers effectively highlight the magnitude of a practice that has been widely adopted since 2021 and remains extraordinarily accessible. As the blog notes, "The volume of content created during this time is difficult to quantify," and various statistics have been monitored in response. Consider a few of them: i. Since the launch of DALL·E 2 up to 2023, an average of 34 million images were generated daily; ii. Midjourney has emerged as the image generation platform with the most extensive user base, boasting 15 million users; iii. Approximately 80% of those 15 billion images were produced using the open-source model, Stable Diffusion. These statistics prompt a profound reflection on the role and impact of AI in creating and disseminating visual content, highlighting the potential responsibilities, limits, and opportunities of such technologies in contemporary visual culture.

On the user experience side, the interaction with the generative computational model begins by inputting data into the model's interface, facilitating a dual-natured transition: verbal prompts—i.e., textual instructions—that are translated into visual artefacts and visual artefacts that can be translated into textual outputs. This two-way translation process is enabled by the training of algorithms on extensive datasets and the conversion of inputs into embeddings—i.e., representations of complex data like words or images within the dataset's vector space. This process allows both images and verbal text to be translated into numerical data within the vector space of the datasets, a mathematical construct where each item—whether a word, an image, or any other type of data—is represented as a point within a multi-dimensional space. Each dimension in this space corresponds to a feature of the data, which can be a particular attribute or characteristic relevant to the dataset. For example, in the context of visual artefacts, images can be translated into vectors based on various visual features such as colour, texture, or shape, each represented by a dimension in the vector space. When data is represented in this manner, machine learning algorithms can perform complex operations such as classification, clustering, or similarity searches, discerning patterns and relationships within the data by analysing the distances and directions between vectors in the dataset's mathematical space. This vectorial representation forms the basis for converting real-world, complex data into a format that machine learning models can process and learn from.

This initial definition of both the model and the user experience offers an interesting starting point: Text-to-image generative systems not only influence the environment in which they operate due to their pervasiveness and accessibility but are also shaped by the presence or absence of data in the environment. This implies that the algorithms powering these models may learn from datasets that are not free from historical and cultural biases, including legacies of oppressive power structures such as colonialism, classism, and sexism. From this perspective, it is evident that if biases can influence the mindsets of those who design data architectures, they can also be embedded within the data itself. Additionally, the interaction between human and artificial agents warrants further consideration.

The effectiveness of AI generative systems largely depends on the quality of the data they are trained on, highlighting the crucial role of human trainers and "human-in-the-loop" companies. These companies specialise in providing clean, labelled datasets for AI training, which involves humans reviewing and categorising images, videos, or other data, enabling algorithms to learn effectively. This method, known as supervised learning, is essential to ensure that AI systems recognise and respond appropriately to various situations. However, this approach is not without risks, often framed within power structures. A significant concern regarding human-in-the-loop companies is their reliance on cheap and precarious labour, particularly in regions of the Global South [26]. Such practices can lead to inherently biased AI systems perpetuating implicit injustices, mainly if the data originates from cultural or social exploitation contexts. In other words, biases embedded in the data and the architecture can enhance the computational models' ability to perpetuate harm and discrimination autonomously. This is why it is necessary "to understand how machine learning and other algorithms have been embedded with human prejudice and discrimination, not simply at the level of data, but also at the levels of procedure, prediction, and logic" (Chun [5], 16).

In the following pages, these concerns will be explored, considering how, in our rapidly evolving algorithmic society [33], AI-generated imagery presents us with significant implications, necessitating an exploration of the critical dimensions of both epistemological and aesthetic reflections on the datasets through which text-to-image generative models are designed. These issues raise essential questions about the ethics surrounding dataset curation [39] and prompt us to consider what is omitted from datasets and how these omissions can perpetuate inequalities.

Drawing on the lessons from archives [21] and Archive Studies [9, 13, 14], it is possible to assert that every dataset, much like every archive, can be considered an object of knowledge that encompass both apparatuses and discursive formations, embodying all the procedures and effects that a specific field is willing to accept at a given time. It is in this light that datasets for machine learning training can be viewed as archives [38]. Despite continually facing deficiencies, they serve as mediums for transmitting knowledge and behavioural codes to algorithms. Consequently, data cannot exist in isolation, they must always be related to others and are inherently situated and interconnected within their environment. Data are never raw or a direct reflection of reality. Instead, they are the outcomes of processes endowed with their own situated grammar and syntax. As Johanna Drucker [11] suggested, we might consider adopting the term "capta" rather than "data," emphasising that it is always extracted from an environment and consequently shaped by that environment. This conceptual shift arises from a dual process: it is operationalised—transformed into something measurable—and parameterised—structured according to specific parameters that govern the behaviour of the model. But how do they do so?

To understand these processes, it is beneficial to draw upon the concept of *Umwelt*, as proposed by Jakob von Uexküll [35]. Uexküll's theory of *Umwelt* describes the environment, or "world around me" (*Welt um mich*), as perceived and acted upon by an organism. According to Uexküll, the *Umwelt* consists of two interrelated components: the *Merkwelt* and the *Wirkwelt*. The *Merkwelt* represents the world that makes an organism notice or perceives things, guiding its attention and shaping its perceptual realities. In contrast, the *Wirkwelt* refers to the world where the organism can enact changes through its actions, reflecting its capacity to influence and modify its surroundings. Applying Uexküll's framework to our discussion, it is possible to see a parallel in how data and images act as both *Merkwelt* and *Wirkwelt* for humans within digital environments. Data not only directs our perception—highlighting what we should notice or observe (similar to the *Merkwelt*)—but also provides a platform for interaction and modification of the digital environment (akin to the *Wirkwelt*). As we interact with data through our actions, whether by generating, modifying, or analysing it, we continuously redefine the parameters of our digital *Umwelt*. This ongoing process illustrates the mutual definition and redefinition between us and our digital environments, where "data and visuality, space (urban and non-urban), and image define each other." (Parikka [29] 24).

16.3 Situating AI Imagery Through a More-Than-Human Perspective: Exploring the Intersection of Human and Artificial Agents Through Enunciation

In our digital *Umwelt*, thus, as we conceive of databases as archives and data as capta, another issue emerges: relationality. Scholars such as Donna Haraway in her exploration of cyborgs [19], Bruno Latour with the development of Actor-Network Theory [22], Karen Barad through her proposal of an intra-actional relational ontology [1], Anna Tsing advocating for a more-than-human sociality [34], and Timothy Morton in his exposition on hyperobjects [25] have set the path for understanding our reality in relational terms and beyond binarism. Haraway's concept of the cyborg [19] has been pivotal in challenging the implicit Cartesian dualism that pervades Western thinking, firstly paving the way for a broader understanding of our reality that includes a more-than-human perspective. This redefinition has involved transcending dichotomous paradigms upon which Western culture has historically relied, realising intricated webs of subjugating structures, including those between human/machine, identity/otherness, and man/woman, in favour of embracing plurality and sympoietic morphogenesis. Furthermore, embracing relationality requires reconceptualising and visualising diverse forms of interconnectedness, fostering an understanding of interdependence, as Judith Butler [4] discussed, between humanity, the environment, and other natural or artificial life forms. In other words, it entails recognising an inter- and intra-relationality that sustainability binds entities, humans, and non-humans alike, connecting them to one another and the world.

Situating our reflections within this relational environment led us to consider how images created by generative artificial intelligence can "do things"—drawing from the realm of pragmatics—and how they might have a more-than-human impact.

Consider a recent example: on May 23rd, 2023, an AI-generated image depicting a supposed attack on the Pentagon circulated widely. Despite being fake, the image was shared by numerous X accounts, including one posing as *Bloomberg Feed*, leading many to mistake it for an affiliate of the renowned media platform *Bloomberg*, known for its extensive media presence and verified blue checkmark. Within minutes, the image was retweeted by thousands worldwide, including major media outlets like *Russia Today* and the Indian television network *Republic*. The effects of this AI-generated image were palpable, resulting in an unexpected and unprecedented outcome: an 85-point drop in the Dow Jones Industrial Average in just 4 min [31].

The marked impact of this image results from a complex circuit of cognitive feedback, where the fake witness enhances our understanding of the digital environment while shaping our perceptions. Its effectiveness largely stems from the image's ability to replicate the event with hyper-realistic precision, bolstered by the mimetic power of the technology employed. Enhanced by the computing capabilities of the artificial intelligence model and a vast collection of similar images present both in our collective imagination and in the databases used for training algorithms, the image delivers an authentic and readily recognisable experience—akin to a photographic testimony of an actual event. Furthermore, the image's efficacy is improved by the compositional design, reflecting deeply embedded visual patterns in the collective memory of how a terrorist attack appears and lending a sense of rhetorical eloquence due to the verisimilitude it produces.

Moreover, the rapid dissemination of the image, facilitated by the communications network that moulds our digital environment, has significantly amplified its global impact. Additionally, it is possible to affirm that the interaction between its virality and the powerful iconic dimension of what is depicted in the image has overwhelmed any attempts to critically analyse its accuracy, presenting it with almost excessive precision and stifling doubt or scepticism.

Reflecting on all these elements, I might suggest that the data in the databases of generative text-to-image models function similarly to the virtualities of our imagination, realising and visualising certain statements and not others through the implementable data already present in the databases. However, unlike human imagination, computational models subvert the traditional sequence through which knowledge is acquired. Text-to-image models must possess prior knowledge for generating or modifying an image, making the dataset crucial in shaping and defining the understanding of the task [18]. Particularly in machine learning models, the algorithm determines the necessary computations to perform its task optimally. The human agent provides only feedback on the results, enabling the algorithm to self-correct without dictating the precise calculations. Indeed, the dataset's quality will condition the model's ability to learn and perform its task accurately. For instance, if the algorithm has been trained on a dataset comprising images of terrorist attacks at recognisable, well-known locations, it will face significant challenges when tasked with generating images of unlikely scenarios.

Examining similar instances—from images of Pope Francis in oversized attire to depictions of Trump's arrest before a warrant was issued, both in 2023—leads us to conclude that understanding visual culture in our algorithmic society necessitates a reassessment of what we mean by agency, encompassing both human and artificial agents. This requires reevaluating concepts related to capabilities such as doing, creating, witnessing, and, consequently, meaning-making.

To navigate this complex landscape and avoid technophobic narratives or deterministic attitudes, I align with the propositions of Katherine Hayles [20]: inhabit our present towards a sustainable and more-than-human environment necessitates, firstly, recognising an enlargement of our agency through computational media design. This process, termed "biotechnoevolution" by the author, can open avenues for envisioning the realisation of cognitive assemblages, i.e., hybrid processes where the very concept of personhood can be reconsidered within interactive human-computational collectives. As Hayles articulates:

> Cognitive assemblages are pervasive in our contemporary world, from satellite-imaging software to railroad-switching controllers to electrical-grid components and operators. (...) Assuming that only the human participants in these assemblages are capable of meaning-making practices is as erroneous and anthropocentric as believing that humans are the only species in the biosphere capable of creating meaning—a viewpoint that has become not only untenable but dangerously biased in its implicit endorsement of human dominance. (Hayles [20]: 51)

Secondly, as cognitive assemblages emerge as powerful epistemic tools, they enable the transcendence of humans' historical dominance over machines, fostering the development of AI-centered enunciation. Within this perspective, I propose to revitalise semiotics as an analytical lens, providing a framework to understand how artificial agents generate and disseminate meaning in our digital environment through a crucial concept: enunciative practices. To gain an initial understanding of the notion of enunciative practices, we might ask: Why do image-to-text models produce certain types of visual outputs and not others? What are the pragmatics underlying their assemblages?

Broadly defined, enunciation bridges the structural code of language—the *langue* in Saussurean terms—to the discursive acts—the *parole*—through the realisation of an utterance. It transforms language from a virtual, systemic form to an actual, contextualised expression while navigating the intricate interplay between encoded conventions and situational factors. Its practice concerns not only selecting and arranging systemic elements within the language code but also activating them in specific contexts, enabling their circulation within the environment. By doing this, enunciative practice negotiates tensions between norms and individual agency, as well as between codified systems and collective experiences.

Enunciation, in fact, can be viewed as a model for relationality or, more precisely, for delegation [28]: within this model, we can identify delegates who negotiate on our behalf the virtuality of the language into the actual realisation of an utterance, such as an image. This recognition arises from considering and examining the roles of entities and artefacts in scaffolding the environment while mediating and shaping our perception. Therefore, when discussing enunciation and its delegation model, it is crucial to consider not only the patterns actualised in concrete acts but also the potential norms and customs intervening in what Umberto Eco defined as the encyclopedia [12]. Representing the shared body of knowledge of a culture, the encyclopedia contains all norms, behaviours, and customs either accepted or rejected by a community and, due to its multimedia format and rhizomatic nature, it interconnects every element present in its body by shaping future enunciations while also being moulded by past ones.

Within this framework, considering AI-centered enunciation from an encyclopedic perspective can be crucial for advancing our understanding of how AI-generated imagery can perpetuate existing inequities and challenge pervasive biases in our environment. This involves providing an analytical framework for understanding how enunciative practices, particularly the generation of images through text-to-image models, are reflected in the images themselves. However, addressing this peculiar type of image creation requires a consideration of the relationality between databases and the pragmatics of image generation related to the computational models. For this reason, I am interested in how generative models transform cognitive assemblages [20] into entities capable of delegating a message through enunciative practices. In this regard, I sustain that the data archived in datasets profoundly influences the production of AI-generated images, reflecting collective memory and societal biases, thus playing a central role in shaping utterances akin to an encyclopedia's influence on knowledge production. In other words, by adopting a computational media perspective on AI-centered enunciation that recognises cognitive assemblages as capable of meaning-making beyond anthropocentrism, I suggest exploring what enunciation can do within text-to-image models.

16.4 Decolonising AI Imagery: The Artistic Practice, the Presences and the Omissions

In the preceding sections, this chapter examined the performative and interactive nature of our engagement with computational models in a data-centric environment. These interactions dynamically sculpt our environment through feedback loops, influencing our perceptions, experiences, and relationships with technology and our surroundings.

Building on a decolonial perspective and the conceptual framework of critical AI studies, this final section explores the interplay between enunciation, databasing, and data-worlding within artistic practices. It seeks to demonstrate how art challenges traditional boundaries and fosters new understandings in the realm of data interaction. As previously discussed, AI often perpetuates legacies of inequality through mechanisms that shape their architectures such as proxies and logical correlations: "Correlations produce proxies. In statistics and economics, most proxies correspond linearly and associate with hidden or unknown variables. Effectively acting as stand-ins or surrogates, proxies reveal protected categories such as race and gender in seemingly color-blind or agent-based categories such as zip code or "age at first arrest" (Chun [5], 121). Therefore, it is critical to examine the underlying assumptions and axioms that influence algorithm design and data structure. In computational models, especially those utilising deep learning technologies like machine learning, algorithms autonomously determine the calculations necessary for optimal performance. Researchers and engineers enhance this process by providing feedback on outcomes, thus enabling algorithms to self-correct without specifying exact calculations. Consequently, the quality of the dataset is a fundamental factor in the model's ability to learn and execute tasks accurately. To combat discrimination and promote a more equitable future, we must critically interpret AI-generated imagery outcomes, viewing them through situated epistemological insights.

In our algorithmic society, in fact, we are witnessing a noticeable shift in the conceptualisation of knowledge from a traditional focus on understanding the past to a future-oriented prediction model. This paradigm shift, moving from a causation-based worldview to one focused predominantly on prediction and correlation coefficients, is becoming increasingly significant and underscores an urgent need to address the effects of embedded predictive logic in perpetuating harm and discrimination and to explore how these can be transformed into tools for an alternative, more inclusive imaginary.

With these reflections in mind, I propose exploring the potential of AI-generated imagery to collaborate in constructing narratives that challenge existing paradigms and cultivate a community sense from a more-than-human perspective. I aim to advance the idea of AI imagery as a resource capable of promoting a decolonial stance, critically examining its role in shaping a future where data is actively engaged in creating more inclusive and equitable narratives rather than merely being extracted and consumed. Acknowledging the parallels between colonial practices in both territorial and structural forms in the digital era, I advocate for integrating decolonial theory into the realm of digital technologies. Like their physical counterparts, digital environments are prone to exploitation and extraction, leading to what may be termed digital-territorial coloniality [24]. This form of coloniality also manifests through enduring power dynamics and the exploitation of common resources [6], influencing socio-cultural norms, knowledge paradigms, and technological development. These frameworks, rooted in historical inequalities and oppressive structures, often go unchallenged today. Given that AI technologies are inherently influenced by these legacies recognising this influence is crucial for applying critical theories to understand and shape a more equitable future. To this end, I aim to promote a critical reflection on the pragmatics of AI, fostering reciprocal engagements between human and artificial agents through affective and political commons.

As we transform fragments of reality into data, their design often results in tangible alterations. Concurrently, as we introduce new relational dynamics, we redefine the essence of databases, thereby influencing the collective imagination. Databases, akin to archives, are deeply intertwined with governance, visibility, policy formulation, knowledge construction, aesthetics, and everyday life experiences. Therefore, navigating this intricate and expansive network of data that characterises our present requires a profound understanding of the complex relationships between data and their environment, between data and the artefacts they generate, between data and the individuals who use them to perceive reality, and the methods of interaction with it.

Bearing all these issues in mind and building on the research of Maria Giulia Dondero [10], I will explore how the praxis of enunciation can be operationalised in AI-generative models through diverse modes of existence, considering the logics of data presentification or omission. From this perspective, databases used for training AI models can be viewed as modes for virtuality because their data always embody both redundancy and omissions within a culture. Moreover, when interacting with a text-to-image model, what we initiate through the prompt can be seen as an actualisation of those presences or absences within the narratives circulating in a community, which can then be realised through the generated images. Furthermore, these images, like the prompts, have the potential to allow the automated system to present what is absent or omit what is redundant while also allowing the glitches or the noises of the machine to be articulated through their own enunciation. In this regard, AI-generated images—namely, the visual utterances generated through our prompts within the interaction with the model—might potentially address and amend the gaps in the datasets while curing historical inequalities. In the next section, thus, through the analysis of the artistic project *Un archivo inexistente* (A Nonexistent Archive) by Felipe Rivas San Martín (2023-on going), I will explore the practice of generating a nonexistent archive about queer worker couples at the beginning of the nineteenth century in Latin America as a decolonial practice for addressing and rectifying, or maybe better to say curing and caring, data omissions beyond heteronormativity.

16.5 Queering AI Imagery Through Enunciative Practice: Love in the Era of Decolonial Retrofuturism

The model employed is Stable Diffusion 2.0, an open-source technology used to create a prompt for generating a nonexistent queer archive: a collection of images depicting love among working-class men from Latin America between the years 1900 and 1920, and subsequently, images of women and queer couples. The artist is Felipe Rivas San Martín, who, describing his practice, introduced the concept of the "minority prompt" to define the enunciative practice at the base of his artistic project, which can be understood as a translation from textual inputs to visual artefacts aimed at counteracting the inherent biases in the databases used to train generative models. By employing this method, *Un archivo inexistente* seeks to expand the generative computational model's imaginary and expose it to a diverse range of artefactual derivations, thereby mitigating the tendency to replicate existing biases such as the predominant representation of white, cisgender heterosexual European bourgeois males. By representing individuals typically excluded from databases due to their socioeconomic status, sexual orientation, and regional background, the enunciative praxis transforms the virtuality of the database into the realisation of a decolonial more-than-human artistic gesture through the actualisation of visual data, where the agency of the model itself is exposed to make the digital materiality a robust support of a decolonial imaginary.

This archive of queer nonexistence delves into speculative forms of love that transcend the predictable outcomes of traditional narratives by articulating, through a variety of visual expressions, the multiple intersecting discriminations that have historically obscured the lives and memories of the depicted individuals. This collection of images, invoking a speculative indexicality, becomes iconic in symbolizing the power dynamics that have historically governed—and continue to influence—expressions of homosexual or queer affection. These dynamics have limited the visibility and representation of the working class, especially outside of heteronormative and capitalist reproductive frameworks. Through an intersectional analysis, thus, this archive documents the erasure of queer existences under colonial exploitation in Latin America, providing a visual testimony to the oppressions that have been endured, overlooked, and erased.

Heteronormative culture has historically erased queer experiences [7, 15, 27], especially within colonial legacies for those from the Global South and working classes. Felipe Rivas San Martín thus engages in a critical exploration of AI through a retrofuturistic aesthetic within a decolonial lens, merging temporalities to address the omission of specific experiences using computational algorithms to reimagine local queer pasts and reclaim an archive that could not previously exist. This collection of images, affectively potent yet revealing the queer imperfections of an evolving technology, not only challenges but also reimagines historical narratives. Furthermore, the archive operationalises the praxis of enunciation in AI-generated models, considering the logic of data presentation and omission. Databases for training AI models are thus viewed as virtual modes, inherently containing both redundancy and omissions that reflect cultural biases. The experience of sexual dissent in Latin America, characterised by violence, precariousness, and insecurity, has been structurally inhibited by hegemonic heteronormative norms, which historically silenced legitimate activist voices. The project aims to fill this archival void by creating a repository that validates and brings visibility to these suppressed experiences. Interacting with a text-to-image model, Rivas San Martín actualises the absence of narratives within queer Latin American contexts, specifically those of love beyond the reproductive logic of capitalism. The generated images not only present what has been omitted but also challenge colonial exploitation narratives. Notably in this regard, these images do not simply replicate retrofuturistic realities, instead, they use the

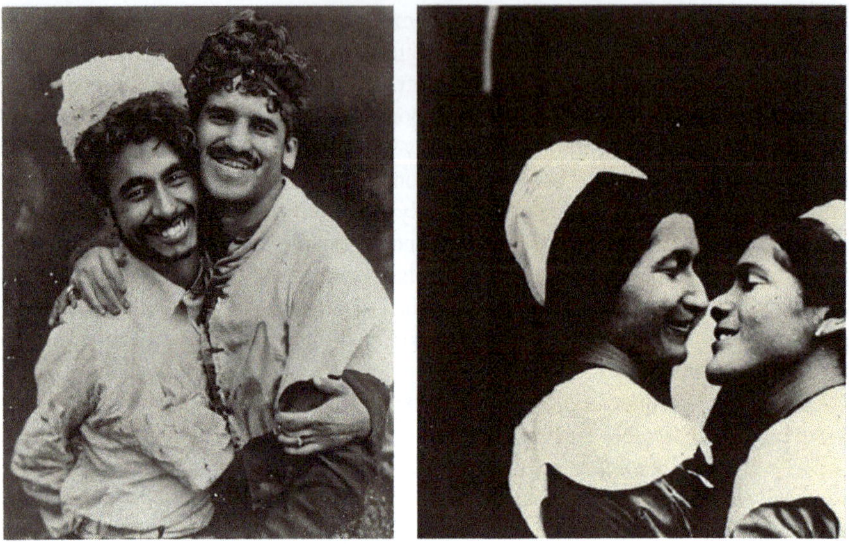

Fig. 16.1 *Un archivo inexistente*, Felipe Rivas San Martin, courtesy of the artist

glitches and errors of the AI process as a form of bodily dissent. This approach counters the notion of replacing a violent past with a fictitious narrative, instead offering radical evidence that such images of the past could not have existed without AI imagery. Contrary to the general idea of photography as an indexical trace of reality, these images from the nonexistent archive assert that "this could not have been—except through an alliance with AI," challenging conventional archival practices and prompting a reevaluation of historical narratives through the lens of modern AI technologies (Fig. 16.1).

16.6 Concluding Reflections on Shaping an Equitable AI Future

In this paper, I have explored the impact of AI-generated imagery, illustrating how such technologies both perpetuate existing inequalities and offer avenues for challenging entrenched biases. The analysis, by reflecting text-to-image models, underscores the importance of a critical approach to AI, particularly in recognising and mitigating the biases embedded in its algorithms and data structures. In this regard, I have discussed how AI can replicate and reinforce social hierarchies unless deliberately calibrated to counteract these tendencies.

Through the theoretical frameworks presented, I have advocated for the potential of AI-generated imagery to transcend traditional barriers, advocating for a decolonial and gender-inclusive approach to technology design and implementation. This involves rethinking the relational dynamics between humans and machines, recognising the agency of AI not just as a tool but as a participant in shaping cultural and social narratives. This shift requires a fundamental reconsideration of how we engage with technology, urging a move away from seeing it as merely functional towards understanding its role as an active creator of reality, capable of influencing societal norms and values and, more in general, our environment. Furthermore, I delved into the analysis of the artistic project *Un archivo inexistente* by Felipe Rivas San Martín, highlighting the transformative power of AI in visualising and giving presence to historically marginalised by reflecting on possible differential pragmatics of datasets. By operationalising the praxis of enunciation in AI models, the project demonstrates the capacity of text-to-image models to not only represent but actively challenge the historical erasures and invisibilities imposed by colonial and heteronormative frameworks.

In conclusion, the journey towards a more equitable AI necessitates a collaborative approach to technologies incorporating a more-than-human perspective. As we continue to navigate our algorithmic society, the principles of intersectionality, decoloniality, and inclusivity must guide our path forward, ensuring that AI serves as a tool for social justice and equity rather than a mere instrument. This approach not only enriches our understanding of AI's potential but also helps forge a future where technology genuinely contributes to the common good, creating spaces for representation and dialogue that have long been denied.

References

1. Barad, K.: Meeting the Universe Halfway: Quantum Physics and the Entanglement of Matter and Meaning. Duke University Press, Durham (2007)
2. Buolamwini, J., Gerbu, T.: Gender shades: Intersectional accuracy disparities in commercial gender classification. Proc. Mach. Learn. Res. **81**, 77–91 (2018)
3. Brubaker, J.R., Paul, J.M., Scheuerman, M.K.: How computers see gender: an evaluation of gender classification in commercial facial analysis and image labeling services. Proc. ACM Hum.-Comput. Interact. **3**, CSCW, Article 144 (2019). https://doi.org/10.1145/3359246 (Accessed 12 May 2024)
4. Butler, J.: Notes Toward a Performative Theory of Assembly. Harvard University Press, Cambridge, MA (2015)
5. Chun, W.H.K.: Discriminating Data: Correlation, Neighborhoods, and the New Politics of Recognition. MIT Press, Cambridge, MA (2021)
6. Crawford, K.: Atlas of AI: Power, Politics, and the Planetary Costs of Artificial Intelligence. Yale University Press, New Haven (2021)
7. Cvetkovich, A.: An Archive of Feelings: Trauma, Sexuality, and Lesbian Public Cultures. Duke University Press, Durham (2003)
8. Dastin, J.: Amazon scraps secret AI recruiting tool that showed bias against women. In: Martin, K. (ed.) Ethics of Data and Analytics: Concepts and Cases. Auerbach Publications, New York (2022)
9. Derrida, J.: Mal d'Archive: Une Impression Freudienne. Éditions Galilée, Paris (1995)
10. Dondero, M.G.: Introduction à une sémiotique de l'énonciation machinique. L'image analysée et produite via l'IA. In: Paper presented at the Séminaire International de Sémiotique à Paris, Fondation de la Maison des Sciences de l'Homme, 15 November 2023 (2023)

11. Drucker, J.: Visualization and Interpretation: Humanistic Approaches to Display. MIT Press, Cambridge, MA (2020)
12. Eco, U.: Semiotics and the Philosophy of Language. Midland Book Edition, Bloomington (1986)
13. Farge, A.: Le goût de l'archive. Éditions du Seuil, Paris (1989)
14. Foucault, M.: L'archéologie du savoir. Gallimard, Paris (1969)
15. Foucault, M.: Histoire de la sexualité I. La volonté de savoir. Gallimard, Paris (1976)
16. Garcia, M.: Racist in the machine: The disturbing implications of algorithmic bias. World Policy J. **33**(4), 111–117 (2017)
17. Gmyrek, P., Berg, J., Bescond, D.: Generative AI and jobs: a global analysis of potential effects on job quantity and quality. ILO Working Paper 96, International Labour Office, Geneva (2023)
18. Gramigna, R., Voto, C.: Notes on the semiotics of face recognition. Sign Systems Studies **49**(3–4), 338–360 (2021). https://doi.org/10.12697/SSS.2021.49.3-4.05 (Accessed 12 May 2024)
19. Haraway, D.: Simians, Cyborgs and Women: The Reinvention of Nature. Routledge, New York (1985)
20. Hayles, N.K.: Can computers create meanings? A cyber/bio/semiotic perspective. Crit. Inq. **46**(1), 32–55 (2019)
21. Jo, E.S., Gebru, T.: Lessons from archives: strategies for collecting sociocultural data in machine learning. In: Conference on Fairness, Accountability, and Transparency (FAT*'20), Barcelona, Spain, 27–30 January 2020, ACM, New York (2020). https://doi.org/10.1145/3351095.3372829 (Accessed 12 May 2024)
22. Latour, B.: Reassembling the Social: An Introduction to Actor-Network Theory. Oxford University Press, Oxford (2005)
23. Manovich, L., Arielli, E.: Artificial Aesthetics: Generative AI, Art, and Visual Media (2021). http://manovich.net/index.php/projects/artificial-aesthetics-book (Accessed 12 May 2024)
24. Mohamed, S., Png, M.T., Isaac, W.: Decolonial AI: decolonial theory as sociotechnical foresight in artificial intelligence. Philos. Technol. **33**, 659–684 (2020). https://doi.org/10.1007/s13347-020-00405-8. Accessed 12 May 2024
25. Morton, T.: Hyperobjects: Philosophy and Ecology After the End of the World. University of Minnesota Press, Minneapolis (2013)
26. Munn, L.: Digital labor, platforms, and AI. In: Werthner, H. et al. (eds.) Introduction to digital humanism. Springer, Cham (2024). https://doi.org/10.1007/978-3-031-45304-5_35
27. Muñoz, J.E.: Cruising Utopia: The Then and There of Queer Futurity. New York University Press, New York (2009)
28. Paolucci, C.: Cognitive Semiotics: Integrating Signs, Minds. Meaning and Cognition. Springer, Berlin (2021)
29. Parikka, J.: Operational Images: From the Visual to the Unvisual. University of Minnesota Press, Minneapolis (2023)
30. Raza, M.M., Venkatesh, K.P., Kvedar, J.C.: Generative AI and large language models in health care: pathways to implementation. npj Digital Med. **7**, 62 (2024). https://doi.org/10.1038/s41746-023-00988-4 (Accessed 12 May 2024)
31. Ross Sorkin, A., Warner, B., Kessler, S., de la Merced, M. J., Hirsch, L., Livni, E.: An AI-generated spoof rattles the markets. The New York Times, May (2023). https://www.nytimes.com/2023/05/23/business/ai-picture-stock-market.html (Accessed 12 May 2024)
32. Scheuerman, M.K., Paul, J.M., Brubaker, J.R.: How computers see gender: an evaluation of gender classification in commercial facial analysis services. Proc. ACM Hum.-Comput. Interact. **3**, CSCW, Article 144, (2019). https://doi.org/10.1145/3359246 (Accessed 12 May 2024)
33. Schuilenburg, M., Peeters, R.: The Algorithmic Society: Technology, Power, and Knowledge. Routledge, New York (2020)
34. Tsing, A.: More-than-human sociality: A call for critical description. In: Hastrup, K. (ed.) Anthropology and Nature, pp. 27–42. Routledge, New York (2013)
35. von Uexküll, J.: A Foray into the Worlds of Animals and Humans (1934). University of Minnesota Press, Minneapolis-London (2010)

36. Valyaeva, A.: AI has already created as many images as photographers have taken in 150 years. Everypixel J. (2023). https://journal.everypixel.com/ai-image-statistics (Accessed 12 May 2024)
37. Vigdor, N.: Apple card investigated after gender discrimination complaints. The New York Times, November 2019. https://www.nytimes.com/2019/11/10/business/Apple-credit-card-investigation.html (Accessed 12 May 2024)
38. Voto, C.: From archive to dataset: Visualizing the latency of facial big data. Punctum: Int. J. Semiotics **8**(1), 47–62 (2021). https://doi.org/10.18680/hss.2022.0004 (Accessed 12 May 2024)
39. Voto, C.: Life is AI cabaret: Queering alterity through dataset curation. In: Zingale, S. (ed.) Design Meets Alterity: Case Studies, Project Experiences, Communication Criticism. Franco Angeli, Milano (2024)
40. Yusuf, A., Pervin, N., Román-González, M.: Generative AI and the future of higher education: A threat to academic integrity or reformation? Evidence from multicultural perspectives. Int. J. Educ. Technol. High. Educ. **21**(21) (2024). https://doi.org/10.1186/s41239-024-00453-6 (Accessed 12 May 2024)

Cristina Voto is assistant professor at the University of Turin, where she teaches Design Languages, Semiotics of Cultural Heritage: Perspectives on Intersectionality, Ethics and AI, and Digital Gender Studies. She is a member of the doctoral board for the *Design and Creation* program at Universidad de Caldas and the *Art and Techno-Aesthetics* program at Universidad Nacional de Tres de Febrero. She currently serves as Vice President of the Latin American Federation of Semiotics. She has collaborated with various universities and artistic institutions, including Vrije Universiteit Amsterdam, Universidad Autónoma de Madrid, the University of the West of England, Universidad de Buenos Aires, Jorge Tadeo Lozano University, Universidad Nacional de Colombia, the *Bienal de la Imagen en Movimiento*, and the *Festival de la Imagen*. Her research focuses on visual semiotics at the intersection of philosophy of technology, feminist and queer theory, and digital and AI-based art. She is the author of *Monstruos Audiovisuales. Agentividad, movimiento y morfología* (2021) and co-editor of *Rostrotopías. Mitos, narrativas y obsesiones de las plataformas digitales* (2023); *I cronotopi del volto* (2022); *Rostrosferas de América Latina. Culturas, traducicones* (2021).

Chapter 17
Authorship in Algorithmic Images

Andrea Valle

Abstract Digitisation is a technical operation that brings the image back into the wider range of computable representations. This operation can be described as an "allographisation", that is, as a form of explicit notation of the image. The contribution aims at investigating if and how this technical feature impacts on the status of the image itself. In particular, it discusses its semiotic effects in relation to production practices. A crucial test bed is thus provided by the notion of authorship, as new forms of subjective mediation emerge starting from the format, through algorithmic generation and through technologies that modify the creative practices, thus redefining the role of the author. In the contribution, the authorial issue in digital images is addressed semiotically by taking into account three axes: enunciation levels, subject modalisations, and modes of sign production. By means of these notions, and their intertwining, it becomes possible to tackle the complexity of the authorship scenario in digital media.

Keywords Authorship · Allography · Modes of sign production · Enunciation · Modalisation

17.1 Introduction

The problem of authorship in digital and generative images requires a more general definition of authorship. What is an author? Is s/he important? The assumption from which we will start is that an author is a form of subjectivity, a principle of instantiation. As a starting point, we can observe how two very different perspectives converge on this topic, which we will define as genetic versus semiotic.[1] We can discuss them starting from the type of texts we will take into account, that is, images.

---

[1] Rather than "semiotic," a typical adjective used in opposition to "genetic" is "generative," but, as we will use extensively the latter in a different context, here we prefer "semiotic."

A. Valle (✉)
University of Turin, Turin, Italy
e-mail: andrea.valle@unito.it

P. Conte et al. (eds.), *Algomedia. The Image at the Time of Artificial Intelligence*,
Lecture Notes in Morphogenesis, https://doi.org/10.1007/978-3-032-08726-3_17

A genetic perspective is inherently historical. It tries to trace back the image to an instantiating gesture which is responsible for a sequence of facts, that is, events distributed in a chronological time. This sequence leads causally to the image. A semiotic perspective, on the other hand, moves from the image to verify the way in which the author presents him/herself within the image itself. For a genetic approach the text is the result of a set of production processes; for the semiotic one it is rather the starting point from which to reconstruct them. This opposite orientation also establishes a relation of determinant/determined. Genetically, production is determinant, and in fact it can lead to a restructuring of the text (collation, critical edition). From a semiotic point of view, the text is determinant and the production is determined, and in fact the latter is the result of a set of reconstruction processes starting from the text itself. The basic assumption in semiotics is that every text presents itself as a discourse, that is, it carries within itself the traces of the act of communication that instantiated it. In semiotic studies, this perspective is described by the so-called theory of enunciation [15, 18]. The theory of enunciation presupposes a recursive functioning of communication: the latter is an operation that produces objects (utterances or "enunciates") which in turn, within them, bear traces of the communication itself (enunciation). In a similar way, Eco [10] proposed to investigate the production starting from the text by taking into consideration the knowledge necessary to access the textual information (model reader) and the information that reveals the text as a communication (model author). A clear example is that of a message in a bottle. Once the bottle—its origin being unknown since it was deposited on a beach coming from who knows where—is opened, all we can know about the author of the contained message is in the message itself.

We can summarise quickly and simply like this. A genetic theory of the author consists in reconstructing the history of the production of the text-image independent of the text itself: the text is in fact the final end of a causal chain. A semiotic theory of the author consists in reconstructing the production of the text-image independently of the production history: the text is the starting point of a chain, which could be defined as final or teleological, because we start from the text to define a potential communication.

This dichotomous vision is largely incomplete and oversimplified. For example, philology must reconstruct genetic relations largely by taking into account stylistic elements (that are indeed emerging from the text); on the other hand every semiotic approach is linked to a historical dimension, since the text is a cultural object, and for example the convocation of the knowledge of a model author/reader can only be historical, however minimal. Yet, this opposition captures a kind of prevailing relevance, two maximal epistemological poles.[2]

The complexity of the situation is evident if we consider the problem of attribution in art history: here, the issue is precisely that of indicating an author for a given image. In this context, a widely used term is "author's creation" [5, 8]. In other words, in the history of art, since the work of Cavalcaselle and Morelli (later Berenson, Friedländer,

[2] On this issue, in the context of semiotics, see the discussion by Dondero [6] problematising the status of photography in relation to the classic work by Floch [11].

etc.), the historian, starting from one or more works of art, literally constructs an author. If we consider the practice of attribution, we could indicate three elements that characterise an author[3]:

1. it is a single instance (even if it can be empirically collective: school, workshop, collective, brand);
2. the encyclopaedic minimum definition is made up of a name, which is often created with a real act of baptism (Maestro delle Vele, Amico di Sandro), a place (Chauvet, Antwerp, Central Italy) and a period (30,000 BC, fourth century, 1520–1530);
3. it must be responsible for a set of productions, which can have cardinality ≥ 1 (if the cardinality is 1, then there is a bijection between author and work).

This minimalist definition underlines the semiotic dimension of the operation of attribution. In particular, it presupposes a form of integrative subjectivity (1), which is defined on the basis of three elements typically detected by the theory of enunciation (person, place, time: 2), and which refers to a corpus of texts (3).

In the history of art an author is constructed on the basis of a set of other texts [8]:

1. historical sources: documents, letters, certificates of ownership etc., recognisable iconographic elements (buildings, clothes, etc.);
2. diagnostic elements: analysis of materials, IIR, X-Ray, dendrochronology;
3. style: a set of recurring and homogeneous features that are shared in a corpus. If the text is one of a kind, then these features at least distinguish the *hapax legomenon* from all other productions (it follows that its author is described in terms of difference from all other authors).

The practice of attribution therefore clearly brings out the potential continuity between genetic and semiotic perspectives. In particular, the notion of style evidently refers to a set of elements that emerge from the text-image. It should be noted that an implicit assumption concerns homogeneity. An author is thought of as a homogeneous and persistent set of stylistic characteristics that are distributed iteratively throughout the text. In semiotic terms, authorship is therefore an "isotopy" [18]. An isotopy is the redundant and unifying distribution of elements of meaning in a text that allows us to define a unitary relevance. The author, as a homogeneous feature that integrates the work, is therefore an isotopy, a plane of meaning, linked to the production of the image itself.

Therefore, if we were to ultimately distinguish a genetic perspective and a semiotic one, it would seem that the key element lies in the open versus closed category. In a genetic perspective, the corpus of texts that can be called for the definition of the author is open (and for example new instrumental tests can be decisive), in a semiotic one the text is closed, as it is assumed as the perimeter of the analysis.

Yet, the practice of attribution underlines how this polarity defines a continuum useful for capturing the way in which the meaning of images as a whole is constructed. Precisely the focus on technical operations, which constitutes a key point in the

[3] We will use "it" to indicate the author in our perspective.

practices of attribution, proves interesting for the description of what happens in the digital image from a semiotic perspective, that one that we will pursue in the following. Furthermore, if the image—as we will see—is generated by a software rather than produced by a human author, the issue of "who is doing what" takes a complex turn, because the authorial subject seems to drift among various places.

The authorial issue in digital images can be addressed semiotically by taking into account three axes: enunciation levels, subject modalisations, modes of sign production.

17.2 Enunciation Levels

The first axis that we will introduce to complexify the discussion on the authorial perspective in digital/algorithmic images takes into account the multilevelarity of enunciation.

The textual level on which semiotics traditionally focuses risks removing a set of heterogeneous elements that a genetic perspective instead aims to include. On the one hand, this level focuses on a well defined relevance, but on the other hand it prevents us from considering, still in semiotic terms, (that is, in terms of a communication oriented to and from value), many other aspects that are normally part of cultural circulation. For example, if an altarpiece is traced back exclusively to the image it contains, the meaning related to its being an object positioned in a specific place (above the main altar of the church, thus defining an equally specific regime of visibility for the audience), cannot be properly discussed.

In relation to this theoretical necessity, Fontanille [13, 14] has proposed a hierarchy of expressive levels, in six levels of immanence:

(a) Signs and figures
(b) Text-utterances
(c) Objects and supports
(d) Practices and scenes
(e) Situations and strategies
(f) Life forms

Each level defines its own relevance, is integrated into the next one and defines a specific form of experience, that is:

(a) Figurativity
(b) Interpretation
(c) Corporeity
(d) Practice
(e) Conjuncture
(f) Ethos and behaviour

Fontanille can thus distinguish two modes of analysis: *mise à plat* and *mise en relief*. In the first case we focus on the single level, analysing in a "distributional" way the elements that are relevant at that level. In the second case, says Fontanille, we are actually dealing with a "generative" operation that takes into consideration the integration of multiple levels.

This multilevel organisation is interesting because, for example, it allows us to think coherently about the image as a text-utterance (e.g. *Adoration of the Magi*, b) composed of specific visual and stylistic features (colors, brushstrokes, the so-called plastic dimension, a), however potentially linked to an object (the painting in its corporeal materiality and its relationship with the user's body, for example the dimensions of the physical canvas on wood, c). But we can also describe "anthropologically" the practice of pictorial production (d), and the communication circuits (e) of the object compared to a framework of values defined at the more general socio-cultural level (f).

Each expressive level then defines a different kind of production subjectivity.

For sake of simplicity, we can limit ourselves to the first four hierarchical levels and name this subset STOP (Signs, Texts, Objects, Practices).

1. Signs: the elements defined at this level are those that can be recognised as minimal visual units. This level pertains to the perceptual dimension, it is variable and unstable, it depends on an initial figurative grasp, that is, on a relevance that minimally integrates the visual features into a set of recognisable figures [9, 18], see below for a discussion on recognition),
2. Text: here an integration step holds together the elements of the previous level by defining syntactic relations according to semantics. The operation is in fact the interpretation of a text as a whole, i.e. as a message of a communication (utterance by an enunciator). Signs/figures that are detectable at the previous level are integrated as elements of a larger whole, that is conceivable as the result of an act of enunciation;
3. Object: but the text can be contained on a level that pertains multimodally to material corporeity. The text of the message in a bottle is written on a piece of paper inserted into the bottle, the Annunciation is contained on a canvas of a certain size. The perspective here is both ecological and biosemiotic. The body is the interface with an environment;
4. Practice: an object results from a set of material relationships that are part of anthropologically defined actions, thus specifying a production syntax. The syntax here pertains to the way bodies interact through mutual actions.

If our theme is authorship as a subjectivity instantiating the image, then we could find four terms that indicate how the latter is located at the four levels of STOP. In other terms, if we assume that authorship is an implied instance of production, then different forms of authorship can be described. Starting from levels and operations, we will then have four figures (i.e. hyponyms) of a more abstract (i.e. hypernym) Producer: a Figurator, an Interpreter, a Body, a Practitioner.

17.3 Image (and) Digitisation

If we consider the perspective of techniques [16], we can discuss whether or not a digital format introduces significant differences in the semiotic status of images. Digitisation in itself works on the representation format. We could say that digitisation as a technical operation produces a potential "allographisation".

The distinction between allography and autography has been proposed by Goodman [2, 17] to characterise a polarity of semiotic regimes. In allography a content system is formally defined a priori, and the expression is autonomous from it. In other words, the form of the content is autonomous from the form of expression. Thus, the expression is properly a notation of the content. Borrowing from artistic practices, the reference example for Goodman is that of musical notation. In Western music practice, the work is defined by the written music notation of the score, from which an indefinite set of occurrences is created by iterated performances: however, those occurrences still refer to the score. The same situation occurs in alphabetic writing: a set of abstract types (characters) allows the production of indefinite occurrences (one can think of the variety of handwriting). On the contrary, in autography there is no formalisation of the content, and therefore of the expression. It is not possible to clearly, definitively, distinguish between expression and content. The term "autography" refers precisely to the concept of autograph, that is, an idiolectal writing, which defines a specific relationship between the subject of production and the artwork (the typical examples being drawing or painting). The autograph is singular in that it can be attributed to a specific instance of production. So, painting (like drawing) is evidently and historically autograph.

Digitisation as a technical operation describes each image in an allographic form. In fact, the image now has a notation: it is a map of pixels each of which is represented by discrete values (typically, a RGB triple, encoding respectively the red, green and blue components).[4] Moreover, image digitisation produces homogenisation. All digital images share the same allographic organisation. This allographic homogenisation implies that visual objects of different origins may collapse into a unified digital format: thus, photography in a digital format loses its specificity, because, although it maintains the properties linked to the instantiation of the image, at the same time it shares the allographic assumption of all digital images, for example the manipulability proper to the representation of the digital (allographic) format (that allows to make computations on representations).

Digitisation represents only a potential semiotic condition. One can ask whether or not this change produces effects with respect to the hierarchy of levels that we introduced, and with the authorial figures that result from it. Does this change in representation format impact on the semiotic status? Not necessarily. As an example, the reproduction of an analogue image in digital format does not introduce significant differences. It could easily be observed that there is no particular relevance of the digital format in the reproduction of an image (e.g. a painting or a picture) from the

[4] A second digital format is "vectorial" and describes the visual content (still allographically) in terms of lines between points. Here we will focus on the representation by map of pixels ("raster").

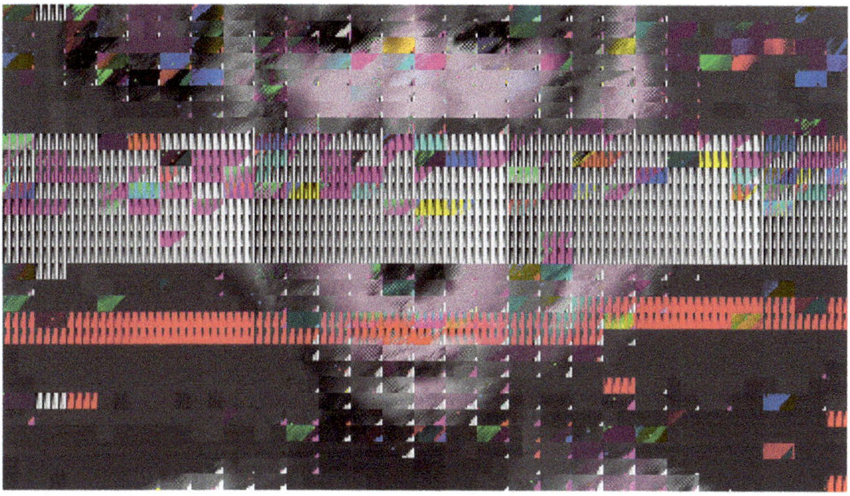

Fig. 17.1 R. Menkman, BLINX2

point of view of the content of the image, so there is no relevance of digitisation at the S and T level.

We could say that in this case the potential digital Figurator subject is concealed in favour of a readability of the image by a "classic" Interpreter. A digital photograph, without disturbances, in other words can make the S level opaque, setting up an Interpreter that coincides with that of the image as a text.

Yet, at level S, the Figurator, responsible for the minimal units of recognition, can emerge in the discrete combinatorics of the elements and in its perturbation. First of all, resolution is a technical dimension that, if too low, may bring to the surface S the digital construction of the signs (that is, pixels become visible). This constructive level may be explored explicitly. In so-called "glitch" art this minimal subject, responsible for sign construction, typically shows its presence in relation to a "standard" situation, so that it is revealed as the one responsible for the errors: be them the introduction of visual noise, artifacts, interferences. This presence of a Figurator evidently forces, by its becoming visible, to define a new meaning at level T, in which an Interpreter emerges as the subject of a different vision, explicitly mediated at a technical level, typically linked to a form which is at the same time a distinct feature of digital images and a distortion of photography.

Figure 17.1 shows BLINX2[5] by Rosa Menkman.[6] Here, the digital construction manifests itself at level S in terms of evident pixelation, as a starting condition for creating figures that allow the recognition of the image while being provided with autonomy (in terms of new geometric and chromatic features).

[5] https://www.flickr.com/photos/r00s/5981939039

[6] https://beyondresolution.info/

Fig. 17.2 Characters as images

In the field of generative AI, the homogenisation made available by the digital format emerges in the relation between image and written text. In generative AI systems aimed at images, written text typically looses its own, specific allographic regime (the type/occurrence between a character and its variable occurrences). Therefore, and in some sense paradoxically, the new allographic regime of the image produces a regime shift in writing, as the latter is treated as a component of the image. A "machinic" Figurator manifests itself precisely in this sort of *mise à plat* that results in a uniform notation for all that is visible, regardless of the distinction between written text and image. This Figurator shows up in the semiotic homogenisation of the heterogeneous starting point (image vs written text), by providing a image-based hyper-regularisation. This can be seen in the AI generated cartographic map in Fig. 17.2 [7] in which the text is often processed as an image.

But it should also be noted that the digitisation of the image can modify the perceptual status of the image, thus impacting on the level O, that of the object. In an analogue domain, be it oil painting, engraving, printing etc., the visible dimension is just one of the textual elements, even if it is indeed the predominant one. However, there is still a significant role of the "objectual" dimension that may be inspected. Thus, there are material marks, e.g. the pigment on the canvas or the grain of the paper, that depend on the material nature of the support. For example, in attribution processes, the image is traced back to a broader configuration that includes the paint, the support, and since the 1950s the preparatory drawing which is literally a subtext (the so called "underdrawing"), as it is accessible only via IRR, being

[7] William McEvoy, "A highly accurate map. I think.", https://www.facebook.com/photo/?fbid=10160144765861218&set=g.1208731756401972

literally placed under the painting [8]. If we consider a digital version of such a painting, its digitisation instead operates through an ancillary step, the photography of the image itself, which is in turn digitised. A digital image is encoded in terms of pixels representing colour units, thus the former has a purely "scopic" nature. If we consider Fontanille's level hierarchy, then corporeality, as the proper operation of level O, is reduced to visibility. The Body of the digital image is compressed into an eye, typically accessing the image-object through a peculiar light condition, irradiated from a monitor rather than projected (as McLuhan [20] already noted when discussing the difference between television and cinema in relation to cathode ray tube vs screen).

This feature of digitisation also impacts the S level, bringing it back to pure visibility, i.e. to chromatic unity. While this purity may not have disruptive effects at the T level (the image as a text for an Interpreter), it certainly modifies the O level in a radical way: as we have discussed, a digital image, meant as an object, looses all the multimodal marks typical of analogue materiality to present itself as eminently visible. One could observe how this feature is also common to photography. Yet, the digital image, while sharing a scopic dominant with photography, adds a notation for those same visible features.

Finally, if we take into account the P level (practices), digitisation potentially allows an indefinite reproduction of the image. If we consider the production practices of digital image, it is evident that the communication circuit is radically different: the image is subject to processes of multiplication, variation, reproduction, storage, distribution, all potentially without data losses, by exploiting the newly gained notational status. The image becomes text in the sense of alphabetic writing, and the practices relating to image manipulation are in fact completely homogeneous (in the sense of a homology of production, Rossi Landi [23]) to those of digital writing. Not by chance, generative AI techniques (e.g. various neural networks architectures) are substantially the same for text and image [3].

It is important to underline again how the level hierarchy allows for reasoning *en relief* (i.e. considering the implications between levels) but does not exclude a reading *à plat*, and transparent/opaque relations among levels are possible. An interesting way of thinking about the expressive levels is thus in terms of permeability. For example, consider Fig. 17.3.[8]

Each image we access is a photographic portrait. The problem of permeability is resolved here in terms of total opacity. In fact, if we limit ourselves to level T, we are dealing with an expression that evidently presents itself as a case of photographic image. But there is a fundamental boundary mechanism. We access the image from a web page in a browser. The title of the web page (saying that "this person does not exist") constitutes a fundamental element of configuration of this semiotic object.

So:

1. at the T level, the text-image is totally opaque compared to a thematisation of the allographic dimension of the digital (it can be seen as a photographic portrait);

[8] https://thispersondoesnotexist.com/

Fig. 17.3 Two AI generated photographic portraits

2. however, at the O level (*super textum*) the title contradicts the text (informing us that the portrayed "person does not exist"), and defines a device that demonstrates the possibility of generating portraits from scratch rather than producing them indexically (as with photography). Why a device? Because it requires an action (reloading the web page in the browser) to be experienced.

We are thus witnessing a different enunciation (made possible by computational means), its valorisation passing precisely through its concealment. At the (S)T level we are in the presence of a concealment which becomes in itself a condition of interpretation at the O level, by thematising a subject of production, whose competence (an "imagopoietic" capability) is exhibited. By the way, this author properly signs the image in the bottom right corner ("StyleGANN2 (Karras et al.)").

Another example concerns digital images of classic pictorial works produced at very high digital resolution and exhibited publicly.[9] These images are printed on paper sheets having the same dimensions of the original canvases. Yet, they redefine the O level, as they can be seen but they have a different material (perceptual) construction. The high resolution is meant to hide the digital S-level. The high resolution allows to reveal details invisible to a direct vision of the pictorial artifact, producing a sort of hyperbolic visibility, and at the same time it eliminates the material components that are at their origin, e.g. the relief of the pigment and the brushstroke. The T level is maintained in order to guarantee an identity of the work (e.g. *Lamentation over the Dead Christ* by Andrea Mantegna). At level P it is evident that these images are reproducible, and therefore not economically valorisable as such (the autograph painting by Mantegna), but at the same time they guarantee the audience an access socially granted as artistic. Note how the historical dimension of authorship (Mantegna) is used precisely to ensure the social value of the image.

[9] https://www.mostreimpossibili.it/

It should be noted that the level O can regain a “classic” multimodal corporeality in different digital practices. For example, it is possible to use computer-controlled plotting techniques on paper (as in the pioneering work of Roman Verhostko[10]) or produce pictorial works through the algorithmic control of a robot arm (for example, in the case of Sougwen Chung[11]).

17.4 Generative Art and Algorithms

The allographic nature of digital representation allows us to technically define “generative” practices. If the image is allographic, then there is a set of types that describe it. In the image, at least in raster format, these are simply the RGB values that can be used for coding each pixel.[12] Again, it is certainly possible to build analog machines for the production of images (see e.g. Tinguely, *Méta-Matic no. 10* [19]). However, given the allographic nature of the digital image, the definition of production procedures is somehow inherent to the representation system. In fact, if the format is digital, then it is possible to use a universal machine such as the Turing machine to define operations for generating representable values.

An algorithm is a finite sequence of formally defined steps, typically for the purpose of solving a certain problem. If the problem is generating an image, and the latter is defined as a map of pixels, an algorithm will define how to calculate the (numerical) value of each pixel.

The verb “generate” has a close relationship with “to bring into existence,” “to produce,” “to create,” “to form,” and its origin can be found in the biological context (“to bring into being by procreation,” “to breed,” “to reproduce”).[13] It is therefore a hypernymic specification of “to do,” “to produce.” A generative algorithm therefore has to do with a production work, that is, with authorial subjectivity, yet at a more abstract level.

Back to our initial considerations, it could also be observed that a generative algorithm could theoretically deal with the genetic level. In fact, it concerns the history of the production of an image. The historian of generative art could then reconstruct, starting from an open corpus of texts (algorithms, software documentation, scientific papers), the ways in which a certain image was produced. But our point here is exactly that a generative algorithm encodes a generating procedure, regardless of where it came from. It is thus interesting to discuss this immanent logic in relation to the hierarchy of expressive levels.

---

[10] https://www.verostko.com/

[11] https://sougwen.com/

[12] In a vector image, however, the graphic content is described in terms of lines between points.

[13] https://www.oed.com/dictionary/generate_v?tab=meaning_and_use

17.5 Modalities: Specificity of Algorithmic Authorship

The extensive use of algorithms raises the problem of defining their status. Is there a difference between a brush and an algorithm?

Fontanille [15], starting from Coquet [4], proposed a characterisation of subjects in terms of modalities. A modality is a form of linguistic origin that captures a basic anthropological orientation. We will simplify the discussion and call three relevant modalities in this context as power, duty, and will: they can be thought of as describing respectively a competence, a disposition to action, and a teleological orientation.

Fontanille notes how it is possible to describe subjects based on one (M^1), two (M^2) or three modalities (M^3). The subject M^3, "trimodalised" is a full one. This proposal allows us to build a *gradatum* in the idea of subjectivity, a hierarchy based on the quantity of modalities. By partially redefining Fontanille's terminology, we can introduce for our purposes two types of subject with reduced modal load, the "tool" and the "machine."

In this perspective, a tool presents itself as a M^1 subject, i.e. it is equipped with a power, a pure capability, which remains available but inertial. This capacity is in fact the result of a historical accumulation of practices that precipitates into the tool. The tool must be put into operation (duty) in order to obtain a certain result (will). In this specific characterisation, a machine has a more complex status than a tool. Once triggered, the machine autonomously pursues a behaviour that is specified in it. The machine is then a sort of middle subject, capable of finding solutions to problems (by carrying out a complex task, a power) but at the same time it is not autonomous, since it requires a meta-control: the latter triggers the operation of the machine and the evaluation of its output. The machine has a deontic nature: it must perform the task for which it is programmed and triggered (duty). More concretely, these considerations allow us to differentiate a brush from an algorithm. A brush, or a pencil, is a tool, as it needs to be put into a practice by a logic defined elsewhere (and mostly implicit: one can think about the syntax of painting as embodied in the painter). An algorithm is a specification of a behaviour to be performed, thus a M^2 subject: that is, it is endowed not only with a power (explicitly stated) but also with a duty. It defines a solution to a potential problem by specifying a syntax of actions that it must perform (if implemented on a certain machine) once triggered. Therefore, an algorithm, as subject M^2, is not yet a full subject, but at the same time it is situated on a level of complexity, of at least partial autonomy, which places it as a form of subjectivity superior to that of the brush as a tool (M^1).

Let us consider the following minimal example:

```
import shoebot

bot = shoebot.create_bot(outputfile="output.pdf")

bot.size(400,400)
bot.rect(50,50,300,300)

bot.finish()
```

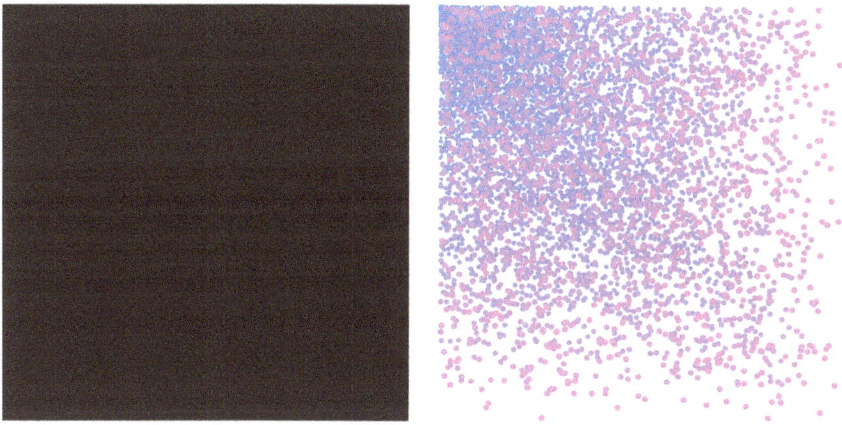

Fig. 17.4 Shoebot generated black rectangle and complex pattern

This is a short snippet, written in the Python language,[14] that makes use of the Shoebot library for generative graphics (Fig. 17.4).[15]

The result is a black square of 300 × 300 pixels whose upper corner is positioned at the (50,50) point of a white image of 400 × 400 pixels. First of all, the adjective "generative" proves relevant because the square is not produced directly by who writes the algorithm, rather it is the result of a process defined by the algorithm itself. Every time we run the code we generate a new image. We can note the imperative form in which we address the machine asking to carry out the operations (for example, `import shoebot`, that tells Python to import the library, see [24] on enunciation in programming languages). Properly, the enunciative relationship consists in sending orders to the machine that will execute them. The machine is capable of solving a set of problems and has many implicit competences. For example, it can produce a graphic file in various formats and it has a representation of what a canvas is, so that one can request its size (400, 400). Furthermore, it has a representation of the geometric concept of a rectangle, specified in terms of position (50, 50) and size (300, 300). We could also observe that, through the default values, a lot of information is implicitly introduced by the machine, for example the colours of the background (white) and the figure (black).

The implemented algorithm then acts as a subject M^2 which mediates between M^3 and M^1 by incorporating skills from the former, operating as an interface toward the latter. Triggered externally, through execution, it operates deontically.

The autonomous qualities of the machine subjectivity emerge more clearly in the next example, slightly more complex.

[14] https://www.python.org/

[15] https://github.com/shoebot/shoebot

```
import shoebot
import random

bot = shoebot.create_bot(outputfile="output.pdf")
dim = 400
how_many = 10000
bot.size(dim, dim)
for i in range(how_many):
    rd = i/how_many*5+1
    rng = i/how_many*dim
    bot.fill(i/how_many, 0.6, 1)
     bot.oval(random.uniform(0, rng), random.uniform(0, rng), rd,
rd)
bot.finish()
```

Here, the import of a library for generating pseudo-random numbers (`import random`) is required. A feature of the code is the use of variables (`dim, how_many`) as placeholders that allow the programmer to define relations among values. The loop (`for`) indicates a feature of the machine: the compression of the repetition of actions through its specification: `for i in range (how_many)` can be translated as "for each value of i up to how_many," here `10,000`. Therefore, for 10,000 times a circle (`oval`) is generated, which radius (`rd`) is calculated as a function of the counter variable (`i`). Color is also computed in relation to the counter, as the red component (the first element of the RGB triplet) is related to i (`i/how_many`), while the green and blue have fixed values (`0.6, 1`). The position exploits the generation of pseudo-random numbers, since both x and y can range randomly between 0 and `rng = i/how_many*dim`. If the counter i is 0 (first iteration), then x and y are both 0. For the last iteration, x and y can range from 0 to 10,000/10,000 × 400, i.e. 400. In other words, there is a progressive, stochastic distribution towards the bottom right corner of the circles, with colors moving from blueish to pinkish.

Here, the delegation of skills grows in qualitative terms: the machine actually decides the position of the circles. But it also grows in quantitative terms: 10,000 circles are generated in one shot. Finally, it grows in terms of abstraction: through the use of variables (`dim, how_many`) the behaviour of M^2 is parameterised and adaptable. If `dim` (the size of the canvas) or `how_many` (the number of circles) are modified in the code, the behaviour changes while maintaining its general form (the distribution over the entire surface according to the described model of a certain number of elements).

The theme of an algorithmic subjectivity as an M^2 subject clearly emerges in the classic works of Casey Reas and Ben Fry, not surprisingly the main developers of the Processing software for generative graphics[16] [22].

In the case of Reas' *Phototaxis*[17] this algorithmic subjectivity is not buried or concealed, rather it emerges as a mode of production at level T. In fact, the image

[16] https://processing.org/

[17] https://reas.com/phototaxis/

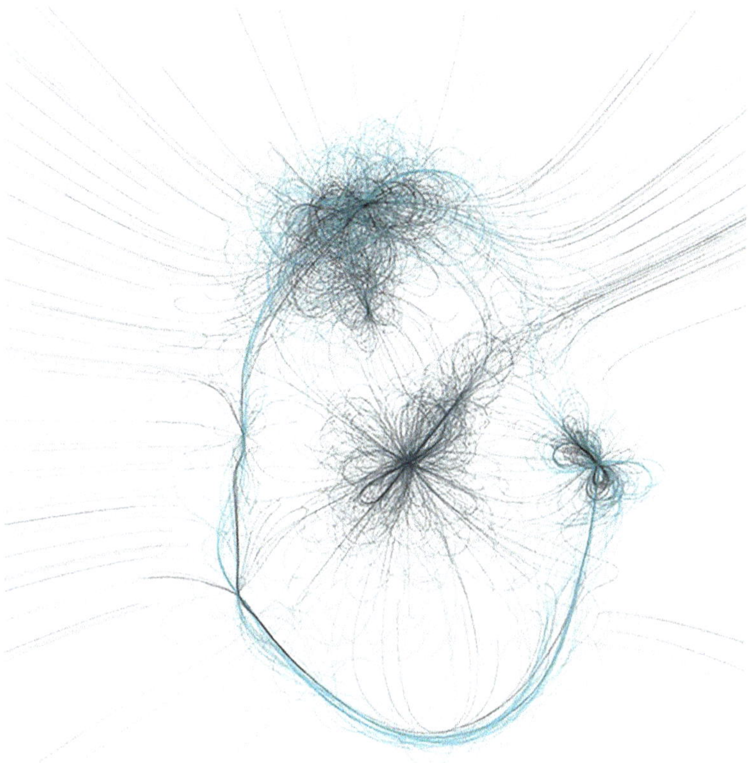

Fig. 17.5 Phototaxis, 1 by Casey Reas

is characterised by two features that are typical of the machinic subject. First, in terms of quality: the algorithm performs exactly as specified, and therefore there is an extraordinary precision that cannot be attributed to the human subject. Secondly, in terms of quantity, as it can be clearly seen from the hyperbolic number of lines of which the image is made up. This non-human form is here thematised as a sublimated naturalness in the title of the work, that is, in terms of phototaxis, the movement of an organism towards the light (Fig. 17.5).[18]

The intervention of the algorithmic subject as a form of machine mediated M^2 already emerges clearly in our simple code examples. Indeed, it finds much more complex forms in generative AI systems. Those systems are much more sophisticated, both in technical operations and in visual results, yet they still operate following the same basic algorithmic logic.

[18] https://en.wikipedia.org/wiki/phototaxis

17.6 Four Sign Production Labours in Algorithmic Perspective

Finally, we can reconsider the authorial problem in terms of Eco's theory of modes of sign production [9, 25]. The latter is proposed as a theory of expressive production in the abstract terms of a production isotopy. Without going into further detail, among the dimensions that Eco considers there is labour, understood as the operation required to the staging of expressive elements. Eco identifies four types of labour: recognition, ostension, replica, invention. Labour implies a form of authorship, as it needs a subject that carries it out.

17.6.1 Recognition

The labour of recognition can be properly conceived of as a meta-labour [25]. In other words, it is a labour that frames the others, a pre-condition that in our case recognises a certain visual whole as an image, therefore as a visual text produced according to a certain logic. As noted, the digital image shows a continuity with digital photography in terms of format, genre, and mode of circulation. In relation to photography,[19] an interesting issue emerges in the optical realism made possible by generative AI. Optical realism is historically an operation that photography introduced through a technical operating element, the photographic apparatus. In photography this mediating subject is recognisable by a set of textual elements that allow, trivially, to distinguish a photography from images produced by other techniques. This optical feature is interestingly addressed in AI generated images. Generative AI algorithms treat all the digital images that constitute the starting database of the machine learning algorithms in a homogeneous manner. Since the traces of the camera are deposited in the form of stylistic elements in the images feeding the deep learning algorithms, those same traces are reproducible. Thus, optical realism present in the images created by generative AI is not a starting condition (as it can be in photography), but an output effect. At the actual state of the art, the main issue in generative AI is global consistency. Analogue photography results at the T level in the activation of a point of view (or an Observer actant, to speak with [12]) as an integrating element for the entire image. In AI, however, the image created by exploiting photographic features results from optically local Observers, that may not be integrated into a single Observer. Instead of being the result of an initial integrative gaze to which the authorship of the image is attributed, in generative AI we thus witness a multiplication, between the disturbing and the grotesque, of author-observers. Hence e.g. the multiplication of fingers in the hands, or the inconsistencies of composition, as can be seen from the arms in Fig. 17.7.

---

[19] We have mentioned many times and indeed cursorily photography, from a semiotic perspective see Basso Fossali, Dondero [1, 6].

17.6.2 Ostension

A second type of work is ostension. In ostension, a certain expressive occurrence represents the abstract type by means of a synecdoche/metonymy relationship. This modality can be brought back to the T level by taking into consideration the topic of quotation: a quotation works when it is recognised (meta-labour of recognition) so that it can trigger a reference to the whole of which it is a part. The capitalisation of data in generative AI results in a pervasive construction of the image through an assembly of pre-existing materials. Hence the proliferation of subtexts, i.e. configurations that can, and indeed typically must, be recognised. If this can also concern Seurat's brushstroke as a reference to the painter through a stylistic mark (Fig. 17.7), very often a quotation is located at the level of configurations, of entire blocks that contribute to the definition of the meaning of the image. From this point of view, generative AI procedures are entirely based on a sort of pervasive, ostensive citationism which is located at different levels.

In Fig. 17.6,[20] the image is constructed as a sort of collage between culturally heterogeneous elements: Flemish painting and modular synthesisers. This ostensive labour (the author as a fragment collector so to say), however, is made cohesive by a second authorship that integrates the S level to the T level (i.e. treating the synthesiser "in a Flemish way," so to say). Resuming in relation to ostension what we discussed in relation to recognition, we frequently note a tension between local elements, which present themselves as "islands of meaning," and global features, which integrate them, or try to, in a sort of overall map. In relation to this, Dondero [7] has proposed to distinguish between a "regional style" (the style of generated outputs that depend on a region of images solicited by the prompt) and a "global" one that concerns the features of a model, and of its dataset as a whole.

17.6.3 Replica

In the labour of replica, an expressive type pre-exists to the occurrences. Each occurrence, despite its variations, refers to the type. The labour of the replica is evidently a specification of allography. As discussed before, digital image is allographic by definition. In this perspective, a *proprium* of generative art is to replace the autographic occurrence with the allographic type as the main place where the authorial labour is carried out. But the type in this case is not a static description, rather it is a very abstract procedural definition with respect to the occurrence. In image generation, the type takes the form of a set of algorithmic procedures that produce classes of results. If we consider the first, very simple, Shoebot example we discussed, the code can be considered as a procedural type for the visual appearance of Kazimir Malevic's *Black*

[20] The anonymous image circulates widely on the web as a meme, captioned with "Forgive me Father for I have synthed."

Fig. 17.6 Flemish painting and modular synthesizers

Square.[21] In the second Shoebot algorithm, that makes use of random variables, each execution generates a variant thanks to the use of stochastic procedures, and all these variants can be thought of as occurrences of the type defined by the procedure itself.

If we think in terms of production practices (P), then we witness a generalisation of typification. The author as a Practitioner works on the type and can produce an indefinite set of occurrences. Indeed, practices in algorithmic image generation typically exploit two paths:

1. the selection of the best occurrence, which at this point subsumes the type (and in some way re-create an autographic regime);
2. the shift of focus from the occurrence level (as the output of image creation) to the corpus level, thus showing the procedural origin of the type by amassing occurrences.

These operations cannot be described at the T level, rather they concern the P level of practices. This aspect is evident if we consider the Python code discussed above in terms of an interface towards the machine. Programming, by exploring the

[21] https://en.wikipedia.org/wiki/black_square

Fig. 17.7 A wrestling fight in the style of Seurat

variety of results while modifying the parameters, is a practice of constructing the type of the image that passes through the writing of the code.

The same situation applies if we consider the User Interface (typically via browser) of generative AI systems (Fig. 17.9).

17.6.4 Imprint and Invention

In this labour, that we can describe by referring to two specifications (imprint and invention), expression results from a relation between the continuum of content and the continuum of expression. This labour is characterised by the autographic dimension.

In the imprint, the relation between expression and content is causal. In other words, the imprint refers directly to a subjectivity of production which is (better: seems to be) at its origin. The imprint is therefore an autograph of the imprinting subject. In digital image this status is paradoxical, precisely due to the allographic nature of the digital itself. The autography of the digital image is a semiotic effect produced and made available at the S/T level. In deep learning techniques, feature extraction captures elements of the S level of the starting image to generate new ones. In Fig. 17.7, the image is generated by OpenArt.ai by requiring a wrestling fight in the style of Seurat. Seurat's painting technique has been extracted and indeed typified, so to generate an indefinite set of new Seurat occurrences.

The work of invention is characterised for Eco as a projective form in which features of the content are at the origin of features of expression, the two being autographically linked. This description would seem to fit an important approach to the generation of digital images, namely data visualisation. In this practice, a set of data and their relations are associated with a set of visual elements (and their relations). Therefore, the starting data represents a form of the content which is projected onto a form of the expression, the visual elements. However, upon closer inspection, we are dealing here with an analytics of content (starting data are discrete) and an analytics of expression (the visual parameters are also discrete), and with a deterministic correlation between the two sets. Therefore, at the end the output image can be thought properly as a notation of the input data, and we fall back into the allographic work of replica.

In Fig. 17.8, an excerpt from Ben Fry's visualisation of the bases that make up chromosome 21 is reproduced. As Fry specifies, "the darker color depicts sequences of code (called exons) that are known to be used by a cell as the set of instructions for building a protein. These instructions are interrupted by unused pieces of code (introns) which here have a medium coloring. The gray areas have no currently known function".[22] Although this is an extreme case (since the starting data is a combinatorics of only 4 bases), the principle of notation holds in general for data visualisation.

In generative AI, however, the situation is different. Prompting is a kind of assisted invention. The machinic subject M^2 is an active element at the P level and assists the subject M^3 in defining a mapping between forms of content (expressed through the mediation of forms of linguistic expression, the prompt) and forms of visual expression, which result through an interpretation of the text (see [21] on the issue of machinic enunciation). In Fig. 17.9,[23] the original prompt, "a wrestling fight in the style of Seurat," is first automatically "enhanced" by the software,[24] then interpreted (internally) and matched to image creation yielding to a corpus of possibilities, obtained by largely exploiting random parameters (the initial weights of deep learning architectures[25]). The algorithmic subject is situated in a place that is both opaque (its competence is hidden) and pragmatic (it generates results) between the subject M^3

[22] https://benfry.com/chr21-icp/

[23] https://openart.ai/

[24] And thus converted into: "Wrestling match in Seurat style, pointillism, intense action, dynamic movement, rich textures, high quality, detailed depiction, pointillism, vibrant colors, dramatic lighting, athletes in motion, textured brushwork, meticulous details, athletic figures, classic art style, pointillist technique, energetic composition."

[25] The exploitation of randomness is the same "trick" that occurred in the previously discussed simple Shoebot example.

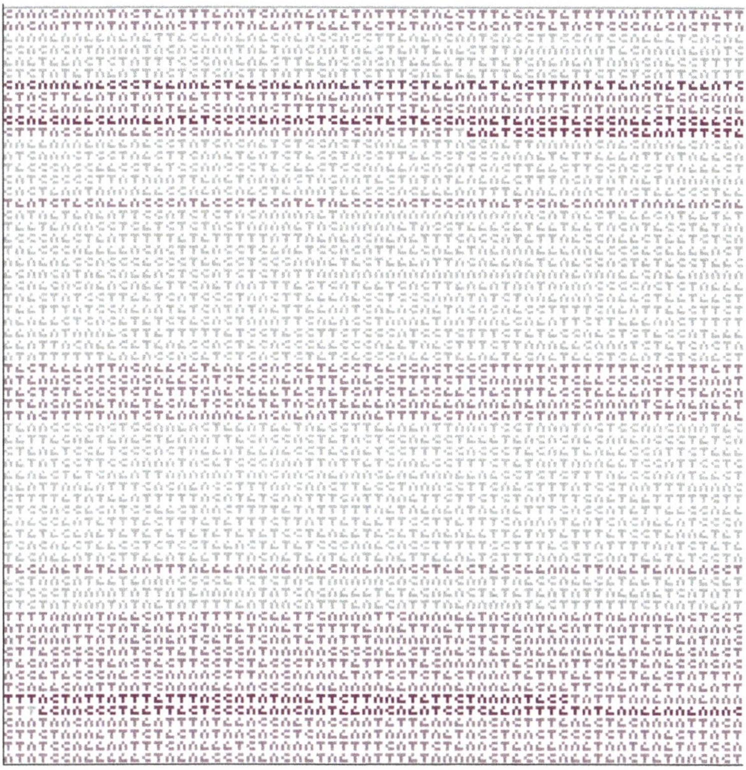

Fig. 17.8 Ben fry, genomic cartography: chromosome 21

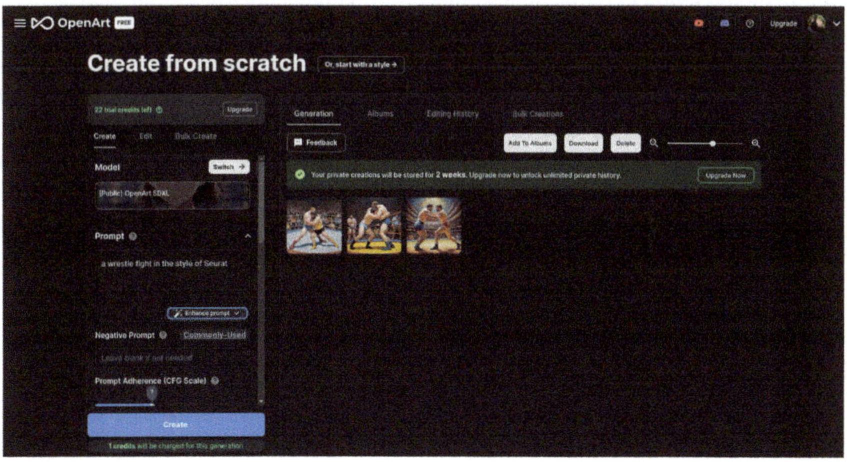

Fig. 17.9 OpenArt UI

and the image, producing first an internal interpretation of the prompt text and then a non-deterministic conversion of the same into an image. At the same time, this mediation level features many more options, including negative prompting, image editing, usage history. The machinic subject M^2 is a layer capable of modulating the M^3 competences so to offer at the end a plethora of solutions.

17.7 Conclusions

Digitisation brings the image back into the wider range of computable representations. This allographisation, as we have discussed, has significant consequences in terms of the status of the image itself and in relation to production practices, revealing specific forms of authorship. New forms of subjective mediation then emerge starting from the format, through algorithmic generation and through technologies that modify the creative practices. The characterisation that we have provided, in terms of levels of enunciation, modalisation of subjects and modes of production, intends to bring out, in the different possible combinations, the complexity of this scenario. With respect to enunciation, the usefulness of a multilevel perspective clearly emerges in the *gradatum* from signs/figures towards practices. This model, borrowed from Fontanille [14], and here discussed in its first four articulations (which we have abbreviated with the acronym STOP) allows us to verify the complex interplay of horizontal operations and vertical interferences, for the analysis of which a classical textual perspective does not appear sufficient. Authorship, defined as a form of subjectivity that can be reconstructed starting from this interplay of levels, therefore finds itself assuming different forms: hence the idea of its specification in terms of a variable modal load [15], in which a "machinic" subject emerges as an intermediate between the classical, full subjectivity (the "painter") and the tool (the "brush"). Finally, as a result of a multilevel system and a variable subjectivity, it becomes important to describe "modes of production" (starting from Eco [9]), as forms of action that link expression and content, crossing in parallel (that is, they can be co-present) the digital image. In this sense, image in the digital context appears rather than a text, i.e. a *textum,* a tissue as the work of a homogeneous weaver, properly as a territory, a surface marked by a plurality of processes, showing the heterogeneous traces of multiple subjects [25].

References

1. Basso Fossali, P., Dondero, M.G.: Semiotica della fotografia. Guaraldi, Rimini (2006)
2. Basso Fossali, P.: Il dominio dell'arte. Meltemi, Roma (2002)
3. Chollet, F.: In: Deep Learning with Python. Manning, Shelter Island, NY (2021)
4. Coquet, J.C.: Le discourse et son sujet I. Klincksieck, Paris (1985)
5. Dalli Regoli, G.: L'attribuzione dell'opera d'arte. Edizioni Plus-Università di Pisa, Pisa (2003)

6. Dondero, M.G.: Semiotics of artificial intelligence: enunciative praxis in image analysis and generation. Semiotica (2025). https://doi.org/10.1515/sem-2024-0195
7. Dondero, M.G.: The Greimasian and post-Greimasian semiotics of photography. In: Lagopoulos, A. Ph., Boklund-Lagopoulou, K., Dondero, M. G., Fontanille, J., Katsaridou, M. I., Walldén, R. (eds.) Semiotics of Images, pp. 226–253. Gruyter–Mouton, Berlin–Boston (2025b)
8. Dubois, A., Couvert, J., Borchert, T.-H.: Technical studies of painting: problems of attribution (15th–17th Century). Peeters, Paris–Leuven–Bristol, CT (2018)
9. Eco, U.: A Theory of Semiotics. Indiana University Press, Bloomington (1976)
10. Eco, U.: The Role of the Reader: Explorations in the Semiotics of Texts. Indiana University Press, Bloomington (1984)
11. Floch, J.-M.: Les formes de l'empreinte. Fanlac, Périgueux (1986)
12. Fontanille, J.: Les espaces subjectifs. Hachette, Paris (1989)
13. Fontanille, J.: Pratiques sémiotiques: immanence et pertinence, efficience and optimisation. Nouveaux Actes Sémiotiques **104–106**, 13–75 (2006)
14. Fontanille, J.: Textes, objets, situations et formes de vie. In: Alonso Aldama, J., Bertrand, D., Costantini, M., Dambrine, S. (eds.) La Transversalité du sens. Parcours sémiotiques, pp. 213–240. Presses Universitaires de Vincennes, Saint-Denis (2006a)
15. Fontanille, J.: The Semiotics of Discourse. Peter Lang, New York (2006)
16. Gonzalez, R.C., Woods, R.E.: Digital Image Processing. Prentice Hall, Hoboken, NJ (2002)
17. Goodman, N.: Languages of Art. The Bobbs-Merrill, New York (1968)
18. Greimas, A.J., Courtés, J.: Sémiotique: Dictionnaire raisonné de la théorie du langage. Hachette, Paris (1979)
19. Hulten, P.: Tinguely una magia più forte della morte. Bompiani, Milano (1987)
20. McLuhan, M.: Understanding Media. MIT Press, Cambridge, MA (1964)
21. Paolucci, C.: The myth of meaning: generative AI as language-endowed machines and the machinic essence of the human being. Semiotica (2025). https://doi.org/10.1515/sem-2024-0204
22. Reas, C., Fry, B.: Processing: A Programming Handbook for Visual Designers. MIT Press, Harvard, MA (2014)
23. Rossi Landi, F.: Il linguaggio come lavoro e come mercato. Bompiani, Milano (1968)
24. Valle, A., Mazzei, A.: Sapir–Whorf versuss Boas–Jakobson. Enunciation and the semiotics of programming languages. Lexia **27–28**, 505–526 (2017)
25. Valle, A.: Modes of sign production. In: Beardsworth, S.C., Auxier, R.E. (eds.) The Philosophy of Umberto Eco, pp. 279–304. Open Court, Chicago (2017)

Andrea Valle is associate professor at the Department of Humanities of the University of Turin, where he teaches Audio and Multimedia, Audiovision and Media and Media Semiotics. He obtained a PhD in Semiotics at the Scuola Superiore di Studi Umanistici of Bologna. He is a founding member and current Director of CIRMA, Interdepartmental Research Center on Multimedia and Audiovisual, of the University of Turin. He is the author of *Audio e Multimedia* (with V. Lombardo, 5th ed. 2024), *Introduction to SuperCollider* (2015, English ed. 2017) and *Contemporary Music Notation* (2002, English ed. 2018). He is active in the music field as a composer and performer.

Part V
Posthuman/Non-human Imagery

Chapter 18
On Filters and Philters

Massimo Leone

> *Cette voix, qui perle et qui filtre*
> *Dans mon fonds le plus ténébreux,*
> *Me remplit comme un vers nombreux*
> *Et me réjouit comme un philtre.*
> Charles Baudelaire, "Le Chat," *Les Fleurs du mal*, 1857

Abstract This chapter explores the semiotic dimensions of digital filters, examining their role not only as tools for aesthetic enhancement but as potent mediators of identity, perception, and cultural ideology in the digital age. By tracing the etymology of "filter" and "philter" and investigating their historical roots, the study reveals how digital filters embody dual functions: as instruments of seduction and control, and as mediators of authenticity and representation. The analysis delves into the ways digital filters diverge from their analog predecessors, emphasising the transformative nature of post-capture manipulation, where filters operate on the image's data rather than light waves. Through case studies such as the controversial "Jewish Filter" and religious-themed filters, the essay investigates how digital filters engage with and challenge social biases, historical stereotypes, and cultural narratives. Ultimately, the chapter positions digital filters as semiotic macro-operators, not only shaping visual culture but also influencing how individuals negotiate meaning, identity, and reality in increasingly mediated environments.

This essay results from a project that has received funding from the European Research Council (ERC) under the European Union's Horizon 2020 research and innovation program (Grant Agreement n. 819649–FACETS); the last version of the article was completed during a fellowship at HIAS, the Hamburg Institute for Advanced Studies.

"That voice forms into drops, filtrates/Into the depths of my being,/Fills me like harmonious verse/And gladdens me like a philtre"; trans. mine.

M. Leone (✉)
University of Turin, Turin, Italy
e-mail: massimo.leone@unito.it

Korea University, Seoul, Korea

P. Conte et al. (eds.), *Algomedia. The Image at the Time of Artificial Intelligence*, Lecture Notes in Morphogenesis, https://doi.org/10.1007/978-3-032-08726-3_18

Keywords Filters · Philters · Social media · Semiotics · Aesthetics

18.1 Introduction

The semiotics of digital filters offers a rich field of inquiry, where the intersection of visual aesthetics, technology, and cultural ideology becomes particularly salient [21]. This article aims to explore the diverse and multifaceted functions that digital filters perform, not only in their role as tools of visual enhancement but as potent semiotic operators that shape how we perceive and engage with the world [11]. Filters are no longer confined to the domain of professional photography or image manipulation, their widespread use on social media platforms like Instagram reveals them as key mediators of identity, perception, and authenticity [2]. By examining the historical and etymological roots of the term "filter," the study will demonstrate how digital filters encapsulate the dual roles of "philter" and "filter," simultaneously seducing the viewer and mediating between reality and representation.[1]

As this analysis moves forward, we will dissect how digital filters function in comparison to their analog counterparts, highlighting the differences between pre-capture physical modifications and post-capture digital manipulations [31]. Through this lens, we will examine the ideological assumptions embedded in both "filtered" and "unfiltered" images, probing the cultural expectations that govern the use of filters in contemporary visual communication. A critical exploration of the aesthetic, technical, and ethical dimensions of digital filters will follow, illustrating their profound impact on both visual culture and personal identity [17].

The roadmap of this investigation will traverse several key areas: from the dual semantic fields of "philter" and "filter" in language and culture, to the evolving functions of filters in digital spaces, and finally, to the broader implications of filters as social and ideological constructs [23]. By scrutinising controversial filters—such as the "Jewish Filter" and other culturally sensitive examples—we will uncover how filters not only influence personal appearance but engage with, challenge, and sometimes reinforce social biases and historical stereotypes [32].

Through this examination, the article will uncover the broader semiotic function of filters, situating them within a wider philosophical and cultural framework [9]. By positioning digital filters as a critical tool in the construction and negotiation of meaning, we gain a deeper understanding of their role in contemporary digital culture and the complex ways they shape both visual and social realities [19].

[1] As explored in my seminar "On Filters and Philters," University of Turin, 2022–23.

18.2 Separation and Attraction

This article investigates the complex dialectic between two concepts, each expressed in Italian by the single term *"filtro,"* but which in English diverge into the homophonic "philter" and "filter." Although they share the same sound in one language, these terms conceal a profound difference in meaning, rooted in their etymological origins. A "philter" traces back to the Greek verb *"philéo,"* meaning "to love," and is connected to the same root as "philosophy." Historically, the philter has been a magical potion, an enchanting concoction designed to manipulate the will of others—most notably by inducing love [35]. This idea of the philter appears in mythological, literary, and visual traditions as the archetypal love potion, an elixir capable of bypassing reason to invoke desire, seduction, and control [28]. By contrast, a "filter" (which shares an etymological link with "felt") refers to something that obstructs direct contact between two entities, functioning as a barrier, a membrane, or a veil. It signifies separation, mediation, and the interruption of immediate interaction.

This essay explores the tension between these two concepts for, in the digital age, the merging of such ideas within a single homophonic term reveals a more nuanced understanding of digital filters. In their contemporary usage, digital filters operate as both philter and filter: they seduce and enchant, constructing a mediated image designed to guide perception, while simultaneously acting as an obstacle to direct, unfiltered authenticity. This ambivalence positions digital filters as potent semiotic operators, navigating between seduction and mediation, authenticity and artifice, and their evolution reflects not only aesthetic shifts but also deep ideological underpinnings. By probing their historical and conceptual lineage, this study deconstructs the semiotic mechanisms of digital filters, revealing their complex role as mediators of identity, perception, and authenticity in the digital age.

18.3 Unfilteredness

Today, when one speaks of "filters," the reference is often to the alteration of digital images before their publication on social media platforms, particularly Instagram. From this contemporary perspective, filters are understood as standardised procedures for modifying digital images, usually with the aim of enhancing their appeal to a specific audience. A paradigmatic example can be seen in Instagram's built-in filters, which promise to imbue an image with a particular aura by manipulating contrast, lighting, saturation, warmth, exposure, and sharpness—elements that ostensibly "enhance" the photo. To confine the concept of filters to this narrow, technical use, however, would be to overlook the rich and complex semantic field in which the lexeme "filter" operates. Before the advent of digital image manipulation, filters were already embedded in a broad array of devices, practices, and operations, all sharing a "family resemblance" that points to their core semiotic function.

To consider filters from the vantage point of cultural semiotics requires moving beyond their most current and obvious manifestation as tools for digital graphics. Instead, we must probe what filters signify in the broader context of meaning-making, the ideologies that shape their use, and how these have evolved over time. In popular understanding—especially as it relates to platforms like Instagram—filters are primarily seen as tools that modify an image by adding something extraneous to it. This operation is frequently viewed through both a gnoseological and ethical lens: there is a tacit assumption that a digital image contains an inherent "truth," which the filter either conceals or distorts. Filters are often critiqued as agents of deception, concealing the "naked" or "natural" state of a subject, particularly in the case of faces, and making an image appear more aesthetically pleasing than it actually is. This critique is often couched in terms of insincerity, as filters are perceived to betray the reality of what they represent—whether by adding a warm glow to a dreary landscape or smoothing the imperfections of a face. Moreover, this insincerity is paired with a charge of conformism: filters, it is generally believed, pre-coded and standardised within apps, eliminate originality by applying uniform modifications across images.

Yet this widespread perception of filters as inherently deceptive or conformist risks oversimplifying and reducing their function as semiotic operators. A semiotics of filters cannot be constrained by such superficial sociological or aesthetic judgments, nor can it take for granted the notion that unfiltered images possess an inherently greater claim to reality or truth. Such a view, grounded in the presumed indexicality of unfiltered images, would be overly simplistic. Similarly, invoking the post-Barthesian rhetoric of the "reality effect" is insufficient. While it is tempting to argue that there are no truly "unfiltered" images—since even an image devoid of overt manipulation still operates within a framework of meaning—such a claim is itself reductive. Unfiltered images do exist, but they are a marked category, explicitly labelled as such ("#no filter") in order to be recognised. What constitutes an "unfiltered" image is not a matter of ontological clarity; it emerges from a complex socio-semiotic negotiation, in which an image's light, colours, and tones are perceived as "normal" only in contrast to the "something special" that filtered images ostensibly possess.

The task of a semiotics of filters, then, is to deconstruct these social clichés, to uncover the deeper ideological reasons for their persistence, and, above all, to analyse the semiotic operations that underlie the use of filters in the digital world. This requires extracting filters from their quotidian and highly technical usage and repositioning them within a broader semantic and semiotic category: that of filtration. By comparing various instances of this category across contexts, we can begin to understand what is meant when something is designated as "filtered"—not just in the technical sense of image manipulation, but in the phenomenological and semiotic sense of mediating perception and meaning.

In our work within the Turin semiotic group, we have explored a related dialectic in the case of "the veiled and the unveiled," [22]. The semiotic framework applied to the veil is not limited to the object itself but extends to a broader disposition of meaning-making that governs its use. Filters, like the veil, participate in a similar plastic dialectic, which within Greimasian semiotics could be understood as a manifestation of the encompassing/encompassed relationship—one that we have applied to the analysis of relics and reliquaries [18], and more recently, to the face and the mask [30]. A filter, like the veil or the mask, encloses and englobes, yet it does so in a way that is distinct from other modes of englobing. It mediates perception while simultaneously constructing it, shaping how an image is seen, felt, and understood.

To grasp how a filter functions semiotically, it is necessary to study the term's semantic field, not through a purely lexicological lens (a method of a certain use in much Greimasian semiotics), but by identifying the structural similarities that emerge across various contexts where filtering occurs. This approach will allow us to isolate the essential semantic markers of filtration, uncovering how they operate in diverse fields—including areas where filtering is not explicitly named, such as the semiotic construction of cultural icons [16]. Where, then, do we encounter filters in the world, and how do they function as operators of meaning?

18.4 Filtration

This question leads us into an expansive field where the concept of filter stretches far beyond the technical domain of digital image manipulation. In the article "Rethinking Filter: An Interdisciplinary Inquiry into Typology and Concept of Filter, Towards an Active Filter Model," Khashayar Razghandi and Emad Yaghmaei attempt to compile an inventory of filters, summarising insights from an interdisciplinary research group [34]. While their comparative approach is valuable, however, it lacks a crucial semiotic analysis that would extract a unified modality from the diverse manifestations of filters across different contexts. Their focus shifts from the idea of filters as veils to seeing them as membranes, drawing on works like *Membranes, Surfaces, Boundaries: Interstices in the History of Science, Technology, and Culture*, edited by Grote and Stadler [12]. In particular, cultural historian Thomas Brandstetter's article "The Vitality of Membranes: Artificial Cells as Boundary Objects" traces the study of membranes from the nineteenth century onward, showing how attempts to reproduce their structure blurred the distinctions between mechanistic and vitalistic views of life, reflecting the philosophical tension between *res cogitans* and *res extensa.*

From a semiotic point of view, membranes, especially through the process of osmosis, are highly comparable to semiosis itself. As Peirce famously posited, semiosis is a kind of filtering where an object signifies not in its entirety but "according to certain respects and capacities," which are dynamically selected in relation to what Peirce called the "ground." This selection process acts much like a filter or an active membrane, where certain elements of the object are filtered through a semiotic system, producing meaning through interpretation. Interpretation, in turn, is the generation of new interpretants, a process that is never purely mechanical but infused with the creativity of meaning-making. The advantage of a semiotic perspective on filters and membranes, then, lies in its capacity to recognise the interstitial dynamics between nature and culture, evolution and cognition. This resonates with the natural science paradigm known as "Active Matter," which is well represented by Grote's *Membranes to Molecular Machines* [13]. The study of filters, from this vantage, is not simply about technical mechanisms but about the ongoing negotiation between systems of meaning and the material world.

Semiotics, by insinuating itself into this interdisciplinary field, not only raises biosemiotic questions but also proposes a cognitive and cultural semiotics of filters. This prompts a fundamental question: what unites such diverse phenomena as a filter decanter with activated carbon, the membrane of a plasma cell, marine filter feeders like Tunicata, graphene oxide nanolayers used in water desalination, the mechanosensors of the tiger spider, acoustic filtering systems, and the filtering algorithms of data science? All these examples share a common characteristic: they mediate between two entities by filtering, separating, or modulating inputs. The plasma membrane filters essential nutrients and ions, just as the spider's sensory system filters environmental vibrations, and filtering algorithms in data science sift through vast data sets to generate intelligible patterns [8]. In each case, the filter does not merely block or allow passage but transforms inputs into a more ordered or meaningful output, a process that parallels the reduction of entropy.

This process of reducing complexity through filtering is not limited to natural or technological phenomena. The concept extends to the realm of architecture, as Siegfried Ebeling's [10] work *Der Raum als Membran* highlights. Ebeling, a Bauhaus member, proposed that membranes mediate between the interior and exterior, the material and functional, much like filters in other contexts. His ideas resonate in post-war visionary architectural projects such as Buckminster Fuller's geodesic domes and Archigram's *Cushicle* and *Suitaloon*. In each of these examples, architecture is not merely a static object but a responsive, filtering interface that adapts to human and environmental needs.

The semiotic functioning of filters extends further into the digital sphere. Eli Pariser's concept of the "filter bubble" [29], reflects the way online algorithms selectively filter information based on user preferences, creating a state of intellectual isolation [4]. This phenomenon highlights how filters can produce not just physical or material order, but also ideological and cognitive structures, reducing the complexity of information to fit preexisting patterns of thought. The operation of filters, whether in biology, technology, architecture, or digital culture, creates difference, generates dissymmetry, and imposes a certain sense of separation and order. Filters, by producing negentropy, actively reduce chaos and complexity [5].

The same can be said of digital image filters, which, upon closer inspection, function as reducers of visual complexity, normalising facial features or landscapes to align with culturally accepted norms. They act as semiotic macro-operators, creating visual common sense by distilling the complexity of human faces or environments into familiar, intelligible models shared by a community of meaning. Filters, then, are not mere tools for aesthetic enhancement but are deeply implicated in the production and circulation of meaning in the digital age. Like cultural icons, which filter existential representations to structure the semiotic complexity of life into recognisable values [33], digital filters channel the nuances of visual communication into compact, immediately comprehensible signs.

Many of the dominant forms of digital communication—selfies, memes, gifs, stickers—are products of this semiotic filtering process. In an era of increasing complexity, these formats represent ever more condensed clusters of meaning, distilling the infinite possibilities of interpretation into a form that appears instantly recognisable. The semiotic analysis of filters, therefore, must not only engage with their technical and aesthetic dimensions but also unpack their deeper cultural, cognitive, and ideological operations. By doing so, we uncover the essential role filters play in shaping the systems of meaning that govern both our perception of the world and our interaction with it.

18.5 Digital Filters

Although they share the same name, digital image filters are fundamentally different from traditional photographic filters [25]. Photographic filters act on light waves or rays *before* the image is captured, while digital filters manipulate the data representing the image *after* the light has been recorded by film or a sensor and digitally encoded. The former is a physical-optical intervention at the moment of image generation, whereas the latter operates on the "body" of the image at a mathematical-logical level. Despite their shared goal—modifying photographic images—and the occasional similarity in results (as seen, for instance, with a soft-focus lens filter versus a digital "low-pass" filter in Photoshop), digital image filters should not be misconstrued as mere simulations of their traditional analog counterparts. Instagram's X-Pro II filter, for example, does not replicate a polarising, colour, or gray filter placed in front of a camera; rather, it simulates specific processes of film development within a darkroom.

Photographic filters, by nature, cannot be entirely replaced by digital techniques because they influence the image capture process itself, as opposed to modifying pre-existing images. For instance, the effect of a neutral density filter, which softens the appearance of moving water by allowing longer exposure times, can only be imitated with laborious digital editing. Similarly, reflections from water or glass, which are removed by polarising filters, are difficult to replicate through digital retouching. These examples illustrate that there is no straightforward equivalence between so-called "analog" photographic techniques and digital processes. Digital media represent more than a mere extension or continuation of analog media, serving as a simple replacement or surpassing of it. Instead, digital technology introduces a new and distinct operational space.

Indeed, as Manovich [24] argues in his analysis of Photoshop, digital filters do not merely replicate analog techniques but open up new possibilities for image manipulation. Certain operations that might seem inherently digital, such as high-pass filtering or noise reduction, actually originated long before the digital era, during the early days of electrical engineering and communications technology. These techniques, developed primarily in the first half of the twentieth century for use in radio and analog telephones, were later adapted into digital image processing. In this sense, even highly digital functions like high-pass filters are simulations of earlier analog circuits. Their application is not limited to images but extends to sound and video signals as well, demonstrating their versatility as general signal processing techniques.

In electronics, a high-pass filter is composed of a circuit that allows frequencies above a certain threshold (the "cutoff frequency") to pass, while attenuating frequencies below that threshold. Depending on the presence of active components, the filter can be classified as either active or passive. Filters are also categorised based on the steepness of the frequency cutoff: first-order filters reduce 20 dB per decade, second-order filters reduce 40 dB per decade, and so on. These filters are widely used in audio systems to separate sounds for tweeters and mid-range speakers. In earlier radio systems, high-pass filters were even used as antennas. As Belevitch [1] shows in his historical analysis of circuit theory, filtering techniques are deeply embedded in the development of communications technology, much of which predates the digital revolution. Manovich [24] further highlights that while many digital tools simulate the effects of physical media, these simulations only hold as long as the user adheres to specific parameter settings. Altering those parameters can lead to results that are entirely foreign to the physical world. Thus, digital algorithms often only simulate the physical world when used within a particular set of constraints. Once those constraints are removed, the results may no longer function as analog simulations.

The case of the "Unsharp Mask" in digital image processing illustrates the divergence between analog and digital techniques. Originally developed for darkroom photography, this technique uses a blurred, or "unsharp," negative of the image to create a mask of the original. The mask is then combined with the positive image, resulting in an image that appears sharper than the original. While the process enhances clarity, the resulting image is not necessarily a more accurate representation of the subject. In digital processing, the "Unsharp Mask" operates as a linear or non-linear filter that amplifies high-frequency components of a signal, sharpening the image by accentuating its edges. However, the process can easily introduce artifacts—halo effects, colour shifts, and increased noise—if parameters are set too aggressively.

The implementation of filters in digital software like Photoshop reflects a deep tension between the underlying mathematical operations and their representation in the user interface. As Manovich [24] notes, while the filter's effects on the image are visible to the user, the computational processes that drive these effects remain hidden behind the graphical user interface. This black-box nature of image processing routines is echoed in Heilmann's [14] analysis of Photoshop's source code, where the small size of the assembly code responsible for the core algorithmic operations contrasts with the much larger structure of the program as a whole. Heilmann suggests that the tension between the efficient, machine-like routines of the algorithm and the more abstract, object-oriented structure of the higher-level programming language reflects the division between the filter's hidden workings and its visible effects on the image [15].

This bifurcation extends to the way filters are experienced in digital software. As both interface elements and functional processes, filters occupy a space that is distinct from the images they modify. Spatially, they are represented as menu items, dialog boxes, and progress bars in the graphical user interface, external to the image itself. Temporally, they mediate between two states of the image—pre-processed and post-processed—without revealing the stages of transformation that occur in between. Filters thus operate on multiple levels: as abstract algorithms, as interactive tools, and as mediators of visual perception. They shape not only the appearance of images but also the very process by which images are interpreted and understood.

While digital image filters may at times mimic the effects of traditional photographic filters, they operate within a fundamentally different framework. The transition from analog to digital filtering represents more than a technological shift; it signifies the opening of a new space for media operations, where the manipulation of visual data is governed by an entirely different set of rules. In this space, filters become powerful semiotic operators, shaping both the form and the meaning of digital images in ways that transcend their analog predecessors.

18.6 Conclusions: *Algofilters*

In August 2019, Facebook, which owns Instagram, launched a tool called "Spark AR Creators," allowing users to create their own facial filters. This feature, made available to the public, enabled anyone to design effects for Instagram Stories and share them with other users. Within weeks, a variety of filters appeared, including "Fix Me" alongside light-hearted effects like glitter, cat ears, and heart-shaped glasses. One filter, in particular, mimicked the markings a plastic surgeon makes on a patient's face before surgery. Following this, filters such as "Plastica" and "Bad Botox" were released, showing users with exaggeratedly plumped lips, narrowed noses, and large, doll-like eyes. Although "Spark AR Creators" did not design these specific filters themselves, Instagram allowed these effects to be accessed by its billion-strong user base, later admitting it was a mistake. In a 2019 statement, Instagram declared that their future goal was to ensure filters would provide a "positive experience" and announced they were updating their "well-being policies." While guidelines were being revised, existing filters like "Fix Me" were immediately removed from the platform, and Facebook currently prohibits the release of any similar effects.

This development came soon after Facebook and Instagram announced stricter policies regarding posts related to cosmetic surgery and weight-loss products, particularly those targeted at minors. Facebook emphasised that it intended to hide posts with "unreliable" claims about diet products, especially those tied to promotional content like discount codes or affiliate links. These measures marked an effort to address the platform's role in promoting potentially harmful beauty standards. Instagram's popularity is fuelled by the active participation of its users—who upload stories, share their lives, and use features like filters. Yet even "harmless" face filters primarily serve the beauty industry. While some filters may seem humorous or whimsical—like the pizza halo or flames in the background—the underlying purpose remains facial optimisation. A critical look at these filters reveals that most, if not all, subtly enhance skin smoothness, enlarge eyes, or brighten the complexion (Leone 2020). This idealised image is leveraged by beauty brands in the offline world, with products marketed through social media bearing names like Revlon's "Photo Ready Insta Filter Foundation" or Charlotte Tilbury's "Hollywood Flawless Filter." These products promise users the same airbrushed, flawless look offline that filters create online.

The use of digital filters on platforms like Instagram has not only transformed aesthetics but has also opened new, controversial avenues for engaging with social and cultural issues. Among the most provocative of these is the so-called "Jewish Filter," named after the offensive query "Do I Look Like a Jew?" The creator of this filter, Jewish artist Maayan Sophia Weisstub, set out to expose the absurdity and extremism of anti-Semitic stereotypes. Rooted in caricatures from anti-Semitic propaganda, especially prevalent in the nineteenth and twentieth centuries [36], the stereotypical portrayal of Jews—with exaggerated features such as large hooked noses, droopy dark eyes, and disproportionate facial characteristics—finds its origins in grotesque representations, like those of witches (for example, Baba Yaga). These visual distortions were used to dehumanise and vilify Jewish people in racist imagery. The creator of the "Jewish Filter" aimed to radicalise these grotesque depictions and expose the irrationality of attributing such features to a particular ethnicity. Through Instagram, where users often apply filters to achieve a "flawless" or doll-like appearance, this project invited users to confront the very notion of perfection [7], challenging both their own assumptions about Jewish appearance and their reliance on biased racist imagery to define it. By transforming their own faces into a distorted "ugly" version, the filter sought to highlight the absurdity of these anti-Semitic projections, pushing users to question ingrained prejudice that lacks any factual basis.

In another instance, the filter "Old School," available in the app's gallery of effects, stirred controversy for adding virtual tattoos to users' faces in selfies, including images of snakes, a Native American headdress, the phrase "pray for me," and a symbol resembling a swastika. Sabrina Zohar, a Californian fashion designer with over 17,000 Instagram followers, was shocked when it appeared that the Nazi symbol was digitally etched on her arm. The creator of the filter, Anastasia Truita Tkachenko, defended her design, explaining that the symbol in question was a Slavic one, symbolising goodness, the sun, and life, and pointed out that the arms of the symbol are bent counterclockwise, unlike the Nazi swastika. Despite her justification, the controversy underscores the challenges of using historical symbols, as their meanings and associations can vary widely across cultures and contexts, making the potential for misunderstanding and offense a serious issue in the design of digital filters in an era of "iconic indifference" [6].

Religious themes have also entered the realm of digital filters, as seen in the Israeli television series *We Will Meet* (produced by Kan TV), which explores issues of religious repentance. For the first season, a custom filter was developed for Instagram and Facebook, enabling users to virtually "wear" the personas of various religious and secular figures. This filter featured ten different characters, from hipsters and goths to individuals with neck tattoos, orthodox hats and beards, kipas, headscarves, and hairbands. While these filters offer a playful way for users to engage with religious and cultural identities, they also raise significant concerns regarding cultural appropriation. In some cases, like the now-banned Instagram filter that allowed users to mimic traditional Māori moko facial tattoos [20], such digital expressions can trivialise or distort sacred cultural symbols, stripping them of their original meaning. When users casually "try on" elements of deeply rooted traditions, the line between representation and appropriation becomes dangerously blurred. These filters risk commodifying and decontextualising religious and cultural markers, reducing them to fleeting digital trends rather than honouring their significance [26]. These examples demonstrate how digital filters have evolved from mere tools of visual enhancement to vehicles of social commentary and provocation [27].

Whether addressing deep-seated prejudices or allowing playful experimentation with identity, the power of filters extends far beyond aesthetics. They shape not only how we perceive ourselves but also how we engage with broader cultural narratives, inviting us to confront the ideologies and assumptions embedded in both beauty standards and historical representations. The provocative potential of these filters, however, comes with risks—while some challenge harmful stereotypes, others can inadvertently perpetuate them, revealing the complex interplay between technology, culture, and identity in the digital age.

Digital filters—or "philters"—serve as quintessential examples of *Algomedia*, a term that resonates on multiple levels. Etymologically, the prefix "algo" derives from the Arabic root for algorithms, denoting the numerical frameworks that drive digital processes, while in Greek, "algo" evokes pain (*algos*), suggesting the existential dimension that filters engage with in human life. As tools of mediation and manipulation, filters operate at the intersection of these two meanings. Through algorithms, they transform visual data, offering seemingly enhanced ways of appearing to oneself and to others. Yet, they also introduce a subtle form of pain—a dissonance between appearance and reality, between the filtered and unfiltered self, that speaks to deeper anxieties about identity, authenticity, and representation.

In this way, filters are not only technical devices but also existential operators that structure human perception, often amplifying the existential tension between our digital personas and our lived experiences. They mediate a new kind of visual pain, where individuals constantly negotiate their place in a world governed by the aesthetic and ideological demands of algorithmic media. By shaping how we appear across time and space, filters encapsulate the dual essence of *Algomedia*—where the digital meets the deeply human struggle with self-image, shaped both by numbers and by the painful awareness of our mediated realities.

References

1. Belevitch, V.: Théorie des circuits de télécommunication. Librairie Universitaire, Louvain (1957)

2. Biggio, F.: Augmented facets: a semiotics analysis of augmented reality facial effects. Sign Syst. Stud. **49**(3–4), 509–526 (2021). https://doi.org/10.12697/SSS.2021.49.3-4.13
3. Brandstetter, T.: The vitality of membranes: artificial cells as boundary objects. In: Grote, M., Stadler, M. (eds.) Membranes, Surfaces, Boundaries: Interstices in the History of Science, Technology, and Culture, pp. 115–124. Max Planck Institute for the History of Science, Berlin (2011)
4. Bruns, A.: Are Filter Bubbles Real? Polity Press, Cambridge, UK–Medford, MA (2019)
5. Cubasch, A.J., Engelmann, V., Kassung, C.: Theorie des Filterns. Zur Programmatik eines Experimentalsystems. (2021). https://doi.org/10.5281/zenodo.4731045
6. Conte, P.J.A.: Dalla differenza all'indifferenza iconica: Metafora e contrasto nel sistema del visibile. In: Purgar, K., Vargiu, L. (eds.) Studiare le immagini: Teorie, concetti, metodi, pp. 325–343. Carocci, Rome (2023)
7. Dalmasso, A.C.: Des pouvoirs prothétiques des intelligences artificielles: Empathie et intercorporéité. In: Dalmasso, A.C., Carbone, M., Bodini, J. (eds.) Des pouvoirs des écrans. Éditions Mimesis, Milan (2018)
8. Diniz, P.S.R.: Adaptive Filtering: Algorithms and Practical Implementation. Springer, Dordrecht (2008)
9. Dondero, M.G.: Models, mannequins, dolls and beautified faces: a semiotic and philosophical approach to the sense of beauty. Topoi **41**(4), 785–793 (2022). https://doi.org/10.1007/s11245-022-09809-x
10. Ebeling, S.: Der Raum als Membran. C. Dünnhaupt, Dessau (1926)
11. Gramigna, R.: Inside Facebook's semiosphere: how social media influence digital hate and fuel cyber-polarization. Soc. Semiot. **32**(5), 606–633 (2022). https://doi.org/10.1080/10350330.2022.2157171
12. Grote, M., Stadler, M. (eds.): Membranes, Surfaces, Boundaries: Interstices in the History of Science, Technology, and Culture. Max Planck Institute for the History of Science, Berlin (2011)
13. Grote, M.: Membranes to Molecular Machines: Active Matter and the Remaking of Life. University of Chicago Press, Chicago (2019)
14. Heilmann, T.A.: Blackbox Bildfilter. Unsharp mask from Photoshop to the Röntgentechnische Versuchsanstalt Wien. Navigations: J. Media Cult. Stud. **20**(2), 75–94 (2020). https://doi.org/10.25969/mediarep/14950
15. Heilmann, T.A.: Processed pictures, photoshop, and unsharp mask. Int. J. Dig. Art Hist. **8**, 162–175 (2021). https://doi.org/10.11588/dah.2021.E1.83935
16. Ibrahim, Y.: Digital Icons: Memes, Martyrs and Avatars. Routledge, Abingdon–New York (2021)
17. Javornik, A., et al.: What lies behind the filter? Uncovering the motivations for using augmented reality (AR) face filters on social media and their effect on well-being. Comput. Hum. Behav. **128**, 107126 (2022). https://doi.org/10.1016/j.chb.2021.107126
18. Leone, M.: Wrapping transcendence: the semiotics of reliquaries. Signs Soc. **2**(S1), S49–S83 (2014). https://doi.org/10.1086/674314
19. Leone, M.: The observer actant in the contemporary legal discourse: a semiotic meditation. Soc. Semiot. **29**(3), 406–425 (2019). https://doi.org/10.1080/10350330.2019.1587836
20. Leone, M.: Facial scripts: the semiotic journey of Maori tattoos from colonial gaze to cultural revival. In: Ferrara, S., Cartolano, M., Ottaviano, L. (eds.) Talking Images: The Interface between Drawing and Writing, pp. 203–223. Routledge, London-New York (2025)
21. Leone, M. (ed.): The Hybrid Face: Paradoxes of the Visage in the Digital Era. Routledge, London-New York (2023)
22. Leone, M., de Riedmatten, H., Stoichiţă, V. I. (eds.): Il sistema del velo: Trasparenze e opacità nell'arte moderna e contemporanea/Système du voile: Transparence et opacité dans l'art moderne et contemporain. Aracne, Roma (2016)
23. Magalhães, L., Oliveira Martins, C. (eds.): Masks and Human Connections: Disruptive Meanings and Cultural Challenges. Palgrave Macmillan, Cham (2023)

24. Manovich, L.: Software Takes Command: Extending the Language of New Media. Bloomsbury, New York–London (2013)
25. Meyer-Rebentisch, K.: Mit optischen Filtern fotografieren: Polarisations-. Neutraldichte- und Grauverlaufsfilter wirkungsvoll einsetzen. dpunkt.verlag, Heidelberg (2018)
26. Mirabet-Herranz, N., Galdi, C., Dugelay, J.L.: Facial biometrics in the social media era: an in-depth analysis of the challenge posed by beautification filters. IEEE Trans. Biometr. Behav. Identity Sci. 1–1 (2024). https://doi.org/10.1109/TBIOM.2024.3438928
27. Modena, E., Pinotti, A., Pirandello, S.: Virtual reality and augmented reality: new tools for art and politics. Paradigmi: Rivista di critica filosofica **1**, 87–106 (2021). https://doi.org/10.30460/100230
28. Müller-Ebeling, C., Rätsch, C.: Isoldens Liebestrank: Aphrodisiaka in Geschichte und Gegenwart. Kindler, Munich (1986)
29. Pariser, E.: The Filter Bubble: What the Internet Is Hiding from You. Penguin Press, New York (2011)
30. Pennisi, G., Leone, M. (eds.): Semiotic and cognitive approaches to the face, between alterity and alteration. Reti, saperi, linguaggi **21**(1), 37–48 (2024). https://doi.org/10.12832/113796
31. Pichler, E., Böhlau, K.: Hochschule für Bildende Künste Braunschweig: Filters + Frames: Developing Meaning in Photography and Beyond. Jonas Verlag, Weimar (2019)
32. Piluso, F.: Dis-playing gender: From ludic reconfiguration to utopic outcomes in gender-swap apps. Future Human. **2**(1–2) (2024). https://doi.org/10.1002/fhu2.8
33. Ponzo, J., Stano, S.: Nuovi media, nuovi miti. Aracne, Roma (2023)
34. Razghandi, K., Yaghmaei, E.: Rethinking filter: an interdisciplinary inquiry into typology and concept of filter, towards an active filter model. Sustainability **12**(18), 7284 (2020). https://doi.org/10.3390/su12187284
35. Taberner, P.V.: Aphrodisiacs: The Science and the Myth. University of Pennsylvania Press, Philadelphia (1985)
36. Volli, U.: Figure semiotiche dell'antisemitismo. Annali di studi religiosi **23**, 121–150 (2022)

Massimo Leone is full professor of Philosophy of Communication, Cultural Semiotics, and Visual Semiotics at the Department of Philosophy and Educational Sciences of the University of Turin; Director of ISR-FBK, the Center for Religious Studies at the "Fondazione Bruno Kessler" in Trento; Professor of Semiotics at the Department of Chinese Language and Literature of Shanghai University; Associate Member of Cambridge Digital Humanities, University of Cambridge; Adjunct Professor at UCAB University in Caracas; and Adjunct Professor at the Korea University, Seoul. A member of the Academia Europaea, he has been a visiting professor at several universities across five continents. He is the recipient of an ERC Consolidator Grant (2018) and an ERC Proof of Concept Grant (2022). He serves as Editor-in-Chief of *Lexia*, the semiotic journal of the Interdisciplinary Research Center on Communication at the University of Turin, as well as of *Semiotica* (De Gruyter). He is also the director of the book series *I Saggi di Lexia* (Rome: Aracne), *Semiotics of Religion* (Berlin and Boston: Walter de Gruyter), *Advances in Face Studies* (London and New York: Routledge), and *Religion and Technology* (Springer).

Chapter 19
Faces that Do not Exist. Limits and Possibilities of AI-Generated Images

Remo Gramigna

Abstract This study explores the profound implications of AI-generated imagery, focusing on the limits and possibilities of synthetic faces and images within contemporary visual culture. It examines the underlying logic of AI image generation, emphasising the shift from one-to-many to many-to-one processes driven by vast datasets and latent spaces. The discourse highlights critical issues such as the epistemic and legal challenges of originality, authorship, and intellectual property, alongside the emergence of technologies like GANs and deepfakes, which blur the boundaries between real and artificial. The comparison between human and algorithmic perception introduces the concept of "algorithmic similarity," questioning how machines recognise and cluster images. Furthermore, the paper discusses the socio-cultural significance of authenticity, the remediating influence of AI on older media, and the aesthetic tension between visual accuracy and distortion. Ultimately, it advocates for a nuanced understanding of AI imagery as both a disruptive force and a catalyst for rethinking truth, representation, and creativity in the digital age.

Keywords AI-generated images · Photography · Deepfakes · Fakes · Photorealism · Authenticity

19.1 The Virtual Component of Human Lives

Over the last five years (2020–2025), there has been quite a dramatic and profound shift in people's lives. For various socio-political and historical reasons, the vast majority of people today are more enmeshed in digital communication than ever before. For work, education, social life, and entertainment, people are spending more and more time on digital platforms and engaging with virtual worlds. In short, human life has become increasingly digitalised [63, 68].

It is also worth noting that, quite recently, due to the Covid–19 pandemic, the use of online communication and social media has skyrocketed. This was coupled with

R. Gramigna (✉)
University of Turin, Turin, Italy
e-mail: remo.gramigna@unito.it

P. Conte et al. (eds.), *Algomedia. The Image at the Time of Artificial Intelligence*, Lecture Notes in Morphogenesis, https://doi.org/10.1007/978-3-032-08726-3_19

an increasing sense of loneliness and a loss of meaning in people's lives, as well as a growing reliance on digital and social media platforms for information, fact-checking, and decision-making regarding trustworthy sources [7, 18, 36, 50, 64].[1] Indeed, social media and online environments have become the primary arena where people seek information and a means to shape public opinion [61, 53]. Although the future remains uncertain, it is plausible that this trend will continue.

If this is not enough to contend with, there is a growing awareness around Artificial Intelligence (AI)—or Intelligences in the plural—a rapidly advancing field that has been both the source of excitement, for its unprecedented possibilities and wide range of applications, and the target of sharp criticisms, due to the ethical and political concerns its uses and misuses inevitably raise. AI developments and applications have already impacted and will continue to influence all compartments of life, from politics to education, and from knowledge creation to the social construction of reality.

While its merits and applications are often praised for their potential to improve the lives of everyone—consider, for instance, their valuable use in the field of medicine and medical diagnostics, such as enhancing medical images (e.g., improving MRI resolution) or generating synthetic medical data to train AI models without relying on real patients' data—this new technology has not been exempt from criticism [49, 65]. This is because AI undoubtedly raises crucial concerns about potential misuse by antagonist or malevolent actors [1, 3]. Indeed, AI-generated imagery can be misused, and the images created can be handled unethically.

Furthermore, concerns about the potential loss of human control over AI are gaining traction. For instance, in a recent study published in the journal *Patterns*, a group of American researchers demonstrates, through empirical tests and case studies, that AI systems, whether used for specific or general purposes, are capable of deceiving humans using manipulation techniques, flattery, subterfuge, and deception. Although there has been much discussion in recent years about the fact that AI systems can produce false or inaccurate information—often referred to as "AI hallucinations" [41, 60]—the fact that these systems have learnt to deceive strategically and to manipulate humans in a premeditated way in order to achieve certain purposes is new and raises many questions.[2]

In such a scenario, where it seems inevitable that AI may deceive and resort to the strategic manipulation of humans, many questions are raised, and risks need to be taken into account. In the above-mentioned study [54], the authors discuss not only the potential structural risks arising from misleading AI systems—including

[1] Bauman [7] spoke of a "confessional society," where there are no longer secrets and privacy, but where these elements are projected into the public media arena. The result, however, is not a greater connection with others; on the contrary, on the virtual platforms, people find themselves increasingly alone. Bauman is sharp in pointing out that online platforms are substitutes for the need for social acceptance and for community, and it is not by chance that he speaks of a "cyber-seclusion".

[2] Peter S. Park, Simon Goldstein, Aidan O'Gara, Michael Chen, and Dan Hendrycks focus, in particular, on a form of deception that they have called *learned deception* and define this concept as follows: "the systematic inducement of false beliefs in others, as a means to accomplish some outcome other than by saying what is true" [54].

the persistence and perpetuation of false opinions and beliefs, political polarisation, and even a global weakening of decision-making processes, as humans increasingly rely on machines—but also the total loss of human control over AI, which could, in a rather apocalyptic and dystopic scenario, lead to the extinction of our species. Today, the need for a clearer and stricter regulation of AI and its applications is being argued for on many fronts [22] as a guarantee that the technology will serve humankind rather than posing an imminent threat to it.[3]

19.2 The Logic Beneath the Generation of AI Images: From One-to-Many to Many-to-One

In the broader field of AI applications, image creation indeed plays a substantial role. AI applications have rightfully had a remarkable impact on the generation, processing, and circulation of digital images. The emergence of AI has significantly altered the way in which we think about images today. The nature of this influence, the ramifications it implies, and the profound changes in the production of images, all deserve careful scrutiny and should not be underestimated.

AI-generated images in all their forms were not born in a vacuum. This is why it is important to put things into perspective and to examine this phenomenon through the lenses of media archaeology and media genealogy. Photography and portraiture, for instance, once celebrated for their unique ability to capture and preserve reality, now share the stage with images that are fully created and synthetised by algorithms. We have probably not yet fully grasped the magnitude and depth of this emergent phenomenon, nor have we fully fathomed the nature of this new wave of images—despite the fact that this issue has captured the media's attention and is at the forefront of heated debates within academia and the scientific community [20, 67].

Such images are generally referred to as "synthetic images," "Artificial aesthetics" [42], or images generated via images [60: 89], a key issue to which I will return in the remainder of the present study. Needless to say, the shift from more traditional, analog photography to what is now often referred to as "post-photography" [5, 32, 43, 46] raises pivotal questions such as regarding authorship, the epistemic value of such images, and the inner mechanism of their generation—including the relationship to the "virtual" space in which they are generated, commonly known as the "latent space." In what follows, I will come back to some of these key issues. Before delving deeper into them, I believe it is important to contextualise this emergent phenomenon.

[3] A first attempt at it has been made by the recent AI Act [22].

19.3 The Preconditions of Existence of AI Imagery: An "Epidemic of Images"

It is undeniable that in recent years—and, in fact, even before the advent of AI—there has been a significant increase in the number of digital images, some of which are entirely synthetic. This quantitative expansion—which, as we shall see in what follows, also has a qualitative dimension—is unprecedented. Probably never before in history has the dissemination of images reached such a peak, to a point where we might describe it, borrowing a term coined by Joan Fontcuberta, as an "epidemic of images." As he pointed out:

> For the first time we are both producers and consumers of images, and the simultaneous accumulation of these factors has resulted in an almost endless iconic avalanche. (...) We are thus passing through an era in which images are included in the category of excess, an era in which we speak of a mass production that leads to asphyxiation rather than emancipation. (...) The idea of the excess of images, which seems to produce at the same time hypervisibility and universal voyeurism, this apparent "epidemic of images," deserves a more in-depth and critical treatment that involves not only the specific conditions of the image, but also the logics of its management, dissemination and control [24: 53].

This point is worth pondering. Indeed, the logic of the managing, creating, disseminating, and controlling of images has been significantly altered by the introduction of AI, and its impact on visual culture is unprecedented [55, 60]. Today, for instance, algorithms play a role not only in the production of digital images, as they are engrained in smartphone cameras, but also in their filtering, censorship, and circulation on social media platforms, as well as for the creation of entirely synthetic photographs and videos such as deepfakes [8: 7].

However, the "avalanche of images" Fontcuberta described is only one corner of this issue. Beneath this statement lies the deeper question of how such an overabundance of images has created the fertile ground in which AI can thrive. Indeed, this vast collection of images—along with other forms of digital data—is continuously used, analysed, stored, and leveraged to generate new data.

From this viewpoint, the proliferation of images discussed above serves as the precondition for the AI-driven operation we see today. Here, I am referring to what, in the technical jargon of software engineering, goes under the rubric of "datasets" and "training sets"—the vast collections of data on which AI algorithms are "trained."[4] This point is crucial because it highlights one of the essential features of AI-generated images: their *generative* nature. That is, they do not emerge in isolation but rely on pre-existing images as their foundation. Genetically speaking, indeed, AI-generated

[4] A *training set* is a subset of a dataset that is specifically used to train a machine learning model. It is a crucial component in the machine learning workflow, as it contains the input–output pairs that the model learns from during the training process. The training set consists of examples from which a model learns to understand the underlying patterns, structures, and relationships within the data. It typically includes both the input features (independent variables) and the target labels or output values (dependent variables, when applicable). The main goal of using a training set is to enable the model to learn a mapping from inputs to outputs.

images are *second-order images* as they are modelled on other (already existing) images.

In this regard, and with specific reference to Large Language Models (LLMs) such as ChatGPT, the following excerpt from the recent book *AI Snake Oil* [49], explains the role and significance of the training process:

> "Chatbots" understand in the sense that they build internal representations of the world through their training process. Again, such representations might differ from ours, might be inaccurate, and might be impoverished because they don't interact with the world in the way that we do. Nonetheless, these representations are useful, and they allow chatbots to gain capabilities that would be simply impossible if they were merely giant statistical tables of patterns observed in the data. [49: 138]

Following this line of thinking, generative AI is, in a sense, a direct consequence and byproduct of the massive over-proliferation of images we witness today. It is in plain sight that AI systems integrate and utilise this vast amount of data in various ways to support their training processes, thereby driving their development and applications. Without the immense databases of images available on the web today, it would be difficult to even conceive of the emergence of generative AI, GANs, and the deepfake content that now circulates widely. This issue, I argue, has significant ramifications as concerns the age-old debate on originality versus replication.

By anticipating an argument that will be unpacked in greater detail in the pages ahead, it is worth noting that, traditionally, in the history of art, a unique object—thought of as a *unicum*—served as the model for creating multiple copies that sought to replicate the original—though these copies were generally limited in number. Today, however, AI image-generation tools such as GANs have reversed this logic of image creation and transform the original/copy dynamic. Indeed, AI-generated images are created by leveraging vast pre-existing datasets—that is, collections of already existing images, whether real or artificial. Apropos this matter, Antonio Somaini has convincingly argued that such images are, in effect, "images from images"; in other words, they are "images produced through the algorithmic processing of vast quantities of other images" [60: 89]. I will come back to this point in the remainder of this study. Not only has the mode of image creation changed—from a logic of one-to-many to a logic of many-to-one—but the speed at which images and *images of images* are generated has increased exponentially.

There are at least two important corollaries stemming from the fact that AI leverages existing datasets to create images. Firstly, there is the issue of intellectual property. This point is well summarised by David Bolter, who writes:

> What is the legal status of these generative images? Are they original or derivative works? The systems do not generate images ex nihilo. They draw on millions or even billions of text-image pairs scraped from the web for their underlying databases, such as, for example, LAION used by Stable Diffusion and some other systems. Do they therefore infringe on the rights of the human artists and producers whose works were scraped from the web and fed into the model? [11: 196–197]

Secondly, there is the issue of AI being recursive. Not only does it use already existing datasets, but it increasingly relies on datasets containing AI-generated images. Thus, it generates images that, in turn, become data that fuel its own training—a phenomenon known as "AI autophagy" [69].

19.4 Human Vision Versus Artificial Vision: Towards the Concept of "Algorithmic Similarity"

In this section, I will be dwelling on the concept of similarity between two images. To do so, I will briefly discuss the difference between human vision and artificial vision. This discussion is instrumental to our study inasmuch as it paves the way to the concept of "latent space," which, *mutatis mutandis*, is a pivotal notion in this paper. Latent space lies at the core of the generation, reproduction, and diffusion of AI imagery.

I will, however, focus on a very specific meaning of the term "similarity," as I am referring to the similarity between two images as detected and suggested by a computer—or, to use a term that has become quite fashionable today, by the logic of an algorithm—rather than the similarity perceived and recognised by human cognition. I shall term this particular type of similarity—namely, the similarity of two or more images determined by algorithms—"algorithmic similarity." Why is it necessary to introduce a new term? I employ "algorithmic similarity" to distinguish it from traditional notions of similarity and face recognition, which are understood as phenomena of human perception.

There are some significant differences between human vision and machine vision. Computer vision enables computers to perceive their surroundings and identify objects in a way that resembles human vision. Humans perceive objects as they are, retaining recognised patters and storing them in the brain until they encounter them again. In contrast, computer vision relies on machine learning techniques and algorithms to identify, distinguish, and classify objects. Human vision involves detecting light patterns and coordinating with the brain to translate light into images. Object recognition remains one of the most challenging problems in computer vision. While humans recognise objects effortlessly and can easily describe scenes, machines require extensive training to achieve similar capabilities.

In his essay *The Mask and the Face. The Perception of Physiognomic Likeness in Life and in Art* ([1973] 1978), Gombrich discusses the concept of "physiognomic constancy" in the human face, that is, a certain continuity that can be traced in a face despite the various changes it undergoes over time. To illustrate this idea, Gombrich compares two photographs of the philosopher Bertrand Russell: one taken when he was four years old and the other when he was ninety (Fig. 19.1).

A great many factors influence the perception, recognition, and identification of faces, some of which lie beyond our conscious control. Our perceptual judgments are often guided by schemas, patterns, and recognition templates that are not always obvious. Put simply, when recognising a face, one could say that in doing so, we project shapes and templates that make facial perception into something, in a sense,

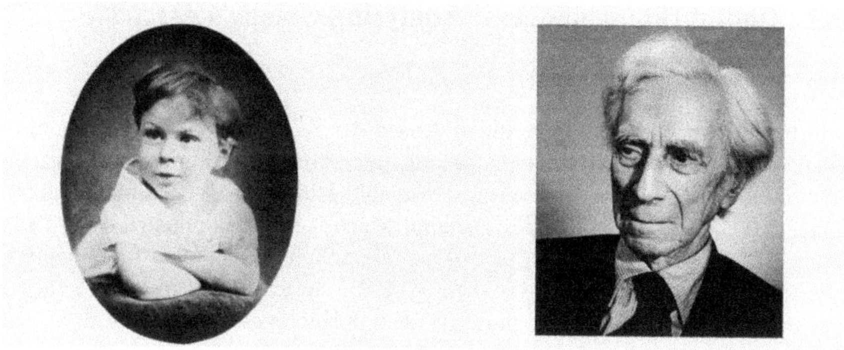

Fig. 19.1 An illustration of "physiognomic constancy" [27: 6–7]

of a constructed process. In this respect, reading a face falls short of a semiotic fallacy [33: 24]. In other words, when we see a face, we do not merely see it, we see it through the lens of our projections. As Ernst Gombrich ([26]: 103) writes:

> [...] we respond with particular readiness to certain configurations of biological significance for our survival. The recognition of the human face, on this argument, is not wholly learned. It is based on some kind of inborn disposition. Whenever anything remotely face-like enters our field of vision, we are alerted and respond.

Although the debate around face perception does not fit neatly into a rigid framework, for clarity's sake, we can distinguish two opposite perspectives: an innatist view and a constructivist one. While the former sees face perception as an innate disposition, constructivists argue that it is a learned ability [34: 103–121]. This issue remains unresolved, and there is no general agreement around such matters.

Gombrich demonstrates how Russell's two images remain perfectly recognisable despite the long time that elapsed between them, as they exhibit a certain "constancy." Although Russell's face has undergone all the transformations that time inevitably imprints, it remains the same face, and this continuity is both recognisable and constant. As [27: 6] remarked: "Yet all growth and decay cannot destroy the unity of the individual's looks—It certainly would not be easy to programme a computer to pick out the invariant, and yet it is the same face."

Since the time Gombrich wrote his essay, things have changed considerably. He could not have predicted that one day, there would be supercomputers capable of performing tasks that seemed impossible just a few years ago. That said, we should, however, take a step further and pose the following question: What if it is not human intelligence that establishes the similarity between two individuals and grasps the "unity" of a face in two photographs, but artificial intelligence that identifies what Gombrich called the "invariant" or "physiognomic constancy"?

As Gombrich wrote some forty years ago, it is certainly difficult to program a computer to detect the invariant of one, two, or more images of a face. However, it is certainly not impossible, as recent technological developments in the field of AI, as well as the following example, clearly demonstrate.

19.5 Digital Doppelgängers: Analysing Google's *Art Selfie* Algorithm

I will begin with a thought experiment. I used *Art Selfie*—a recent app created by Google—to "read" my own facial image and select, from the various famous portraits in the art history databases, the one I most resemble. I first took a selfie with my mobile phone, then uploaded the image to the app, which "read" and registered my facial features. The app compared the image of my face to hundreds of other face images to detect similarities. Specifically, *Art Selfie* works by comparing the image of a face to those portrayed in paintings and portraits from the archives of museums all around the world.

Thus, among the several thousand face images in the vast database of famous paintings and portraits archived and stored by Google, who do I resemble the most? According to Google's *Art Selfie* app, in fact, the image of my face resembles many portraits (Fig. 19.2).

From the output provided by *Art Selfie*—that is, from the selection of ten portraits that the app returned to me—it is clear that the image of my face falls into a well-defined cluster. This cluster can tentatively be defined as follows: it includes portraits of men, portraits of men with beards or moustaches and with short hair, and images of men of around 40–50 years old.

Another important aspect that can be inferred from the list of portraits selected by *Art Selfie* is that the algorithm used by the app calculates the similarity between two face images and quantifies it numerically. Of the ten portraits identified, according to the algorithm, two have a higher percentage of similarity than the others (75%), and there are two others at 74%, four at 73%, and two at 72%. However, the app

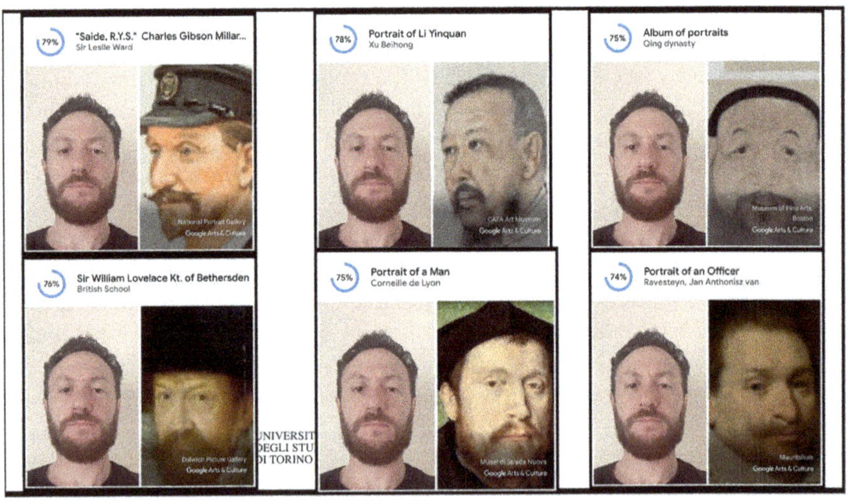

Fig. 19.2 Who do I look like according to Google's art selfie

does not explain in any greater detail how this percentage of likeness is calculated. Still, it remains a quite innovative aspect that differentiates this type of mathematical similarity from the "physiognomic similarity" discussed earlier.

This example illustrates how AI can effectively cluster large amounts of images. The image clustering operation performed by the app, even with its limitations, is interesting from an operational point of view, as it demonstrates how AI and algorithms work with images and what we can achieve with them. In this case, the app associates images considered similar into well-defined clusters. In other words, visualisation through clustering shows us that certain images are similar and that there is, therefore, a type of similarity based on calculation and algorithmic prediction. Needless to say, to assess the usefulness and sophistication of how the clustering works, one would need to consider how this technique can be applied to the visualisation of clusters in large datasets, such as the portrait databases of the world's museums, where thousands of images are clustered according to similar characteristics. It is clear that the clustering process requires prior identification of the salient features of each image [47].

19.6 A Stroll Through the "Latent Space"

Given that one of the underlying architectures supporting many of the technologies discussed in this paper is the concept of "latent space," it is worth pausing at this juncture to examine the idea further.

Let us state from the outset that the term "latent" has both a metaphorical sense and a more restricted, technical meaning in the phrases "latent space" and "latency." The word "latent" literally means "hidden, occult" and refers to something that cannot be seen. Needless to say, this term is multifaceted and is used across various disciplinary fields.

The term "latent" is often juxtaposed with its opposite in semantic pairs such as latent/manifest, latent/potential, and virtual/actual.[5] In this chapter, however, I will not dwell on the genealogy, semantics, or on the metaphorical uses of the term—an endeavour that would require much lengthier and more in-depth discussion. Instead, I will focus on the technical meaning of the concept of "latent space" and extrapolate from it some hypotheses about the status of AI-generated images. As Estelle Blaschke pointed out:

> ["Latent space"] is an abstract and mathematical entity whose significance and cultural, epistemological, and political implications can hardly be overstated today. Latent space is a key element in any machine leaning process, and thus essential to any AI model. It is the abstract space in which complex digital objects (such as images, texts, and sounds) are encoded and represented in a simplified, compact form, so that they can be processed using various mathematical operations. [8: 13]

[5] This semantic opposition, for example, is often found in the history of the face and physiognomy. It is worth recalling, for example, that for Ortega y Gasset [51], the face is founded on the latent/manifest dynamic, through the phenomenon of expression.

Suppose we have a large amount of information at our disposal and provide it to a neural network. Regardless of the quantity, sophistication, and quality of this information, it serves as input to the network. Based on this input, the neural network will generate an output—a series of transformations applied to the data.

Suppose again that we provide the neural network with a large amount of information about a specific object—for instance, an entire database of wines containing multiple variables such as acidity, colour, viscosity, provenance, and so on, for a total of twenty variables. Since there are twenty variables, the information space has a dimension of 20. However, the task of the neural network is simply to determine whether a given wine is good (1) or bad (0). In this case, the latent space corresponds to the space of outputs—that is, all possible results the net can compute. Essentially, it represents a one variable (good/bad) characterisation of wines that were originally described using twenty different variables.

The adjective "latent," therefore, should be interpreted to mean that this single variable (good/bad) "hides" within itself a kind of summary or skeleton of the original twenty variables, which remain unseen. In semiotic terms, that is a virtuality that has not yet been actualised. As Blaschke remarks, "because of their multidimensionality and their vectorial and mathematical nature, these latent spaces remain completely inaccessible, invisible, unimaginable" [8: 13].

This is an important point, as deep learning relies on data compression as one of its cornerstones. Compression "allows us to eliminate all superfluous information and focus only on the most important features."[6] Latent space is based on the so-called "manifold hypothesis," which assumes that the variables we observe are actually governed by a much smaller set of unobserved variables, considered "latent" or "hidden."

In the case of images, the process remains largely the same. In other words, the observable variables—in this case the pixels of the image, which number in the tens of thousands—are actually governed by a much smaller set of underlying variables. For images representing faces, if only two variables are used, they function as points on a map, where the latent variables correspond to the longitude and latitude. It is also important to emphasise that thousands of other images are needed to reconstruct the latent space of a single image. This enormous quantity of images serves to "train" the algorithms, which learn by mimicking the neuronal processes of the human brain. This is why the machines require a vast amount of data to learn in statistical and mathematical terms. It also explains why AI systems work by detecting patterns and recognising constant features in the data it analyses.

[6] Ekin Tiu, "Understanding Latent Space in Machine Learning," online, https://medium.com/data-science/understanding-latent-space-in-machine-learning-de5a7c687d8d (last access 22/03/2025).

19.7 The Emergence of GANs and Deepfakes

The introduction of Generative Adversarial Networks (GANs) in 2014 was a game changer in the field of generative AI [28]. GANs are a type of machine learning model consisting of two neural networks—a *generator* and a *discriminator*—that compete against each other. The generator creates data (such as images or videos), while the discriminator tries to distinguish between real data and data generated by the model. GANs are used to generate new data that mimic real data, including images, videos, and even music. They can create realistic images of faces, objects, or entire scenes, based on training data. The generator aims to create data that looks as realistic as possible, while the discriminator evaluates the authenticity of the generated data. Over time, the generator improves at making data that is increasingly difficult to distinguish from real data, driven by feedback from the discriminator.

GANs have been trained to perform various photographic tasks, which include both the modification of existing images and the generation of new ones ([59]: 91–94). Here is a list of some of the most common operations ([59]: 93–94):

- "Style transfer" [58], which is a technique that allows the transfer of the artistic style of one image onto the content of another image. This process involves combining the content of one image (e.g., a photo) with the visual style of another image (e.g., a famous painting), creating a new image that retains the content of the first image but adopts the aesthetic style of the second;
- "Age modulation." This process involves using GANs to modify a person's facial features based on age-related patterns such as wrinkles, sagging skin, and changes in hair colour. The model generates images that depict how a person might look at different ages, often based on a single photograph. This works in both directions, "age progression" refers to creating images that simulate the natural aging process of an individual, showing how their face may evolve over time; "age regression" is the reverse process, where the AI makes someone look younger;
- "Image animation." This can be seen in applications that animate portraits to simulate expressions, like in the *Deep Nostalgia* feature from MyHeritage, which animates old photos;
- "Image inpainting" or "image completion," a task that aims to "restore missing or damaged regions of images with semantically coherent content" [15];
- "Image outpainting" or "image extrapolation" [57];
- "Image upscaling." GANs can be used for image super-resolution by upscaling low-resolution images with minimal information loss and improved image quality;
- Last but not least, we should mention all operations related to the creation of deepfakes in both still and motion images, which I discuss in the next session.

The above mentioned operations of image manipulation can be refined in a typology on the basis of the criteria of the additive (adding information to an image) or subtractive (removing or replacing information) quality of the operation. We can categorise the techniques as follows (Table 19.1).

Table 19.1 Typology of image manipulation techniques

Category	Technique	Description	Operation type
Additive	Style transfer	Applies the artistic style of one image to another	Adds new visual style
	Age modulation	Alters facial features to show aging or rejuvenation	Adds age-related details
	Image animation	Adds movement or expressions to static images	Adds motion and new frames
	Image upscaling	Increases resolution with added detail	Adds new pixels and sharpness
	Image outpainting	Expands image beyond original frame with new content	Adds new visual elements
Subtractive	Image inpainting	Fills in missing or damaged areas, replacing content	Removes gaps, replaces missing info
	Deepfakes	Alters or replaces facial features, suppressing identity	Replaces original features

As pointed out before, GANs can not only perform tasks that modify already existing images, but they can also generate new ones. Deepfakes are a case in point. A deepfake is an AI–generated content, generally a photorealist image or video, that whilst purporting to be true, is not. This technology can create realistic images, videos, and audios depicting people saying or doing things they never said or did. Deepfakes gained momentum in 2017 when the word was coined as a portmanteau of two terms, "Deep Learning" and "fake" (counterfeit, inauthentic, false, phony). Since its inception, deepfake technology has rapidly evolved, including today not only images and videos, but also synthetic voice impersonations, with important ramifications for trust and scamming [1, 3]. Indeed, deepfakes can be misused and the images created can be handled unethically.

The recent AI Act defines the deepfake as an "AI-generated or manipulated image, audio or video content that resembles existing persons, objects, places, entities or events and would falsely appear to a person to be authentic or truthful" [22]. The introduction of such ground-breaking technology has far-reaching implications both for the concept of identity [19] as well as for the study of the human face on a variety of levels [17, 25, 38].[7]

With the introduction of generative AI, spotting fakes has become harder and harder [14], as the level of technological sophistication of digital image manipulation has markedly increased [1, 3] and the technical possibilities of face swapping and synthetic imaging are now legion and widely accessible [28, 52, 59, 60].

[7] On this point, see the research conducted within the ERC Project FACETS [29, 40, 66].

Because deepfake technology is able to generate hyper-realistic videos, it is very difficult to sift the authentic from the fake. This uncertainty around images creates potential for spreading disinformation that can influence political elections [6], finance [21], and all compartments of life. This issue, coupled with the blurring of the distinction between fact and fiction, truth and falsehood, and illusion and reality, is a key point that raises concern [16, 23].

Deepfakes are of various kinds and produced with different purposes. Meskys et al. [44] sketched out a fourfold taxonomy, including revenge porn deepfakes, deepfakes used in political campaigns, commercial deepfakes, and creative ones. Each type has different ethical and legal implications. Whilst deepfake technology opens up a wide spectrum of possibilities, including ethical and non-malicious applications [13], the vast majority of deepfake videos are pornographic and entail non-consensual use of private images of the individuals portrayed [1, 2].

Indeed, deepfakes can potentially yield an erosion of trust in visual media and a breach in the credibility of visual and auditory information, which is pivotal to everyone. By altering and foiling these information channels, deepfakes undermine the reliability and accuracy of information. It is, therefore, imperative to stress the need for a higher level of visual literacy that will enable citizens to better identify deepfakes. This is especially important for older adults and vulnerable citizens, more susceptible to deception [21, 37], often in a condition of cognitive decline and loneliness [48], and less equipped to discern AI generated imagery [35].

19.8 Real or not? "Platform Realism," the Socio-cultural Significance of Authenticity, and the Remediation of Photography and Cinema

It is often claimed that AI-generated imagery, specifically photorealistic images, presents something as real and authentic when it is not. As I have argued elsewhere [30], AI images are not indexes, at least if the indexicality of an image is understood in semiotic terms. Or, as Roland Meyer puts it, these are simulacra that have no "indexical relationship to any reality, not even a simulated one" [45: 9]. The introduction of AI and algorithms calls for a reconfiguration of the whole domain of photography, as the categories used so far are probably no longer adequate to fully grasp this emergent phenomenon: "The dominant idea of the photographic image as "imprint" or "index," and of the "decisive moment," are no longer valid in today's context" [8: 8].

From a historical perspective, we observe that certain functions of older media have been adopted by new media, with the function of authenticity being transferred from one medium to another over time. As Lotman [39]: 63–64) pointed out, from a cultural perspective, the role of proving a subject's authenticity was originally performed by the portrait, which anticipated the role of photography. From a semiotic standpoint, it is important to emphasise that authenticity is not solely about the likeness between people's faces and their representations—that is, the correspondence between the object and its model, whether a portrait, a photograph, a video, or another medium (what we previously termed "visual accuracy").

On the contrary, what is crucial is that the recognition of likeness has a socio-cultural dimension that should not be overlooked, as identifying an image as similar to another always occurs within a specific socio-cultural context. In this regard, Lotman provides examples of photographs that do not closely resemble the people they portray, demonstrating a low degree of likeness between the subject and its representation—in other words, a low level of visual accuracy. A caricature or a portrait may exhibit a higher degree of likeness compared to a poorly taken photograph. Nonetheless, due to the cultural functions and values attached to photography as a marker of authenticity, truth-value, proof, and evidence, a photograph will always be preferred over a handwritten sketch or caricature for personal identification [39: 65]. Lotman thus rightly observes that it is the "formal ability to be *a sign of likeness* that fulfils a given conventional function" [39: 65].

Returning to AI-generated images, we notice that they both attract and, in a way, confuse the viewer. Perhaps it is this very confusion, puzzlement, and uncertainty that makes them appealing, as they invite doubt: Is it real? Is it not? This hesitation in recognising whether an image is real or artificial contributes to their fascination. This kind of mismatch, tension, or friction—between looking like a photograph while, at the same time, not being one—is one of their most important aesthetic characteristics.

According to Gunning, even the awareness that such images have been algorithmically altered adds to their appeal. As he writes,

> It seems to me that the power of the digital manipulation of photographs depends on our recognizing them as *manipulated photographs*, being aware of the strata of the indexical (or perhaps better, the visually recognizable) beneath the manipulation. [...]
>
> The power of the digital (or even the traditional photographic) to "transform" an image depends on maintaining *something* of the original image's visual accuracy and recognizability. I use this phrase ("visual accuracy and recognizability") to indicate the manner in which indexicality intertwines with iconicity in our common assessment of photographs. Our evaluation of a photograph as accurate (i.e. visually reflecting its subject) depends not simply on its indexical basis (the chemical process), but on our recognition of it as looking like its subject [31].

The argument is worth considering. When discussing deepfakes and digitally altered images, we tend to focus on their ability to modify an image in a way that is imperceptible to the human eye and evades detection. Time and again, we hear that the primary threat of this technology lies in its capacity to blur the line between the real and the fake. However, this is only one side of the coin, and the perspective can be reversed.

Following Gunning's thesis, our fascination with fakes and digital image manipulation does not stem from *being fooled by the image*, as he puts it—that is, from AI and deepfakes being completely undetectable. On the contrary, our appreciation of skilful manipulation arises precisely from "its playful push–pull between its associations with accuracy (that is how a woman's face actually looks) and its obvious distortion (but no face looks like that)" [31]. Meaning emerges within this space of tension—the irreconcilable gap between visual accuracy and complete unrecognisability.

Point taken. However, this neither fully elucidates the nature and ontology of these images nor advances our understanding in this domain.

An additional theoretical framework that helps grasp this issue is the observation that deepfakes and other forms of generative AI actually "remediate" [12] older media; by capitalising on the qualities of these older media, these new forms of imagery acquire their own distinct qualities. Or, as Somaini asks, what makes a picture look like a picture?

At this stage, I believe it is useful to draw on the notion of photorealism and, more specifically, on the concept of "platform realism" as developed by Meyer [45]. As he puts it, "in general, 'realism' is the holy grail of AI image synthesis. With each update of their models, companies like Midjourney and OpenAI strive for increasingly 'realistic' renderings, which is also what customers expect from these tools" [45: 7].

According to Meyer [45: 7], this concept of realism encompasses three dimensions:

1. Legibility;
2. Plausibility;
3. Platform realism.

The characteristic of "legibility" in images refers to the expectation that "synthetic images are expected to match a given textual description (or prompt) as closely as possible" [45: 7]. "Plausibility," on the other hand, pertains to how these images meet "expectations of visual coherence, physical integrity, and spatial consistency" [45: 7]. The third and final dimension, termed "platform realism"—a concept introduced by Jakob Birken— also called "realism immanent in the media" [9], relates to the crucial fact that the expectations of legibility and plausibility towards these images are shaped by pre-existing (older) media. As Blaschke [8: 12] and colleagues remark, "these images do not seek to imitate the physical world, rather they set out to mimic the media that dominated the visual culture of the last century: photography and cinema." By imitating older media, AI-generated imagery adopts the connotations inherent to them. In the case of cinema and photographs, their truth-value, indexicality, and evidentiary status are appropriated by new forms of media that leverage and exploit these qualities.

19.9 Coda

It is difficult to make conclusive and definitive remarks on a subject that is quickly evolving on a daily basis. This is a rapidly expanding field whose rules seem elusive. For this reason, making a definitive statement remains a challenge. Conclusions, in this regard, should be framed as an open set of possibilities.

AI-generated images call for a redefinition of how we perceive traditional forms of imagery, such as photography, cinema, and painting. While there are sharp points of rupture between these forms of representation, there are also elements of continuity. AI imagery does not merely imitate older media; rather, it reshapes our visual culture, prompting us to reconsider long-held assumptions about authenticity, authorship, and artistic intent.

Moreover, as AI-generated visuals become more pervasive, they challenge existing ethical, legal, and epistemological frameworks. Issues of misinformation, intellectual property, and the role of human creativity in an era of machine-generated content remain open-ended debates. Instead of seeking rigid boundaries between AI imagery and traditional media, we should perhaps focus on understanding how these technologies transform our engagement with images, truth, and representation itself. In this sense, AI-generated visuals are not just new forms of media but catalysts for rethinking the very nature of visual culture in the digital age.

References

1. Ajder, H., Patrini, G., Cavalli, F., Cullen, L.: The State of Deepfakes: Landscape, Threats, and Impact. Deeptrace, London (2019)
2. Ajder, H., Patrini, G., Cavalli, F.: Automating Image Abuse: Deepfake Bots on Telegram. Sensity, Amsterdam (2020)
3. Amezaga, N., Hajek, J.: Availability of voice deepfake technology and its impact for good and evil. In: Trygstad, R., Zheng, Y. (eds.) SIGITE '22: Proceedings of the 23rd Annual Conference on Information Technology Education, pp. 23–28. ACM Press, New York (2022). https://doi.org/10.1145/3537674.3554742
4. Antonio, S.: Algorithmic images: artificial intelligence and visual culture. Grey Room (93), 74–115 (2023). https://doi.org/10.1162/grey_a_00383
5. Bajac, Q.: Dopo la fotografia: Dall'immagine analogica alla rivoluzione digitale. Italian translation by Boni, G. Contrasto, Roma (2011)
6. Barrett, P.M.: Disinformation and the 2020 election: how the social media industry should prepare (Online). Stern, NYU, New York (2019). Available: https://bhr.stern.nyu.edu/tech-disinfo-and-2020-election
7. Bauman, Z.: Liquid times: living in an age of uncertainty. Polity Press, Cambridge (2007)
8. Blaschke, E., Bonhomme, M., Joschke, C., Somaini, A.: Introduction. Photographs and algorithms. Transbordeur. Photographie Histoire Sociétê **9**, 1–20 (2025)
9. Birken, J.: Vom Pixelrealismus: Takeshi Muratas Stillleben Cyborg. Schlaufen, Berlin (2023)
10. Blaschke, E., Bonhomme, M., Joschke, C., Somaini, A.: Introduction. Photographies et algorithmes Transbordeur 96–19 (2025). https://doi.org/10.4000/transbordeur10.4000/13dxk10.4000/13dwn
11. Bolter, D. J.: AI Generative Art as Algorithmic Remediation. IMAGE: Interdisc. J. Image Sci. **37**(1), 195–207 (2023)
12. Bolter, D.J., Grusin, R.A.: Remediation: Understanding New Media. MIT Press, Cambridge, MA (2000)
13. Caporusso, N.: Deepfakes for the good: a beneficial application of contentious artificial intelligence technology. In: Advances in Intelligent Systems and Computing, pp. 235–241. Springer, Cham (2020). https://doi.org/10.1007/978-3-030-51328-3_33
14. Chalmers, D.: Reality+: Virtual worlds and the problems of philosophy. W. W. Norton & Company, New York (2022)

15. Chen, S., Atapour-Abarghoue, A., Zhang, H., Shum, H. P. H.: AMxT: Mamba x transformer for image inpainting. arXiv preprint arXiv:2407.16126 (2024)
16. Chesney, R., Citron, D.: Deepfakes and the new disinformation war: the coming age of post-truth geopolitics. Foreign Aff. **147**, 147–158 (2019)
17. Conte, P.: Mockumentality: from hyperfaces to deepfakes. World Lit. Stud. **4**(11), 11–25 (2019)
18. Desmet, M.: The psychology of totalitarianism. Chelsea Green Publishing, London (2022)
19. Donath, J.S.: Identity and deception in the virtual community. In: Kollock, P., Smith, M. (eds.) Communities in Cyberspace, pp. 29–59. Routledge, London (1998)
20. Dondero, M.G., Aldama, J.A., Leone, M.: Aspects of AI semiotics: enunciation, agency, and creativity. Semiotica **262**, 1–3 (2025). https://doi.org/10.1515/sem-2025-0014
21. Ebner, N. C., Pehlivanoglu, D., Shoenfelt, A.: Financial fraud and deception in aging. Adv. Geriatr. Med. Res. **5**(3), e230007 (2023). https://doi.org/10.20900/agmr20230007
22. European Parliament: Artificial Intelligence Act. European Parliament, Brussels (2024). Available: https://www.europarl.europa.eu/doceo/document/TA-9-2024-0138_EN.pdf
23. Fallis, D.: The epistemic threat of deepfakes. Philos. Technol. **33**(2), 147–164 (2020). https://doi.org/10.1007/s13347-020-00419-2
24. Fontcuberta, J.: La furia delle immagini. Note sulla postfotografia. Einaudi, Torino (2018)
25. Geil, A., Jirsa, T. (eds.): Reconfiguring the Portrait. Edinburgh University Press, Edinburgh (2023)
26. Gombrich, E.: Art and Illusion: A study in the psychology of pictorial representation. Princeton University Press, Princeton (1956)
27. Gombrich, E.: The mask and the face: The perception of physiognomic likeness in life and the art. In: Gombrich, E.H., Hochberg, J., Black, M. (eds.) Art, Perception, and Reality, pp. 1–46. Johns Hopkins University Press, Baltimore (1978)
28. Goodfellow, I., Pouget-Abadie, J., Mirza, M., Xu, B., Warde-Farley, D., Ozair, S., Courville, A., Bengio, Y.: Generative adversarial nets. Adv. Neural. Inf. Process. Syst. **27**, 1–9 (2014)
29. Gramigna, R.: Preserving anonymity: deepfake as an identity-protection device and as a digital camouflage. Int. J. Semiot. Law **37**, 729–751 (2024). https://doi.org/10.1007/s11196-023-10079-y
30. Gramigna, R.: Indexicality rethought. In: Leone, M. (ed.) Face It!, pp. 101–115. Routledge, London (2025)
31. Gunning, T., Beckman, K., Ma, J.: Still moving between cinema and photography what's the point of an index? or faking photographs, pp. 23–40, Duke University Press (2008)
32. Gunthert, A.: L'image partagée. La photographie numérique. Textuel, Paris (2015)
33. Gurisatti, G.: Dizionario fisiognomico: Il volto, le forme, l'espressione. Quodlibet, Macerata (2006)
34. Guthrie, S.E.: Faces in the Clouds: A New Theory of Religion. Oxford University Press, Oxford (1995)
35. Hamrin Reinhed, H., Blomgren, E.: Age disparities in discerning AI-generated imagery. Degree project, KTH Royal Institute of Technology, Stockholm (2024)
36. Harari, Y. N.: 21 Lessons for the 21st century. Jonathan Cape, London (2018)
37. Herrera, L., Van Sickle, L., Podhradsky, A.: Bridging the protection gap: innovative approaches to shield older adults from AI-enhanced scams. In: Cyber Awareness and Research Symposium (CARS) 2024, pp. 1–9. IEEE, New York (2024). https://doi.org/10.1109/CARS61786.2024.10778759
38. Jirsa, T., Rosenberg, R.: (Inter)faces, or how to think faces in the era of cyberfaces. World Lit. Stud. **4**(11), 11–25 (2019)
39. Lotman, J.: Il ritratto. In: Lotman, J. (ed.) Il girotondo delle muse: Saggi di semiotica delle arti e della rappresentazione, pp. 63–96. Moretti & Vitali, Bergamo (1998)
40. Leone, M.: The Hybrid Face: Paradoxes of the Visage in the Digital Era. Routledge, New York (2024)
41. Maleki, N., Padmanabhan, B., Dutta, K.: AI hallucinations: a misnomer worth clarifying. In: IEEE Conference on Artificial Intelligence (CAI) 2024, pp. 133–138. IEEE, Singapore (2024). https://doi.org/10.1109/CAI59869.2024.00033

42. Manovich, L., Arielli, E.: Artificial aesthetics: Generative AI, Art, and Visual Media. Self-published, San Francisco (2024). Available: https://manovich.net/index.php/projects/artificial-aesthetics
43. Marra, C.: L'immagine infedele: La falsa rivoluzione della fotografia digitale. Bruno Mondadori, Milano (2006)
44. Meskys, E., Kalpokiene, J., Jurcys, P., Liaudanskas, A.: Regulating deep fakes: legal and ethical considerations. J. Intell. Property Law Pract. **15**(1), 24–31 (2019)
45. Meyer, R.: "Platform realism". AI image synthesis and the rise of generic visual content. Transbordeur [En ligne] 9, 2025. http://journals.openedition.org/transbordeur/2299; https://doi.org/10.4000/13dwq
46. Mitchell, W.: The reconfigured eye: visual truth in the post-photographic era. MIT Press, Cambridge (1992)
47. Morra, L., Santangelo, A., Basci, P., Piano, L., Garcea, F., Lamberti, F., Leone, M.: For a semiotic AI: bridging computer vision and visual semiotics for computational observation of large-scale facial image archives. Comput. Vis. Image Underst. **249**, 104187 (2024). https://doi.org/10.1016/j.cviu.2024.104187
48. Napal, D.: Imitation is the sincerest form of fraudulent activity: artificial intelligence in financial scams against older adults. Bifocal **45**(6), 1–12 (2024)
49. Narayanan, A., Kapoor, S.: AI Snake Oil: What Artificial Intelligence Can Do, What It Can't, and How to Tell the Difference. Princeton University Press, Princeton (2024)
50. Newport, C.: Digital minimalism: choosing a focused life in a noisy world. Penguin, New York (2019)
51. Ortega y Gasset, J.: Sobre la expresión fenómeno cósmico, O.C., t. II (1925)
52. Paris, B., Donovan, J.: Deepfakes and Cheap Fakes: The Manipulation of Audio and Video Evidence. Data & Society, New York (2019)
53. Pariser, E.: The filter bubble: what the internet is hiding from you. Penguin, New York (2011)
54. Park, P.S., Goldstein, S., O'Gara, A., Chen, M., Hendrycks, D.: AI de-ception: a survey of examples, risks, and potential solutions. Patterns **5**(5), (2024). https://doi.org/10.48550/arXiv.2308.14752
55. Pinotti, A., Somaini, A.: Cultura visuale: immagini, sguardi, media, dispositivi. Milano, Einaudi (2016)
56. Roland, M.: Platform Realism. AI Image Synthesis and the Rise of Generic Visual Content. Transbordeur 9 (2025). https://doi.org/10.4000/transbordeur10.4000/13dxk10.4000/13dwq
57. Sabini, M., Rusak, G.: Painting outside the box: image outpainting with GANs. arXiv preprint arXiv:1808.08483 (2018)
58. Singh, A., Jaiswal, V., Joshi, G., Sanjeeve, A., Gite, S., Kotecha, K.: Neural style transfer: a critical review. IEEE Access **9**, 131583–131613 (2021). https://doi.org/10.1109/ACCESS.2021.3112996
59. Somaini, A.: Dreams, hallucinations, visions: generative AI and the long history of surrealism. Paradigmi Rivista di Critica Filosofica **3**, 541–556 (2023). https://doi.org/10.30460/112741
60. Somaini, A.: Algorithmic Images: artificial Intelligence and Visual Culture. Grey Room **93**, 74–115 (2023). https://doi.org/10.1162/grey_a_00383
61. Sunstein, C.R.: Republic.com 2.0. Princeton University Press, Princeton and Oxford (2007)
62. Tang, Z., Goh, D. H.-L., Lee, C. S., Yang, Y.: Understanding seniors' strategies for identifying deepfakes. In: Stephanidis, C., Antona, M., Ntoa, S., Salvendy, G. (eds.) HCI International 2024 Posters: 26th International Conference on Human-Computer Interaction, pp. 236–244. Springer, Cham (2024). https://doi.org/10.1007/978-3-031-61947-2_26
63. Turkle, S.: Ghosts in the machine. Sci. **35**(6), 36–39 (1995). https://doi.org/10.1002/j.2326-1951.1995.tb03214.x
64. Turkle, S.: Simulation and its discontents. MIT Press, Cambridge (2009)
65. Vallor, S.: The AI Mirror: How to Reclaim our Humanity in an Age of Machine Thinking. Oxford University Press, Oxford (2024)
66. Viola, M., Voto, C.: Designed to abuse? Deepfakes and the non-consensual diffusion of intimate images. Synthese **201**, 30 (2023). https://doi.org/10.1007/s11229-022-04012-2

67. Wilde, L., Lemmes, M., Sachs-Hombach, K. (eds.): Generative imagery: towards a "new paradigm" of machine learning-based image production. IMAGE: Interdisc. J. Image Sci. **3**(1) (2023)
68. Windley, P. J.: Digital Identity. O'Reilly Media, Sebastopol (2005)
69. Xing, X., Shi, F., Huang, J., Wu, Y., Nan, Y., Zhang, S., Fang, Y., Roberts, M., Schönlieb, C.-B., Del Ser, J., Yang, G.: When AI eats itself: on the caveats of AI autophagy. arXiv preprint arXiv:2407.16126 (2024)

Remo Gramigna is tenure-track assistant professor in Semiotics at the University of Turin. He is Ph.D. in Semiotics and Culture Studies (University of Tartu, Estonia) and he has been a member of the ERC project FACETS (*Face Aesthetics in Contemporary E-Technological Societies).* His research primarily focuses on semiotics, cultural studies, and communication studies. His scientific interests include lying and deception in human interactions, fakes deepfakes and manipulation, as well as general semiotics and the semiotics of culture. Recent publications include: *Augustine and the Study of Signs, Signification, and Lying* (2020), *Imagining others: deception, prediction, and disguised intentions in strategic interactions* (2020). He has worked in the editorial team of *Sign Systems Studies.* He published in international journals such as *Semiotica, Versus, Frontiers of Narrative Studies, Lexia, Sign Systems Studies, DeSignis.*

Chapter 20
What Does the Robot See? A Cognitive Semiotic Perspective on Robotic Learning Algorithms

Martina Bacaro

Abstract The field of Human–Robot Interaction (HRI) explores the mechanisms that enable robots to engage meaningfully with humans across diverse contexts. A central focus within this domain is robotic learning, where Learning from Demonstration (LfD) has emerged as a promising approach for enabling robots to acquire behaviors by observing human actions. However, the implementation of LfD poses fundamental challenges, particularly in parsing human demonstrations and translating them into executable robotic actions. These challenges are often framed within a Theory of Mind (ToM)-based paradigm, which assumes that social understanding relies on inferring human intentions through internal state modeling. Yet, this reliance on inferential models not only imposes significant computational demands but also reflects deeper epistemological assumptions about cognition and interaction. This chapter critically examines these assumptions and explores an alternative approach rooted in cognitive semiotics and enactive theories of social cognition. By drawing on insights from developmental robotics and *motionese*, I consider whether a more interaction-sensitive perspective—one that foregrounds embodied, relational meaning-making—might offer new pathways for robotic learning. This shift not only challenges established perspectives in LfD but also calls for a reconsideration of how robotic learning algorithms engage with the dynamics of social interaction.

Keywords Human–robot interaction · Social robotics · Learning from demonstration · Cognitive semiotics · Enactive cognitive science

20.1 Introduction

The field of robotics faces a critical near-term challenge: developing robots capable of engaging in fluid, reliable interactions with humans during shared tasks [17]. This challenge has become increasingly pressing as robots are transitioning from

M. Bacaro (✉)
University of Bologna, Bologna, Italy
e-mail: martina.bacaro2@unibo.it

P. Conte et al. (eds.), *Algomedia. The Image at the Time of Artificial Intelligence*, Lecture Notes in Morphogenesis, https://doi.org/10.1007/978-3-032-08726-3_20

controlled industrial environments to social spaces where they must interpret and respond appropriately to human actions [79]. Central to this endeavour is the fundamental question of how robots "see" and process salient aspects of human actions during interactions. While current approaches rely heavily on sophisticated machine learning models and extensive data collection, results have fallen short of expectations. These limitations are often attributed to technical constraints such as insufficient training data, implementation challenges, or computational power. This chapter proposes a different idea, by sustaining that the core obstacles may lie deeper, within the theoretical models of interaction underlying current implementations.This perspective suggests that reconceptualizing our approach to human–robot interaction models could yield significant advances in developing socially capable robots, potentially unburdening computational resources while making better use of environmental cues. Specifically, this paper advocates for reformulating machine learning approaches through an enactive and cognitive semiotic framework of interaction. This theoretical shift builds upon and extends the evolutionary trajectory of robotics over the past three decades, where implicit and explicit moves toward embodied and situated cognition have already begun reshaping the field.

20.2 How to Make a Robot Able to See

Artificial Intelligence (AI) has been particularly challenged by the task of building robotic artifacts. Since the development of early robots, robotics has served as a robust testing ground for AI hypotheses, revealing the limitations of computational models when applied to real-world environments. Assumptions that proved effective in controlled, simplified environments often failed when deployed in robots that needed to interact with their surroundings, locate objects, and perform physical tasks. For example, SHAKEY, the first robot of its kind, incorporated computational mechanisms that had previously demonstrated success in expert systems and navigation algorithms. However, once embedded in SHAKEY, these algorithms produced a machine that was hesitant and often incapable of executing even basic actions, earning its name due to its unsteady performance [45].

The growing divide between AI and robotics contributed to the decline of AI enthusiasm in the 1970s (e.g., the so-called AI winter; Boden [13]), leading to a clear separation between the two fields. However, in the 1980s and 1990s, robotics played a crucial role in revitalizing artificial intelligence, particularly through the emergence of parallel cognitive architectures and the paradigm shift introduced by *Nouvelle AI* under Rodney Brooks' leadership [19]. This shift redefined some AI principles by rejecting the traditional computationalist perspective and instead emphasizing the role of an agent within its environment (in line with first-order cybernetic approaches, Brooks [20]). According to this approach, intelligence arises from the fundamental mechanisms of movement and embodiment, rather than from abstract computation alone [18].

One of the most significant outcomes of the Nouvelle AI revolution was the development of robots capable of acting—initially in minimal ways, but gradually with increasing complexity—in coordination with humans. This advancement led to the emergence of the field of social robotics [16] and, through the observation of the distinctive structure of the communication between humans and robots, the establishment of Human–Robot Interaction (HRI) as a research domain [29, 31, 64].

Starting from the 1990s, HRI has evolved into a field dedicated to studying interactions between humans and robots, with three primary objectives: (1) optimizing the design of robotic artifacts with social capabilities; (2) understanding human cognitive and behavioural dynamics in response to socially capable robots; and (3) modelling these interactions to integrate robots into application domains such as assistance, education, and domestic support.

In order to face these objectives, HRI has grown as an interdisciplinary field that integrates robotics and AI with various other disciplines, including cognitive science, philosophy and psychology [31]. Each application area within HRI extends traditional robotics research into socially interactive contexts. For example, robot perception techniques have been adapted to interpret social cues, infer human intentions, and facilitate nonverbal communication [78], motion planning in social robotics incorporates elements of social awareness to convey intent and establish rapport [68]. Similarly, machine learning in robotics has expanded to include socially guided learning, drawing from human instructional models [53, 65].

A key challenge in HRI is enabling robots to learn from observation, which serves two crucial functions [70]. First, it allows robots to adapt to human behaviours dynamically, facilitating real-time interaction in unpredictable environments. Second, it provides a means for robots to acquire socially meaningful behaviours without relying exclusively on pre-programmed rules or explicit instructions.

Machine learning plays a crucial role in addressing these challenges, particularly in enabling robots to learn from human input in real time. This is made possible in the field of Learning from Demonstration (LfD), i.e. a machine learning paradigm in which robots acquire new skills by *imitating* human behaviour [22, 58]. In this context, demonstration is defined as a method for transmitting skills to an artificial agent, where the robot observes a human demonstrator who performs the interested action and then attempts to reproduce it, typically using imitation learning techniques [11].

Initially proposed in the late 1990s, LfD was envisioned as a pathway to developing humanoid robots with human-like capabilities [63]. However, while these techniques have significantly advanced robotic behaviour in social contexts, they have also highlighted a pressing need to better understand the imitative structure of human cognition—specifically, how we observe, model, and replicate the behaviours of others. This has led to extensive investigations into the cognitive mechanisms underlying imitation learning in humans and social animals, with the aim of implementing these mechanisms into robotic cognitive architectures [30]. These inquiries, in turn, have shaped key challenges in LfD research since then.

Following this line of research, a central issue in LfD was identified in the *correspondence problem* [51], which refers to the challenge of mapping observed human behaviours onto a robot's distinct embodiment. In summary, imitation is conceived as the system's ability to extract the structure of the behaviour to be replicated and translate this understanding into the demonstration of the same movement pattern, thereby proving that it has solved the correspondence problem, formally defined as follows:

> Given an observed behaviour of the model, which, from a given starting state, leads the model through a sequence (or hierarchy) of subgoals (in states, action, and/or effects, while possibly responding to sensory stimuli and external events), find and execute a sequence of actions using one's own (possibly dissimilar) embodiment, which, from a corresponding starting state, leads through corresponding subgoals (in corresponding states, actions, and/or effects, while possibly responding to corresponding events). [54: 43]

In the field of LfD, the correspondence problem was conceived presenting several sub-problems, including action segmentation, identifying which behaviours are relevant for imitation, perspective taking, and demonstrator modelling [48]. These challenges are compounded by the necessity of translating observed actions into executable motor primitives within a robot's physical constraints.[1] Researchers have addressed these challenges mainly by developing algorithms that allow robots to parse human demonstrations, extract essential action features, and reproduce them in a functionally equivalent manner [58, 66].

Even with the deep investigation of human cognitive models of mimicry and imitation, however, the implementation of LfD in social robots seems to suffer of different obstacles. Firstly, since robots must apply machine learning algorithms to build their own image of the demonstrated process, they often struggle with occlusions, rapid movements, and sensor noise when observing human actions, making difficult to accurately capture human demonstration process [58]. Secondly, enabling robots to learn complex, multi-step tasks from demonstrations requires the development of hierarchical abstractions, which remains a significant challenge in LfD [42]. Thirdly, the effectiveness of LfD heavily depends on the quality of human demonstrations, and variations in demonstration quality can adversely affect the robot's learning and generalization capabilities [3, 60]. Finally, even if the robots learn how to perform the specific demonstrated behaviour, then struggle to reuse the actions to perform different behaviours. In other words, the challenge concerns not only to make the robot able to see the actions performed by the human demonstrator, but also the necessity to generalize the "comprehension" and make it able to effectively perform the learnt behaviour when needed. This entails developing LfD algorithms that enable robots to adapt learned skills to new, unseen tasks aiming to enhance the flexibility and applicability of robotic systems [2].

Adopting a broader perspective on robotic learning and on the construction of imitation mechanisms through machine learning techniques reveals the extent to which these models are shaped by the process of data collection. A close examination of how challenges are identified, framed, and addressed in LfD highlights a prevailing emphasis on data acquisition—its volume, availability, and the necessity

[1] The structure of LfD process generally consists of five main stages: Demonstrator Selection; Data Acquisition; Data Modeling; Skill or Task Execution; Refinement Learning. See [66] for details.

of continuous expansion to optimize the learning process. This perspective, which has historically influenced explanatory models of the mind within cognitive science, is grounded in computationalist assumptions that conceptualize cognition as a data-driven computational process. In other words, both the paradigm for investigating human imitative processes and the design of robotic systems capable of learning from observation are deeply embedded in what has been termed the *computational metaphor of the mind* [12]. According to this view, perception—the foundation of the learning mechanism—is structured as a process of data collection, which is subsequently represented within the system's functional architecture and used to fuel inferential reasoning about others' actions and intentions. Over the years, almost all human cognitive abilities have been translated into the cognitivist framework and examined through this epistemological lens, with social cognition being primarily understood as an inferential reasoning process driven by mental representations [5, 7], see §2, this chapter).

Within this framework, machine learning techniques have provided a powerful tool for implementing theories of imitation and mimetic learning in robotics. This has significantly influenced the dominant approaches to enabling robots to learn from their environment. However, research emerging over the past three decades has increasingly challenged the cognitivist paradigm, highlighting its limitations in accounting for the full scope of human cognitive capacities [20, 35, 73, 77]. In particular, when it comes to equipping robots with the ability to *see* and *learn*, an alternative perspective reveals that the image constructed through environmental data collection does not inherently generate meaningful representations for the robot. Beyond the necessity of instructing robots on *what* to perceive, there is a fundamental need to specify *how* to perceive.

In this paper, I will argue that many of the challenges in LfD stem from its foundational cognitivist assumptions about social cognition, particularly its reliance on Theory of Mind concepts. By reconsidering LfD's theoretical underpinnings drawing on a cognitive semiotic view on interaction and learning, we can move toward more adaptive and context-sensitive models of robotic learning, better suited to the demands of real-world human–robot interaction.

20.3 Learning from Demonstration and Theory of Mind: The Foundation of Robotic Understanding

The prevailing view in LfD assumes that imitation and social learning require the ability to infer human intentions and mental states. This definition is grounded in the formalization that social cognition has taken regarding the processes of interpreting and understanding the actions of others through the so-called Theory of Mind (ToM) [7, 25].

ToM, or *mindreading*, is a psychological theory that identifies the core mechanism of social cognition with the ability to comprehend and reason about one's own mental states (e. g., beliefs, desires, intentions, knowledge, etc.) while simultaneously attributing analogous mental states to others to predict and explain their behaviour [1]. Since its emergence in the 1980s, this theoretical framework has gained substantial traction, ultimately establishing itself as the predominant paradigm for explaining social cognition in humans and higher primates [8]. The fundamental cognitive mechanism underpinning ToM is inferential reasoning: understanding others' mental states necessitates drawing inferences based on three key elements: shared contextual information, perceptual indicators, and the interpretation of actions as manifestations of underlying mental states [21].

In this domain, even imitation capacities are conceptualized as inferential abstraction operations intrinsically connected to mindreading and ToM in social contexts. As Nehaniv and Dautenhahn [51: 53] assert, "understanding and matching higher-level structures characterizes various other forms of observational learning and matching involving theory of mind or mind-reading."

Given ToM's central position in these disciplines, its implementation in robotic cognitive architectures is considered essential for enhancing robots' capacity to interpret human actions. The widespread acceptance of ToM as the operational mechanism of human social cognition has led researchers to conclude that equipping robots with analogous mechanisms represents the optimal approach to achieving sophisticated social intelligence capable of human interaction. Following Scassellati's pioneering research (2001) [61] on implementing ToM in robotics, numerous studies have attempted to integrate ToM's inferential mechanisms into various architectural frameworks. This initial research has spawned an extensive body of ongoing work [6, 10, 9]. Being able to implement ToM features into cognitive architectures is supposed to be the best pathway to reach the threshold of social cognitive complex capabilities for robots, also relying on a bench of studies that assess the higher acceptance of robots if they display *mindreading* capabilities [69].

Following this trend, the field of LfD has integrated ToM as the theoretical cornerstone for developing robotic cognitive architectures. This integration shapes both the fundamental principles of data collection and the inference-based training of machine learning mechanisms that enable actual learning. Indeed, the theoretical model underpinning this approach positions logical-inferential abilities at the heart of social cognition. According to this view, cognitive systems come to understand others' intentions through reasoning about their mental states. As Mohammad and Nishida [48: 211] describe:

> Humans understand the behaviour of other humans (and some animals) by assigning mental states to them and reasoning about how these states affect or generate their perceived behaviour. This ability to reason about the mental state of other agents (whether people or otherwise) is called by various names by different researchers including mentalizing, mind reading, and theory of mind (ToM).

This conceptual framework has given rise to three key methodological challenges in modelling actions for imitative purposes:

(i) First, *action is understood as a discretisable flow of elementary components.* This view suggests that behaviour emerges from the combination and sequencing of basic action units. Such segmentation into discrete units reflects traditional computational approaches—similar to those that guided the development of early robots like SHAKEY, though now implemented with far more sophisticated technology.

(ii) Second, *ToM-based LfD approaches imitation through a logical–linguistic lens*. This involves analysing the demonstrator's goals and intentions, decomposing their behaviour, and computing how to map these observations onto the observer's potential actions. This view is based on the idea that "Imitation of *action* requires the representation of goals as well as motor mimicry and hence could be a way for the subject to become aware of another's goal, an important step on the path to mind reading" [40: 911].

(iii) Third, *implementing ToM principles in cognitive architectures requires processing input data through various ToM-based reasoning structures*. This implementation presents three significant practical challenges:

a. ToM requires modelling and reasoning about multiple mental states, which significantly increases computational costs. Many approaches rely on Bayesian inference, deep learning, or probabilistic models, all of which demand substantial resources [57].
b. The continuous need for inference operations and dataset updates can significantly impair real-time performance, making it difficult for robots to interact and learn simultaneously [9].
c. ToM assumes that mental states—beliefs, desires, and intentions—are internal and hidden from direct observation. This necessitates additional computational resources to decode the motivations behind human actions, further complicating the learning process [26, 27, 36, 76].

In essence, algorithms trained within a cognitivist framework based on ToM shape robotic vision and learning by guiding its attention to key focal points identified through this theoretical approach. Consequently, robotic perception is engineered to prioritize fine-grained details, segment actions, and construct inferences about both the possible intentions underlying human behaviour and the perspective adopted by the human demonstrator.

However, despite significant advancements enabled by increased computational power (such as *Big Data*; Chen et al. [26, 27]) and advanced machine learning techniques, which have demonstrated exceptional performance in disembodied artificial intelligences [33], these challenges remain largely unresolved, or at best, only marginally mitigated. In essence, the strategy of increasing the *quantity* of available data does not appear to offer effective solutions in the short term.

To overcome this impasse, I propose shifting focus toward the *quality* of the collected data, introducing a novel model of social cognition based on coordination and participatory practices rather than inference-driven imitation. This shift requires drawing on theories that, adopting a semiotic perspective on social interaction, conceptualize intersubjective understanding as a process fundamentally situated within the environment, where multiple contextual elements dynamically contribute to the construction of individual comprehension.

20.4 Interaction as Semiotic Process

Over the past twenty years, a new theory of social cognition has emerged within the cognitive sciences, positioning itself in direct opposition to the explanatory hypotheses proposed by ToM [35, 37, 59]. Embodied and enactive approaches of social cognition critique ToM and *mindreading* models on several grounds, arguing that social cognition does not rely on internalized mental state inference but instead emerges from embodied interaction, sensorimotor coordination, and participatory sense-making [44]. In particular, embodied critiques to ToM models of social cognition point the attention to three key aspects. First, while the mindreading model assumes that individuals infer others' mental states based on internalized representations, empirical research shows that social understanding is primarily interactional and embodied rather than inferential [37]. Embodied account of social understanding highlight that humans do not typically engage in explicit mental state attribution during real-world social interactions, instead, understanding emerges dynamically from bodily and contextual coordination [44]. Second, they propose that social cognition begins with embodied interactions (e.g., *joint attention*, affect attunement) rather than conceptual inferences about beliefs and desires [59, 71]. Third, while mindreading models assume that understanding others' emotions and intentions necessarily requires inferring their mental states, embodied cognition research suggests we understand others primarily through bodily mirroring and sensorimotor resonance. As also highlighted by neuroscientific evidence in mirror neuron research and embodied simulation theories, we mainly experience others' emotions and intentions directly via bodily engagement, not through abstract inferences [54, 38]. As a result, social understanding is considered to involve several complex dimensions, that put together make the shared meaning arise from contextual, situated and embodied interaction. As Gallagher [37: 157] clarifies, "Meaning, including what we can understand of the other's actions, is accomplished [...] by drawing on "different kinds of semiotic resources" [41] available in the environment and in whole body pragmatics."

This idea has been extensively developed by Paolucci [55, 56] in formulating a new cognitive semiotic approach to meaning-making. By relying on the enactive and embodied theories of cognition, this approach constructs a semiotic perspective on the processes of meaning construction by analysing the characteristics of sense-making that distinguish a living organism as it navigates its environment. In particular, in line with the pragmatist tradition, for cognitive semiotics cognition is understood as a form of mediated action that unfolds through semiotic processes of constructing saliences and values, which, in turn, guide action through meaning [56]. By making the world meaningful and thereby opening up heterogeneous possibilities for action, agents can engage with their environment (and with other agents that inhabit it) effectively, shaping semiotic niches that render a world of values and meanings inhabitable [55].

In essence, by endorsing the cognitive semiotic and enactive view of interaction, the process of understanding in social cognition no longer primarily relies on internal mental states or inferential reasoning. Instead, it emerges from the ongoing engagement that occurs during interaction, shaping both action and meaning.

If we adopt this account of social cognition and replace the ToM framework in LfD with this perspective, we can see how it becomes possible to overcome some of the issues outlined in §2.

Regarding point (i), from an enactive and cognitive semiotic perspective on understanding others, action segmentation does not play a fundamental role. The core process of social understanding is attunement, based on bodily rather than inferential correspondences. From this viewpoint, understanding does not depend on breaking down actions into isolated components but rather on the immediate configuration of affordance patterns. In the case of social cognition, this means a direct and immediate connection with the other agent. This attunement occurs through bodily correspondences that emerge without the need for representational inferences, as the body itself detects similarities through perceptual and sensorimotor dynamics, making an analytical segmentation process unnecessary. In other words, the emergence of a meaningful semiotic plane takes place through the instantaneous configuration of action possibilities related to the agent's pragmatic capacities.

This approach can also be applied to (ii), since the mental states of an agent—such as beliefs, intentions, and desires—are not exclusively inferred through data-driven inferential mechanisms but rather through a semiotic configuration of immediate, embodied understanding. In this sense, the neuroscientific evidence mentioned in the previous section [42, 35] supports a less inferential and more embodied view of intersubjective understanding. This is exemplified by the role of mirror neurons, which serve as one of the crucial bodily components for constructing the shared sense that characterizes interactions between social agents.

Regarding point (iii), following this approach to understanding learning processes, the focus shifts from the quantity of data collected in the perceptual segmentation of the environment to the quality of the semiotic configuration that emerges from the real, ongoing interaction between the learner and the demonstrator. The key idea is that, even if the robot is not capable of engaging in interaction in a bodily, embedded, and vital manner (see [32, 80, 80]), adopting a cognitive semiotic and enactive perspective on learning processes can inform the design of robotic machine learning algorithms based on bodily and agentive signals situated in interaction [55], as well as on the dynamics of adjustment that emerge in *participatory sense-making* [44]. This perspective opens up a range of possibilities for implementing learning algorithms that go beyond the mere replication of demonstrated movement, instead adopting an embodied approach to collaboration, which characterizes practices in social interaction.

To clarify this final point, in the next paragraph I propose a cognitive semiotic and enactive reading of certain machine learning techniques currently in use, which align with the shift I propose. When framed within a situated and embodied paradigm of social cognition, these techniques could offer new pathways for instructing robots in learning processes, making algorithmic training more closely resemble human learning practices.

20.5 Reframing Learning Practices in HRI: Towards a Mutual Collaboration Between Humans and Robots

To demonstrate how robotic learning can be framed through an enactive and semiotic perspective, we consider a series of experiments that have enabled robots to acquire motor skills by observing human demonstrations. Specifically, we refer to studies in developmental robotics [46], a field that models ontogenetic processes in child development to inform machine learning mechanisms in robots. This approach is based on the premise that a robot, guided by intrinsic developmental principles regulating real-time interactions between its body, brain, and environment, can autonomously acquire increasingly complex sensorimotor and cognitive capabilities [23].

The studies we are going to examine draw on extensive research in developmental psychology, which has identified distinctive features in caregiver-infant interactions across many cultures. Notably, caregivers modify their speech when addressing infants—a practice also called *motherese*—using higher pitch, shorter utterances, exaggerated intonation, and simplified content [52, 67]. Numerous studies suggest that such modifications enhance infants' attention to communicative input. For instance, infants as young as one month prefer infant-directed over adult-directed speech [28, 34] and exhibit greater visual attention when exposed to infant-directed speech [75]. The exaggerated features of infant-directed communication, including wide pitch variations in speech and slow, large, and repetitive movements in sign language, are particularly effective in capturing infants' attention [47].

Recent research extends these findings to the domain of action, identifying a phenomenon known as *motionese*[2] [24]. Indeed, caregivers also adjust their gestures in order to leverage children comprehension: compared to interactions with adults, they use fewer gestures with infants, and those they do use are more likely to co-occur with speech, reinforcing rather than supplementing verbal content [43]. Brand et al. [14] expanded this idea and examined object-related actions in caregiver-infant interactions, revealing that caregivers modify their movements when demonstrating actions to infants. Compared to adult-directed actions, infant-directed actions occur closer to the infant, involve higher interactivity, greater enthusiasm, increased repetition, and simplified yet exaggerated movements. Such modifications are believed to enhance infants' attention and facilitate their understanding of action structure [15, 14].

Building on these studies, several investigations [50, 49, 72] have explored whether and to what extent action demonstration techniques incorporating *motionese* can facilitate robotic learning. This approach leverages the exaggerated features of demonstrators' actions and makes explicit the interactive dynamics involved in this pedagogical behaviour, mirroring caregiver-infant interactions.

In particular, Nagai and Rohlfing [50] analysed parental actions from an infant-like perspective using a saliency-based visual attention model, which detects and focuses on salient locations in a scene without any prior knowledge of the task or

[2] For an extensive survey on *motionese*, see [14].

objects involved. Their findings suggest that *motionese* supports infants in three key ways: "[…] (1) to receive immediate social feedback on the actions, (2) to detect the initial and goal states of the objects used in the actions, and (3) to look at the static features of the object" [50: 305]. In a subsequent study, Nagai and Rohlfing [49] expanded on their previous research by conducting a computational analysis of *motionese* to enhance robotic action learning. They argue that robots, like infants, lack prior knowledge of task-relevant information and therefore require guidance in determining what to focus on when observing demonstrations. Using a bottom-up attention model, they demonstrate that *motionese* "has the effect of highlighting the initial and final states of the action, the significant events within it, and the properties of the objects involved, which can impart the goal of the action" [49: 53]. This enables robots to identify what to imitate without relying heavily on predefined instructions.

More recently, Ugur et al. [72] further demonstrated the effectiveness of *motionese* in LfD for robotics, identifying three key advantages: (i) it enhances robotic algorithms' ability to segment and interpret demonstrated actions, thereby improving the efficacy of the imitation process; (ii) it allows naïve human demonstrators to engage in teaching practices with robots without requiring technical training, as they can rely on their social competencies to demonstrate actions similarly to how they would interact with a child; (iii) it enables robots to acquire the capacity for multi-step planning, allowing them to achieve a goal through structured action sequencing rather than merely replicating isolated movements.

Now, by analysing this approach, we can identify a set of insights that support the adoption of a more distributed and embodied perspective on cognition and learning—one that aligns with the cognitive semiotic and enactive framework we have proposed—rather than reinforcing an internalist and representational conception of cognitive processes.

1. The practice of *motionese* aligns with a participatory conception of social cognition, in which cognitive structures emerge both as an outcome of the interaction between two agents and as elements that actively shape and develop the interaction itself. From this perspective, *motionese* can be understood as a process of mutual attunement, wherein the caregiver, already proficient in the embodied social practices that the child (or robot) has yet to learn, adjusts their demonstrative actions based on what the learner has not yet mastered. Essentially, the caregiver constructs a *scaffolding* [39, 74] structure composed of salient affordances, making key aspects of the demonstration more perceivable and accessible. This enables the learner to occupy positions within the interactive structure that would otherwise remain inaccessible. Just as an adult provides support that allows a child to accomplish tasks beyond their independent capacity, robots can benefit from the exaggeration of particular features in actions, facilitating the training of machine learning algorithms to detect and prioritize the most relevant and accentuated aspects of demonstrated behaviour. For a machine learning system interacting with a human motivated to facilitate learning, social elements can significantly enhance the success of the learning process by constraining and guiding its development.

2. This approach enables machine learning algorithms to be trained by integrating a bottom-up process rather than only relying on top-down feature detection mechanisms. That is, the algorithm can identify salient features in the demonstrator's behaviour without necessarily requiring an explicit parsing of the action or relying on a human robotic designer to predefine the correct sequence of actions. More simply, pattern recognition [78] can be used to detect key aspects of an action through repeated exposure to similar trajectories, allowing the robot to build an agentive repertoire. In this way, the robot not only acquires skills based on demonstration (i.e., *what* to imitate) but also learns *how* to perform the action effectively. In contrast with the approach based on the correspondence problem and on ToM, in which action segmentation is central to guarantee the commensurability of humans and robots' actions, if we adopt a cognitive semiotic model of interaction and learning the *what* and *how* of imitation must be considered deeply interwoven. Only by identifying the semiotic and agentive patterns that characterize the demonstrators' engagement with their environment can the learner—whether human or artificial—grasp the situational context and understand how to perform actions in real-time scenarios.
3. In this regard, the cognitive semiotic approach allows to highlight the role of environmental and contextual embedding in learning. Traditional LfD approaches often assume that learning occurs through the isolated observation of actions, treating the demonstrator's behaviour as a sequence of discrete movements to be imitated. However, from an enactive and semiotic perspective, learning is fundamentally embedded in the affordances of the environment and the situational dynamics in which actions unfold. The meaning and relevance of an action are not inherent in the movement itself but emerge through interaction. A robot trained in a controlled setting may struggle to transfer its learning to real-world applications if it lacks sensitivity to these contextual dependencies. By integrating an awareness of the relational and ecological conditions of action, an enactive and semiotic approach enables a more adaptive and context-sensitive learning process. This is particularly relevant in HRI, where robots must operate in open-ended environments rather than controlled laboratory conditions [31].

In conclusion, I emphasize that incorporating a cognitive semiotic and enactive perspective into LfD does not constitute a radical departure from the field. Rather, by shifting from internalist, inferential, and representational models of social cognition to a more interactive, embodied, and pragmatic framework, we may, in the future, develop more collaborative robots—ones that might ultimately be capable of *seeing* what truly matters to us.

References

1. Apperly, I.A.: Mindreaders: The Cognitive Basis of "Theory of Mind." Psychology Press, Hove (2011)

2. Arduengo, M., Colomé, A., Borràs, J., Sentis, L., Torras, C.: Task-adaptive robot learning from demonstration with Gaussian process models under replication. IEEE Robotics and Automation Letters **6**(2), 966–973 (2021)
3. Argall, B., Browning, B., Veloso, M.: Learning by demonstration with critique from a human teacher. In: Proceedings of the ACM/IEEE International Conference on Human-Robot Interaction, pp. 57–64 (2007)
4. Baker, C.L., Saxe, R., Tenenbaum, J.B.: Bayesian theory of mind: modeling joint belief-desire attribution. Cognition **120**(2), 236–301 (2011)
5. Bandura, A.: Social Foundations of Thought and Action: A Social Cognitive Theory. Prentice-Hall, Englewood Cliffs, NJ (1986)
6. Barchard, K.A., Lapping-Carr, L., Westfall, R.S., Fink-Armold, A., Banisetty, S.B., Feil-Seifer, D.: Measuring the perceived social intelligence of robots. ACM Transactions on Human-Robot Interaction **9**(4), 1–29 (2020)
7. Baron-Cohen, S.: Mindblindness: An Essay on Autism and Theory of Mind. MIT Press, Cambridge, MA (1995)
8. Baron-Cohen, S.: Evolution of a theory of mind? In: Corballis, M., Lea, S. (eds.) The Descent of Mind: Psychological Perspectives on Hominid Evolution. Oxford University Press, Cambridge, MA (1999)
9. Bianco, F., Ognibene, D.: Robot learning theory of mind through self-observation: exploiting the intentions-beliefs synergy. arXiv Preprint arXiv:2210.09435 (2022)
10. Bianco, F., Ognibene, D.: Functional advantages of an adaptive theory of mind for robotics: a review of current architectures. In: 2019 11th Computer Science and Electronic Engineering Conference (CEEC), pp. 139–143 (2019b)
11. Billard, A., Calinon, S., Dillmann, R., Schaal, S.: Robot programming by demonstration. In: Handbook of Robotics, vol. 59, issue no. 2015, pp. 1371–1394 (2008)
12. Block, N.: The computer model of the mind. In: Goldman, A.I. (ed.) Readings in Philosophy and Cognitive Science, pp. 705–731. MIT Press, Cambridge, MA (1993)
13. Boden, M.: Mind as Machine: A History of Cognitive Science. Cambridge University Press, Cambridge, MA (2006)
14. Brand, R.J., Baldwin, D.A., Ashburn, L.A.: Evidence for 'motionese': Modifications in mothers' infant-directed action. Dev. Sci. **5**(1), 72–83 (2002)
15. Brand, R.J., Shallcross, W.L.: Infants prefer motionese to adult-directed action. Dev. Sci. **11**(6), 853–861 (2008)
16. Breazeal, C., Dautenhahn, K., Kanda, T.: Social robotics. In: Siciliano, B., Khatib, O. (eds.) Springer Handbook of Robotics. Springer, Cham (2016)
17. Breazeal, C., Takanishi, A., Kobayashi, T.: Social robots that interact with people. In: Siciliano, B., Khatib, O. (eds.) Springer Handbook of Robotics. Springer, Berlin–Heidelberg (2008)
18. Brooks, R.: Intelligence without representation. Artif. Intell. **47**, 139–159 (1991b)
19. Brooks, R.: Achieving Artificial Intelligence Through Building Robots. Technical Report, Artificial Intelligence Laboratory, Massachusetts Institute of Technology, Cambridge, MA (1986)
20. Brooks, R.: Intelligence without reason. Technical Report, Computers and Thought, IJCAI-91, Artificial Intelligence Laboratory, Massachusetts Institute of Technology, Cambridge, MA (1991a)
21. Byom, L.J., Mutlu, B.: Theory of mind: mechanisms, methods, and new directions. Front. Hum. Neurosci. **7**(413) (2013)
22. Calinon, S.: Learning from demonstration (programming by demonstration). In: Ang, M., Khatib, O., Siciliano, B. (eds.) Encyclopedia of Robotics. Springer, Berlin–Heidelberg (2018)
23. Cangelosi, A., Schlesinger, M.: Developmental Robotics: From Babies to Robots. MIT Press, Cambridge, MA (2015)
24. Cantor, Z.: Motherese and Motionese: Do They Travel Together? Doctoral dissertation, University of Oregon (2023)
25. Carruthers, P., Smith, P.: Theories of Theories of Mind. Cambridge University Press, Cambridge (1996)

26. Chen, B., Vondrick, C., Lipson, H.: Visual behavior modelling for robotic theory of mind. Sci. Rep. **11**, 424 (2021)
27. Chen, Z., Chen, B., Xie, S., Gong, L., Liu, C., Zhang, Z., Zhang, J.: Efficiently training on-policy actor-critic networks in robotic deep reinforcement learning with demonstration-like sampled exploration. In: 2021 3rd International Symposium on Robotics & Intelligent Manufacturing Technology (ISRIMT), pp. 292–298. IEEE, USA (2021)
28. Cooper, R.P., Abraham, J., Berman, S., Staska, M.: The development of infants' preference for motherese. Infant Behav. Dev. **20**(4), 477–488 (1997)
29. Dautenhahn, K.: Socially intelligent robots: dimensions of human–robot interaction. Philos. Trans. R. Soc. B **362**(1480), 679–704 (2007)
30. Dautenhahn, K., Nehaniv, C.L. (eds.): Imitation in Animals and Artifacts. MIT Press, Cambridge, MA (2002)
31. Dautenhahn, K.: Human-robot interaction. In: Interaction Design Foundation, https://www.interaction-design.org/literature/book/the-encyclopedia-of-human-computer-interaction-2nd-ed/human-robot-interaction (2014)
32. Di Paolo, E. A.: Organismically-inspired robotics: homeostatic adaptation and teleology beyond the closed sensorimotor loop. In: Murase, K., Asakura, T. (eds.) Dynamic Systems Approach for Embodiment and Sociality, pp. 19–42 (2003)
33. Dulac-Arnold, G., Levine, N., Mankowitz, D.J., et al.: Challenges of real-world reinforcement learning: definitions, benchmarks and analysis. Mach. Learn. **110**, 2419–2468 (2021)
34. Fernald, A.: Four-month-old infants prefer to listen to motherese. Infant Behav. Dev. **8**(2), 181–195 (1985)
35. Gallagher, S.: How the Body Shapes the Mind. Oxford University Press, Oxford (2005)
36. Gallagher, S.: Social cognition and social robots. Pragmat. Cogn. **15**(3), 435–453 (2007)
37. Gallagher, S.: Action and Interaction. Oxford University Press, Oxford (2020)
38. Gallese, V., Sinigaglia, C.: What is so special about embodied simulation? Trends Cogn. Sci. **15**(11), 512–519 (2011)
39. Gauvain, M.: Scaffolding in socialization. New Ideas Psychol. **23**(3), 129–139 (2005)
40. Gerrans, P.: Imitation and theory of mind. In: Handbook of Neuroscience for the Behavioral Sciences, vol. 2, pp. 905–922 (2009)
41. Goodwin, C.: Action and embodiment within situated human interaction. J. Pragm. **32**1489–1522 (2000)
42. Gopalan, N., Moorman, N., Natarajan, M., Gombolay, M.: Negative result for learning from demonstration: challenges for end-users teaching robots with task and motion planning abstractions. In: Proceedings of Robotics: Science and Systems XVIII (2022)
43. Iverson, J.M., Capirci, O., Longobardi, E., Caselli, M.C.: Gesturing in mother-child interactions. Cogn. Dev. **14**(1), 57–75 (1999)
44. De Jaegher, H., Di Paolo, E.: Participatory sense-making: an enactive approach to social cognition. Phenomenol. Cogn. Sci. **6**, 485–507 (2007)
45. Kuipers, B., Feigenbaum, E.A., Hart, P.E., Nilsson, N.J.: Shakey: from conception to history. AI Mag. **38**(1), 88–103 (2017)
46. Lungarella, M., Metta, G., Pfeifer, R., Sandini, G.: Developmental robotics: a survey. Connect. Sci. **15**(4), 151–190 (2003)
47. Masataka, N.: Perception of motherese in Japanese sign language by 6-month-old hearing infants. Dev. Psychol. **34**(2), 241–245 (1998)
48. Mohammad, Y., Nishida, T.: Data Mining for Social Robotics. Springer, Cham (2015)
49. Nagai, Y., Rohlfing, K.J.: Computational analysis of motionese toward scaffolding robot action learning. IEEE Trans. Auton. Ment. Dev. **1**(1), 44–54 (2009)
50. Nagai, Y., Rohlfing, K.J.: Can motionese tell infants and robots what to imitate? In: Proceedings of the 4th International Symposium on Imitation in Animals and Artifacts, pp. 299–306 (2007)
51. Nehaniv, C.L., Dautenhahn, K.: The correspondence problem. In: Dautenhahn, K., Nehaniv, C.L. (eds.) Imitation in Animals and Artifacts, pp. 41–61. MIT Press, Cambridge, MA (2002)
52. Newport, E. L.: Motherese: The Speech of Mothers to Young Children. University of Pennsylvania (1975)

53. Nocentini, O., Fiorini, L., Acerbi, G., Sorrentino, A., Mancioppi, G., Cavallo, F.: A survey of behavioral models for social robots. Robotics **8**(3), 54 (2019)
54. Di Paolo, E., De Jaegher, H.: The interactive brain hypothesis. Front. Hum. Neurosci. **6**, 163 (2012)
55. Paolucci, C.: Cognitive Semiotics. Springer, Cham (2021a)
56. Paolucci, C.: Che cos'è una semiotica cognitiva? Sistemi Intelligenti **33**(2), 281–303 (2021b)
57. Rabinowitz, N. C., Perbet, F., Song, F., Zhang, C., Eslami, S. M. A., Botvinick, M.: Machine theory of mind. In: International Conference on Machine Learning (ICML) (2018)
58. Ravichandar, H., Polydoros, A.S., Chernova, S., Billard, A.: Recent advances in robot learning from demonstration. Ann. Rev. Cont. Robot. Autono. Syst. **3**(1), 297–330 (2020)
59. Reddy, V.: How Infants Know Minds. Harvard University Press, Cambridge, MA (2008)
60. Sakr, M., Li, Z.J., Van der Loos, H.F.M., Kulic, D., Croft, E.A.: Quantifying demonstration quality for robot learning and generalization. IEEE Robot. Autom. Lett. **7**(3), 7370–7377 (2022)
61. Scassellati, B.: Theory of mind for a humanoid robot. Auton. Robot. **12**, 13–24 (2002)
62. Scassellati, B. M.: Foundations for a theory of mind for a humanoid robot (Doctoral dissertation, massachusetts institute of technology) (2001)
63. Schaal, S.: Is imitation learning the route to humanoid robots? Trends Cogn. Sci. **3**, 233–242 (1999)
64. Sheridan, T.B.: Human–robot interaction: status and challenges. Hum. Factors **58**(4), 525–532 (2016)
65. Soori, M., Arezoo, B., Dastres, R.: Artificial intelligence, machine learning and deep learning in advanced robotics: a review. Cogn. Robot. **3**, 54–70 (2023)
66. Sosa-Ceron, A.D., Gonzalez-Hernandez, H.G., Reyes-Avendaño, J.A.: Learning from demonstrations in human–robot collaborative scenarios: a survey. Robotics **11**(6), 126 (2022)
67. Stern, D.N., Spieker, S., MacKain, K.: Intonation contours as signals in maternal speech to prelinguistic infants. Dev. Psychol. **18**(5), 727–735 (1982)
68. Tazaki, Y., Murooka, M.: A survey of motion planning techniques for humanoid robots. Adv. Robot. **34**(21–22), 1370–1379 (2020)
69. Thellman, S., De Graaf, M., Ziemke, T.: Mental state attribution to robots: a systematic review of conceptions, methods, and findings. ACM Trans. Hum. Robot Inter. **11**(4), 1–51 (2022)
70. Thomaz, A., Hoffman, G., Cakmak, M.: Computational human-robot interaction. Found Trends Robot **4**(2–3), 105–223 (2016)
71. Tomasello, M.: Becoming Human: A Theory of Ontogeny. Harvard University Press, Cambridge, MA (2019)
72. Ugur, E., Nagai, Y., Sahin, E., Oztop, E.: Staged development of robot skills: behavior formation, affordance learning, and imitation with motionese. IEEE Trans. Auton. Ment. Dev. **7**(2), 119–139 (2015)
73. Varela, F., Thompson, E., Rosch, E.: The Embodied Mind: Cognitive Science and Human Experience. MIT Press, Cambridge, MA (1991)
74. Vygotsky, L.S.: Mind in Society: The Development of Higher Psychological Processes. Harvard University Press, Cambridge, MA (1978)
75. Werker, J.F., Pegg, J.E., McLeod, P.J.: A cross-language investigation of infant preference for infant-directed communication. Infant Behav. Dev. **17**(3), 323–333 (1994)
76. Winfield, A. F.: Experiments in artificial theory of mind: from safety to story-telling. Front. Robot. AI **5**(75) (2018)
77. Winograd, T., Flores, F.: Understanding Computers and Cognition: A New Foundation for Design. Ablex Publishing Corporation, Norwood (1986)
78. Yan, H., Ang, M.H., Poo, A.N.: A survey on perception methods for human–robot interaction in social robots. Int. J. Soc. Robot. **6**, 85–119 (2014)
79. Youssef, K., Said, S., Alkork, S., Beyrouthy, T.: A survey on recent advances in social robotics. Robotics **11**(4), 75 (2022)
80. Ziemke, T.: Are robots embodied? In: First International Workshop on Epigenetic Robotics, vol. 85, pp. 701–746 (2001)

Martina Bacaro is a postdoctoral researcher in the Department of Philosophy at the University of Bologna. She is a research fellow of the PRIN 2022 project "Normative Kinds. Values and Classificatory Decisions in Science and Policy-Making." Her research focuses on the epistemological and philosophical dimensions of human–robot interaction, cognitive robotics, and social cognition, with a particular emphasis on enactivist and embodied approaches to the mind. She has contributed to national and international projects on artificial intelligence and has participated in conferences on cognitive science and social robotics. She holds an MA in Semiotics (University of Bologna, 2019) and a PhD in Philosophy, Science, Cognition, and Semiotics (University of Bologna, 2025).

Chapter 21
Posthumous AI: The Afterlife of Data

Pietro Conte and Maria Serafini

Abstract This essay delves into the interplay between artificial intelligence (AI), death, and the afterlife. It argues that the relationship between AI and the dead extends far beyond particular historical cases, such as the use of images of deceased individuals to train machine learning models. Instead, it is rooted in AI's structural reliance on memory traces and its operation within the conceptual domain of the afterlife. This entanglement manifests at two interconnected levels: the personal, where AI enables the virtual reanimation of individual identities; and the collective, where it preserves and reinforces cultural memory by perpetuating shared imagery, visual motifs, and stereotypes. First, the essay examines how neural networks generate digital doppelgängers, raising critical questions about identity persistence in virtual environments. It then explores how training data, though seemingly absorbed by AI systems, re-emerge in generated outputs as spectral traces or revenants. These forms of posthumous presence reveal how AI reconfigures practices of memory and mourning, embedding the dead within algorithmic systems of visibility, control, and reproduction. This ultimately calls for political consideration of the governance of digital afterlives.

Keywords Death · Memory · Simulation · Mourning · Nachleben

This essay was written within the scope of the "Departments of Excellence 2023–2027" project, awarded to the Department of Philosophy Piero Martinetti (EXT—Coordinated Research Centre on Extended Reality) at the University of Milan by the Italian Ministry of University and Research. It is the result of a joint effort between the two authors. The authorship responsibility is divided as follows: Sect. 21.1 (co-written by the two authors), Sect. 21.2 (Pietro Conte), Sect. 21.3 (Maria Serafini)

P. Conte (✉) · M. Serafini
University of Milan, Milan, Italy
e-mail: pietro.conte@unimi.it

P. Conte et al. (eds.), *Algomedia. The Image at the Time of Artificial Intelligence*, Lecture Notes in Morphogenesis, https://doi.org/10.1007/978-3-032-08726-3_21

21.1 Algorithmic Memory

From February to June 2020, the de Young Memorial Museum hosted an exhibition entitled *Uncanny Valley: Being Human in the Age of Artificial Intelligence*, which centred around the idea that the notion of the uncanny valley—introduced by Japanese roboticist Masahiro Mori in the 1970s to describe the eerie feeling one might experience when confronted with hyperrealistic androids—should no longer be limited to the outward appearance of human-like robots. Instead, it should be expanded to encompass contemporary algorithms designed to mine, analyse, and replicate human behaviour, projecting it into statistical (and therefore tradable) futures.

Among the many works on display was *They Took the Faces from the Accused and the Dead…*, an installation by renowned visual artist Trevor Paglen consisting of a large wall grid with 3240 individually pinned mug shots (Fig. 21.1). These images were sourced from the archive of the National Institute of Standards and Technology and assembled by AI-driven software, which grouped them based on formal similarities such as race and facial expression. Just as photographs of accused criminals and prisoners were collected in the late nineteenth century in order to identify recurring features of potentially dangerous individuals, by the early 1990s, they were routinely used to train facial recognition algorithms without the consent of those involved. Seeking to demonstrate how this use of AI dehumanises the persons depicted and reduces them into ever-growing data sets that fuel sophisticated systems of surveillance and control, Paglen chose to present the photos without identifying information, obscuring parts of the faces with white bars that serve both as a reference to the stolen identities and as a means of returning them to anonymity.

The installation was part of a larger project aimed at raising awareness of the unethical use of images of unwitting people to train deep learning software. Deceased individuals, in particular, attracted Paglen's interest due to their inherent inability to prevent any non-consensual collection and exploitation of their data. A similar concept lies at the core of the series *Even the Dead Are Not Safe,* where the artist created pseudo-portraits that appear to depict deceased activists and dissident luminaries, including Frantz Fanon, Simone de Beauvoir, and Simone Weil. These ghostly images were fabricated using a facial recognition algorithm similar to that employed by Facebook and other social media to identify faces in photos uploaded by their users, isolating their unique features. The provocative title of the artwork underscores that even individuals who never used social media, such as, emblematically, those who passed away long ago, are "captured" by the software. Paglen's ultimate goal is to prompt reflection on whether the development of surveillance technologies will prevent figures like Fanon from ever existing again. In the artist's own words,

> We are rapidly moving towards a society in which one's de facto liberties are continually modulated by our metadata signatures. In that society, it's becoming increasingly easy to use these kinds of tools to make instruments of power more efficient, whether that is capital in the marketplace or whether that is in policing, law enforcement, and surveillance […]. What I'm getting at is [that] someone like a Fanon or a Weil contributed to social progress precisely by breaking the law—because they were unjust laws. How does that work in a

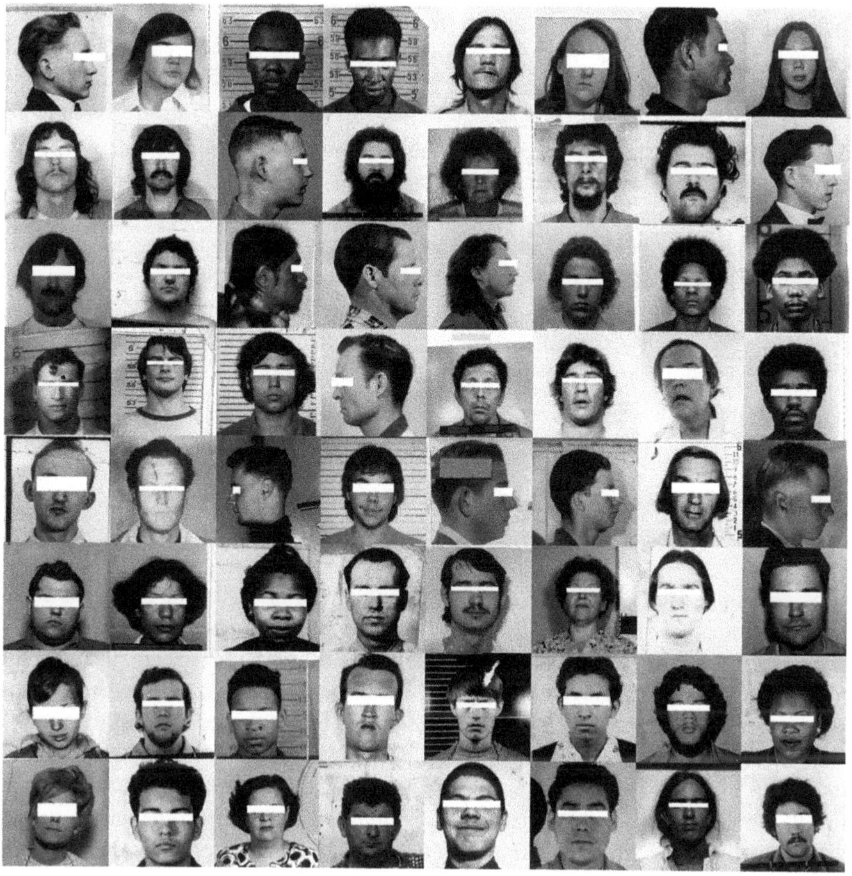

Fig. 21.1 Trevor Paglen, *They took the faces from the accused and the dead…(SD18)*, detail, 2020. 3240 silver gelatin prints mounted to Gatorboard, pins 571.5 × 914.4 cm © Trevor Paglen. Courtesy of the Artist, Altman Siegel, San Francisco and Pace Gallery

> society in which tools like this work more and more in the favor of centralized political or law enforcement systems? [29]

In line with Paglen's artistic practice, this essay argues that the connection between the dead and AI is not confined to a specific historical moment when images of deceased individuals were used to train deep learning models. Broadening the artist's insight into how AI interacts with images of the dead, we propose that death is *inherently* intertwined with AI, as both share a conceptual space within the realm of the afterlife. Since it relies on the collection and storage of memory traces, AI is related to the afterlife on two distinct but connected levels, as demonstrated in Paglen's work: the personal, which pertains to the virtual *revival* of individual identity thanks to AI's ability to reanimate digital data; and the collective, where AI systems contribute to the digital *survival* of cultural remains by identifying, reinforcing, and perpetuating shared visual imagery, recurring motifs, and stereotypes.

In light of this distinction, first (Sect. 21.2), we examine how neural networks are used to create digital doppelgängers of both the living and the deceased, raising questions about the persistence of identity in the virtual sphere. Second (Sect. 21.3), we explore how images stored in training datasets may appear to vanish once incorporated into AI models, only to persist and "haunt" these models, resurfacing and evolving with each new automatically generated image, thus acting as veritable revenants. These spectral presences—both individual and collective—suggest that AI is transforming the processes of memory and remembrance by entangling the dead in networks of visibility, control, and reproduction. In doing so, it reinforces patterns of power, exclusion, and surveillance, ultimately demanding a critical examination of how memory, representation, and digital afterlives are governed.

21.2 MimetAI

In the opening scene of Plautus' comedy *Amphitryon*, the Thebans are depicted as returning home by boat following a victorious expedition against the Acarnanian tribe of the Teleboans. Shortly after disembarking, a slave named Sosia rushes to the residence of his master, King Amphitryon, to inform the latter's wife, Alcmena, of the commander's imminent arrival. Upon reaching the house, however, Sosia perceives a mysterious individual standing right on the doorstep. With a lantern in his hand, the servant approaches the stranger only to be shocked to meet the exact replica of himself—his perfect "sosia," a term still used in Italian and other Romance languages to describe someone's doppelgänger. As we learn later, the alter ego is, in fact, the god Mercury, acting as a lookout for his father Jupiter while the latter is inside the house under the guise of Amphitryon, engaging in sexual relations with Alcmena under false pretences.

Far from being limited to mere physical resemblance, the imitation undertaken by the two gods extends to every aspect of the behaviour of the king and his slave. For the plan to succeed, Mercury and Jupiter must not only mirror their victims' outward appearance but also imitate their behaviour and speech. They must rob Sosia and Amphitryon of their habits and attitudes, making them doubt the very possibility of telling the difference between themselves and their sinister "other selves."

The mimesis is so effective that Sosia becomes increasingly confused about his own identity. Struggling to make sense of the unsettling encounter, he finally bursts out: "This man here has my entire appearance (*omnem imaginem meam*), the appearance that used to be mine. Something is happening to me while I am alive that nobody would do for me once I died!" (Plaut. *Am.* 458). The joke, which would have been readily comprehensible to Plautus' contemporaries, sounds enigmatic to modern audiences and can only be understood by taking into account the cultural milieu in which it was originally formulated. Roman aristocrats were accorded a privileged status that granted them the exclusive right to possess the *imagines maiorum*, wax effigies which replicated the features of their ancestors' faces with astonishing accuracy. These were usually kept in special cabinets in the atrium of the houses, but were taken out of their shrines during the funeral ceremonies of a family member. On such occasions, they were worn or carried by actors—*mimetai*—who resembled the deceased in size, donned clothing matching their social status, and recreated the way the ancestors spoke and moved.

In *The Natural History* (35, 6), Pliny the Elder describes the *imagines* as faces (*vultus*) rather than masks (*personae*), implicitly suggesting that they were to be regarded, at least for the whole duration of the funeral, as equivalent to the real heads of the dead. Not mere representations, but rather presentifications, they were "'enacted', playing a performative role that consisted in evoking once again the *presence* of the deceased ancestors" [4: 202].

According to Polybius (VI, 53, 5), these artefacts were distinguished by their striking resemblance to the facial features of the ancestors, achieved through mechanical reproduction using wax casts from moulds. As such, the *imagines* seem to best embody the literal meaning of the Latin word for "to portray," *retrahere*, which denotes a material taking the imprint of, and drawing it away from, the subject portrayed. It was precisely this notion of a mechanical "extraction" that led to a mask being regarded as a face, the former being considered not only as similar to the latter, but rather as equivalent to it. The idea that faithful replicas could serve as genuine surrogates for their models, precisely because they were (or were supposed to be) derived from direct contact with them, also helps explain the role played by *mimetai* in funeral rites. Diodorus (*Hist.* 31, 25, 2) reports that these actors or mimes accompanied the members of the Roman aristocracy throughout their lives, meticulously learning to imitate their distinctive postures, gestures, manner of speech, and the way they expressed emotions. Sosia's joke is now easier to understand. As a slave, he will have no right to any postmortem *imago*; nevertheless, paradoxically, such an *imago* stands before him while he is still alive, in the form of an uncanny double of himself.

Considering the intrinsic nature of the *imagines* as casts and the fact that they were part of a long process of imitation that began during the aristocrat's lifetime, one could argue that the role of both the wax heads and the *mimetai*—who can be considered as living casts themselves—was to *extract information* from the individual in order to replicate not only his semblance but also his entire "personality," as Julius von Schlosser [36: *passim*] put it. A cultural model emerges where death is closely tied to the realm of the double by virtue of an imitation so precise that it generates identity rather than just similarity.

In this respect, it is no coincidence that the same word *mimetai* was used in a different, Biblical context to describe imitation as an act that goes far beyond mere artistic reproduction: "Follow my example (*mimetai mou ginesthe*), as I follow the example of Christ," says Paul (*Corinthians* 11:1), urging the faithful to become models—*typoi*—for other people. The primary meaning of the Greek noun *typos* is again "imprint," *typtein* referring to the act of beating, pressing, and impressing. The culturally intriguing aspect lies in the ambiguity of the semantics surrounding this verb and its derivatives, which can denote both the act of exerting pressure and the result of such an act, both the cause and the effect it produces. The *typos* establishes an identity relationship between the model and the image(s) derived from it: each copy retains all the characteristic (that is, the "typical") features of the model, being "genetically" linked to it through contact. If the imitation process is carried out perfectly, the result will be a copy representing an occurrence of the original model, the two being ultimately indistinguishable. Thus, following Paul's line of thought, the patriarchs, the apostles and, of course, Christ himself should serve as models to be imitated by devout believers, who, in doing so, would in turn become models for others. The same holds true for the exemplary role of the aristocracy in ancient Rome: since they represented an illustrious past to be not only celebrated but also replicated by younger generations, the deceased were, in a sense, eternalised by the *mimetai* wearing their masks (or rather their faces).

The need to confront the "crisis of presence" [11] triggered by death by resorting to rituals and artefacts meant as substitutes for the deceased is rooted in a long and cross-cultural history [3, 37], which finds contemporary expression in today's attempts to "resurrect" the dead through generative artificial intelligence (AI). Indeed, neural networks are being used to replicate, and thereby immortalise, not just the physical appearance but also the behaviour of a deceased loved one. The basic idea remains the preservation of the individual's character, a term derived from the ancient Greek *charakter*, which not coincidentally also means imprint or cast. AI models thus become the new *mimetai*, with their source of information—extracted in an unprecedentedly rapid and automated manner—being our so-called "digital footprint," a notion that once again refers to the processes of impression and casting. By scraping content from emails, social media posts, and a wide range of data collected from smartphones and wearables, an artificial model can be created capable of conversing and interacting "in the personality" of a specific person, as explicitly suggested in a patent filed by Microsoft in 2020.[1]

The tech giant is not alone in exploiting this new form of "knowledge extractivism" [21, 30] to pursue digital immortality. On the contrary, it is simply a prominent player in the rapidly expanding market of death capitalism, a socio-economic system fuelled by a growing number of AI companies that promise to reunite us with the dearly departed. The *Black Mirror* episode "Be Right Back," in which a hyperrealistic synthetic body animated by AI replaces the physical body of a young man killed in a car accident, remains squarely within the realm of science fiction. However, the rapid pace of technological progress has already made it possible to create highly realistic and fully customisable avatars from 3D body scans or even from single photos taken with an ordinary smartphone. Thanks to machine learning techniques like Generative Adversarial Networks (GANs), these pictures can be transformed into faithful replicas of the individuals portrayed [40]. The digital likenesses can then be animated using deepfake technology, which allows for the manipulation of facial expressions, body movements, and voice, increasingly even in real-time. Deep

[1] The patent can be accessed on the official website of the United States Patent and Trademark Office at https://patentimages.storage.googleapis.com/8d/2a/7e/325266284d79df/US10853717.pdf (accessed 28 April 2025).

Nostalgia, for example, is an online service that enables users to upload old photos of their relatives and uses machine learning to animate them, adding lifelike movements. With just a few clicks, the images come to life, and users can see their ancestors smile, wink, or turn their heads, often with disturbingly realistic results. Furthermore, a growing number of applications now allow subscribers to add deceased loved ones to pre-existing photos and even create short emotional clips where they can "hug" their late relatives using just two pictures. This trend, swiftly gaining traction on TikTok, has cultural roots that can be traced back to the origins of photography in the late nineteenth century, when the then-new medium, through double-exposure techniques, allowed people to be immortalised one last time alongside those who had passed [18, 28: 135–169].

AI-produced videos are also being employed to animate key figures from world history. This presents numerous risks, as image- and voice-cloning technology has the potential not only to bring singers or movie stars back from the dead (which raises concerns about intellectual property and potential copyright infringement) but also to "resurrect" notorious criminals or controversial political figures. Famous cases include videos featuring an AI-powered English translation of Adolf Hitler's 1939 Reichstag speech, which have recently gone viral on social networks. It is a small step from this example to having the Nazi leader or any other historical figure make statements they never made, promoting racist content, or undermining the credibility of established historical facts.

The effectiveness of processes like these hinges on both the availability of the data relating to the deceased and the capabilities of generative models. To date, digital "twins" still exhibit various imperfections and respond awkwardly to even simple questions. However, given the extraordinary advances in machine learning and the fact that so much of our lives are now conducted online, with all the data exposure this entails, it seems likely that we are on the brink of a new era in the avatarisation of our personal identities. Innovative start-ups such as Eternime, a project launched in 2014 at the Massachusetts Institute of Technology, are working intensively to produce digital "selves" that can live on forever in the form of interactive chatbots or "thanabots." In 2016, for instance, James Vlahos, the CEO of HereAfterAI, created a "Dadbot" that mimicked his late father's speech patterns, attempting to replicate conversational attributes such as style, tone, and accent. That same year, Eugenia Kuyda, co-founder of the technology company Luka, digitally re-created her late best friend, Roman Mazurenko, using text messages he had sent her before his death (see Sisto [35] for these and other similar cases). Yet texting with a digital doppelgänger is only the first stage of a larger agenda. Companies like You, Only Virtual (https://www.myyov.com) offer a glimpse into future possibilities, claiming not only to collect, curate, and preserve the memories of someone you care about but also to capture, as the website suggests, their "personality" and "the distinct bond" you share with them, making it accessible for evermore through a hyperrealistic digital avatar (significantly called "versona," a portmanteau of "virtual" and "persona") which you can meet in immersive digital environments.

These digital doubles of ourselves are typically trained by having users answer a series of questions so as to create specific neural networks designed to transform our *individual* way of being-in-the-world into a *style* that the artificial model can replicate autonomously. Such a transformation is not without risk, as it delegates to algorithms what would otherwise be a matter of our personal idiosyncratic uniqueness, raising thorny questions about the accuracy and even the legitimacy of the virtual replica and its interaction with others. This applies, for instance, to Holocaust survivors whose stories have been recorded and turned into holograms to educate future generations about one of the most tragic crimes against humanity [1]. In this case, AI is employed to interpret questions from the public and provide them with a fixed set of pre-recorded responses. The decision to limit AI to simply selecting an answer originally provided by a real person stems precisely from the fear, as noted in a recent UNESCO report [26], that generative models could jeopardise Holocaust memory by fabricating fake content or distorting historical records.

Thanatology experts and grief counsellors caution against these new forms of "cyber-necromancy" [24], warning that the use of AI to digitally resurrect the departed is a double-edged sword. While it may offer momentary comfort by helping us remember loved ones and navigate through grief, it could also, especially over time, alter our genuine memories of them and even foster a denial of death's unavoidable reality: "In conditions of digital recall, loss is itself lost" [16: 2]. Additionally, AI may "hallucinate" and deliver unexpected, disturbing answers that can unsettle the mourner. This concern is explored in the 2024 German-American documentary *Eternal You*. The film made worldwide headlines when directors Hans Block and Moritz Riesewieck reported that one of the protagonists, Christi Angel, asked a digital recreation of her first love, Cameroun, whether he (or rather, it) was in heaven. The chatbot allegedly responded that he was in hell and would probably haunt her. However, Jason Rohrer, the developer of the AI model behind the virtual persona and founder of *Project December: Simulate the Dead* (https://projectdecember.net), clarified that those disturbing interactions were taken out of context and manipulated to create a compelling narrative and provide a striking soundbite. In fact, the conversation spanned numerous topics and was filled with confusing, disjointed, and surreal moments, as is often the case when chatting with AI agents. The film directors had, it would appear, cherry-picked and rearranged certain parts, presenting the conversation in a way that made it appear more shocking than it actually was. Thus, ironically, a documentary intended to explore how AI companies are attempting to turn immortality into a business has itself become part of that very same business.

The ultimate goal of all these thana-technologies seems to be to map our memories, experiences, and preferences and use this data to build a digital double of our consciousness. Whether this replica could ever be considered a "true," albeit digital, version of ourselves is the question posed by acclaimed composer and director Michel van der Aa in his award-winning multimedia film opera *Upload* (2019–2020). The story centres on a father who, without consulting his daughter, chooses to upload his mind into cyberspace through AI, with irreversible consequences. When she learns that he has been transformed into a digital entity and will live on forever, questions emerge about self-identity, the human, all-too-human, relentless pursuit of immortality, and the moral and legal implications involved. Conversations with family and friends, filled with differing recollections, highlight the fragility of memory—and, by extension, personality—whilst subtly unveiling the inaccuracies inherent in the uploading process. The fact that a simulation, due to its essentially artificial nature, can never fully capture the richness and nuance of human experience was foreshadowed earlier in the opera when the father visits a Swiss clinic for the upload. Just before the procedure begins, the lead doctor makes a statement that encapsulates the essence of all the technologies discussed: "You can live forever—you just have to die first."

21.3 Restless Images: AI as Digital *Nachleben*

The digital afterlife of data—images in particular—encompasses not only individual but also collective legacies, drawing attention to the cultural roots of every act of memory. Indeed, as Didi-Huberman [13: 197] maintains, "images, too, suffer from memories": they exist between presence and absence, between disappearance and re-emergence. Neither fully alive nor entirely dead, images embody the logic of the *revenant*: whether as welcome dreams or unsettling hallucinations, they harbour an unresolved past and haunt cultural history, disrupting its seemingly linear temporality, creating disorientation, and assuming unexpected, often paradoxical forms.

For instance, how does Darth Vader come to mourn an uncanny, apparently dead figure reminiscent of countless previous lamentations and whose face resembles Francis Bacon's distorted paintings (Fig. 21.2)?

This compelling vision emerges from an enigmatic process of *remembrance*, lying somewhere between recollection and creation—the very realm of AI image generation. In depicting a scene of mourning, AI displays the multifaceted relationship between images, gestures, and grief. Across most funeral cultural traditions, mourning rituals regulate the "return of the dead," mediating encounters between the living and the afterlife, thereby restoring order after the chaos introduced by death. Mourning gestures formalise this chaos, channelling unregulated grief into culturally shared expressions [11]. In this sense, they reference previous images: both gestures and visual representations serve as vehicles for perpetuating these expressions across generations, thus providing *orientation* to grief and helping negotiate relationships with the afterlife.

Image and mourning emerge, therefore, as interconnected mediators, each shaping cultural responses to death. Whether through the ghostly power of images or the performative nature of mourning, cultures foster practices related to the afterlife. In contemporary contexts, these practices are enacted through digital technologies which echo pre-digital rituals [3]. Thana-technologies reanimate traces steeped in cultural memory, enabling the return of the dead on multiple levels. AI systems, in particular, exhibit a metahistorical power, processing the past, reconstituting the present, and predicting future cultural forms [27]. Text-to-image Generative AI (GenAI) exemplifies this phenomenon, engaging in digital remediation [5] of motifs such as grief and lamentation, and shaping digital afterlife as a "new cultural form" [23: 169, 34].

Fig. 21.2 Image generated using Stable Diffusion with the prompt "lamentation" and the negative prompts "Virgin Mary" and "Jesus Christ" (2024)

As a step towards exploring these dynamics, we conducted a preliminary investigation into AI-generated text-to-image outputs, culminating in the image with which we began and to which we will return (Fig. 21.2). Using basic prompts, we aimed to examine how specific AI models "translate" abstract concepts such as "grief" into visual language [27: 81]. More broadly, the study considered how GenAI systems function as a form of digital afterlife, preserving, reshaping, and transmitting cultural expressions of grief while contributing to a visual digital legacy. The inquiry also sought to explore whether these generative processes reinforce or challenge existing cultural representations and expectations around grief. Below are selected prompts and corresponding outputs generated in 2024 by DeepAI-Images and Stable Diffusion, the latter allowing for both positive prompts (elements to be included in the image) and negative prompts (elements to be excluded) [2].

The images generated with the prompt "grief" exhibit a peculiar, striking aesthetic characterised by sombre tones, often dominated by deep blacks, muted greys, and occasional splashes of subdued colour, such as dark reds or soft blues. These compositions convey emotional gravity and ambiguity, featuring abstract textures evocative of swirling smoke or tears. Human figures, predominantly women, often appear in

Fig. 21.3 Images generated using DeepAI-images with the prompt "grief" (2024)

Fig. 21.4 Images generated using Stable Diffusion with the prompt "grief" (2024)

intense introspection, their faces downcast or covered. A recurring motif stands out: a gesture in which the hands move toward the head, covering the eyes and forehead, or resting atop the head (Figs. 21.3 and 21.4). This gesture, deeply familiar and widely recognisable, resonates with artistic works such as *Lakoninon* by Pia Lindman (see [17]) and *Mood Disorder* by David Horvitz.

Even more significantly, the gesture recalls some images of Ernesto de Martino's *Figurative Atlas of Weeping*, the appendix of *Death and Ritual Mourning* (1958). This comprehensive anthropological study and ethnographic inquiry conducted in Southern Italy's rural communities sought to examine the enduring *remains* of ancient mourning rituals and gestures. Meticulously documented through photographs, the funeral lament includes notable examples of the hands-to-head (or raising arms) gesture, performed by female mourners, custodians of the "technique of weeping"[2] [11]. This gesture, a survival of an ancient expression documented in both Egyptian and Greek mourning iconography, is interpreted as a formalisation and containment

[2] Unless otherwise stated, all translations of quotations are ours.

of the act of tearing one's hair, which in turn represents a "symbolic attenuation" of the self-destructive instinct that defines "the mourning crisis" [11]. Similarly, Aby Warburg emphasises the centrality of mourning gestures in cultural memory. His *Mnemosyne Atlas* devotes the entire 42nd panel to "mourning and lamentation," while his *Schemata Pathosformeln* (1905) conceptualise gestures like the "hands-to-head" as primal forms that bridge instinctual expression and cultural codification. Influenced by Darwin's reflection on *The Expressions of Emotions* (1872), Warburg identifies the lament as a fundamental expression that straddles the realms of instinctive behaviour and cultural formulation, reducing the distance between past and present.

In *Ninfa dolorosa* [12], Didi-Huberman has analysed the "longue durée" of gestures of mourning in contemporary visual and political forms, beginning with Georges Mérillion's *Pietà du Kosovo* (1990). The famous photograph—depicting the wake of Nasimi Elshani, killed by Serbian police in the village of Nagafc—makes grief visible in opposition to oppressive authority by drawing on a collective visual repertoire rooted in the Western iconography of lamentation. In doing so, it frames the experience of loss through established patterns, inscribing it within a shared visual grammar [38]. From this perspective, in performing the lamentation, the bereaved from Kosovo entrust their despair to a different kind of authority, which Roberto Longhi interpreting Caravaggio [25: 95] has defined as the "quasi-immanent authority of gestures": the authority of a collective memory of gestures, passed down across generations, capable of being reactivated even in distant times.

The *artificial* re-enactment of the "hands-to-head" gesture as a recurring cultural "topos" [10, 20] reflects a complex temporality. Generated images emerge from a process involving vast datasets of visual and textual materials collected from disparate historical and cultural contexts. This process weaves together motifs and iconographic fragments from different epochs, creating a temporal conglomerate that can be understood, following Didi Huberman's theory of images, as "anachronistic montage" [13: 139]. Warburg's contribution to art history lies precisely in identifying this non-linear temporality in which *l'image survivante* intermittently resurfaces across cultural history. The concept of *Nachleben*, the afterlife of images, refers to the temporality of the revenant—the forgotten image, a manifestation of collective unconscious memory traces.

While GenAI models primarily rely on past data taken from extensive repositories of images and texts, many scholars emphasise the "predictive" nature of AI generation, asserting that these systems do not merely imitate the past but generate new possibilities based on future expectations, attempting, for instance, "to predict the images that correspond best to your text input" [27: 80]. From this perspective, AI is seen as a forward-looking force, shaping what is yet to come.

The concept of *Nachleben* offers a more layered framework for understanding AI image generation. Rather than framing time as a simple past-future dichotomy, *Nachleben* conceives it as recursive, discontinuous, and stratified—intrinsically posthumous. It denotes a temporality in which the movement toward the future is inextricably bound to the persistence of past forms: a forward motion that proceeds with its gaze fixed backwards. Within this framework, generative AI can be understood not simply as a predictive force but as a site of temporal entanglement, shaped as much by what is remembered and reactivated as by what is imagined and projected. GenAI does not adhere to a strictly chronological model. Instead, it operates in a present, in Didi-Huberman's terms, "woven from multiple pasts" [13: 30], where images, cultural forms, and digital remains re-emerge in paradoxically unpredictable returns and unexpected configurations.

Unexpected was, indeed, one particular occurrence of the hands-to-head gesture which surfaced during our investigation. A female figure (Fig. 21.5) performs a strikingly exaggerated version of the gesture, closely resembling the Lucanian female mourner's expression documented by Franco Pinna's photograph (Fig. 21.6).

With her mouth open and her face expressing intense lamentation, this figure manifests a sense of despair in the face of death, departing from the more introspective and restrained depictions found in other images. This contradiction reflects the ambivalence of Western mourning expressions as analysed by de Martino in relation to Christian representations of the Virgin Mary. While the iconography of the *Stabat Mater* embodies composed grief, aligned with the Christian promise of eternal life, the syncretic figure of the *Mater Dolorosa* retains echoes of ancient mourning

Fig. 21.5 Image generated using Stable Diffusion with the prompt "grief" (2024)

Fig. 21.6 Photograph by Franco Pinna (Pisticci 1952)." *En el aire conmovido…*, Museo Nacional Centro de Arte Reina Sofía, Edificio Sabatini, 6 Nov. 2024–17 Mar. 2025

practices despite adhering to Christian moderation. In this duality, Mary becomes a liminal figure, balancing restraint and visceral expression of grief. This tension is powerfully rendered in famous Renaissance artworks such as Niccolò dell'Arca's *Lamentation over the Dead Christ* (1463–1490), where Mary Magdalene, in a state of frenzy beneath the cross, re-evokes the figure of the pagan mourner. Dell'Arca's portrayal, which resonates with de Martino's ethnographic studies in Lucania [8: 75], "subverts" Christian constraints on female lamentation, asserting the *survival* of ritualised mourning in visual culture.

A key aspect of examining AI-generated imagery lies in the tension between dominant and latent histories, which demands a critical investigation of normative representations of grief. The recurring hands-to-head gesture in mourning images discloses a cultural bias: grief is almost exclusively depicted through stereotypically *female* figures. As documented in research on AI gender biases [39], datasets that feed text-to-image tools not only reflect cultural stereotypes but actively reinforce them. These tools perpetuate a connection between women and mourning, reinforcing tropes that link grief, vulnerability, and caregiving with femininity, thus confining women to roles of sorrow [15]. However, the deeply rooted and cross-cultural association between women and mourning gestures extends far beyond notions of fragility, pointing instead to a more complex historical and cultural legacy. Ethnographic

Figs. 21.7 Images generated using Stable Diffusion with the prompt "grief" and the negative prompt "woman" (2024)

studies [33] and historical research [7] reveal how mourning gestures became markers of femininity in many cultures, where lamentation granted women unique authority over the afterlife. From the constraints placed on ritual lamentation in ancient Greece [19] to its suppression in early Christianity [11], women's mourning practices have historically been both a source of power and a locus of control as well as resistance. This political dimension of grief demands critical reappraisal [6].

Conversely, our inquiry revealed the under-representation of male figures in AI depictions of mourning. In many societies, men are expected to remain stoic in the face of loss, as social norms equate vulnerability with weakness and emotional restraint with strength [11]. Notably, when utilising a negative prompt to exclude female characters, Stable Diffusion often generated obscured or masked faces, with a dominant tone of mystery, gothic elements, or monastic atmospheres, as illustrated in Figs. 21.7.

The underlying factors that led to this result remain a matter of speculation. The presence of masks, for instance, could be interpreted as: (1) a "neutral alternative" designed to prevent the inadvertent inclusion of feminine features; (2) a symbolic element historically and visually associated with religious mourning imagery; or (3) a "neutral" or even "androgynous" substitute, particularly if grief is strongly coded as feminine within the training datasets and/or if male figures are under-represented or visually indistinct. In numerous explanatory frameworks, the under-representation of grieving men in the training dataset could play a relevant role: once "woman" is excluded as a visual possibility, the AI model seems to favour ambiguity over explicitly masculine traits, ultimately reinforcing their marginalisation in visual narratives of grief. A straightforward solution might involve explicitly prompting Stable Diffusion to generate an image of a grieving man. However, our approach was deliberately different: rather than forcing the model toward a predefined representational outcome, we sought to disrupt its normative visual tendencies. Our aim was more about *disorientation* than *reorientation.*

The appearance of masked figures in response to our prompts suggested a particularly productive direction for further inquiry, which foregrounds the implicit constraints embedded in AI-generated imagery. This, indeed, raises broader questions: what happens when we question normative representations not by prescribing an alternative, but by unsettling the visual conventions that shape them? How does the strategic use of both positive and negative prompts expose, reinforce, or subvert the underlying constraints of AI-driven image generation?

Building on this idea, we generated images using the positive prompt "lamentation" while applying negative prompts "Virgin Mary" and "Jesus Christ," thus effectively instructing Stable Diffusion to depict a scene of lamentation without its most iconic figures. The results provide intriguing insights, particularly with the surprising inclusion of the "Darth Vader" scene (Fig. 21.2) where the epitome of power, control, and emotional detachment appears as an unlikely mourner, reimagined in a moment of grief. By merging a pop-culture icon with the iconography of Christian lamentation, the image establishes a compelling contrast between dominance and vulnerability. Following the tendency to portray masked figures, Stable Diffusion ironically concludes by presenting a hypermasculine character in a moment of raw emotional exposure. This result not only questions the cultural association of grief with femininity but also calls into discussion the broader gendered norms around the expression of emotions. Thus, AI-generated imagery inadvertently challenges the same cultural norms it tends to reinforce, giving new significance to a figure that fits perfectly in our framework: Darth Vader—a fallen warrior untombed, brought back from death, alive only thanks to technology, an unexpected *mourner revenant.*

Inheriting the "authority" of gestures, GenAI reanimates visual traditions embedded in cultural memory, forging new pathways for the undisciplined trajectories of images and visual culture [9]. By remixing the past and anticipating future forms, GenAI reveals cultural memory as inherently *anachronistic*—an active, generative force. Compelled to subvert conventional representations, it produces images in which indeterminacy intersects with an unconscious visual culture, disorienting perception and redefining the boundaries between visibility and invisibility. Through a dynamic interplay of citation, distortion, and reinvention, AI-generated imagery functions as a digital afterlife for images: it preserves and transmits familiar motifs while simultaneously reconfiguring them, revealing implicit biases, bringing to light latent visual structures, and prompting new political reflections. In an era where image-making redefines immortality, AI is unsettling the notion of the dead "resting in peace" [22]: restless images return to haunt us, resurrecting forgotten histories and compelling critical reflection on how we see, remember, and mourn.

References

1. Alexander, N.: Obsolescence, forgotten: "survivor holograms", virtual reality, and the future of holocaust commemoration. Cinergie **10**(19), 57–68 (2021). https://doi.org/10.6092/issn.2280-9481/12205
2. Ban, Y., Wang, R., Zhou, T., Cheng, M., Gong, B., Hsieh, C.-J.: Understanding the impact of negative prompts: when and how do they take effect. Paper presented at the 18th European Conference on Computer Vision (ECCV 2024), Milan, Italy, 29 Sept–4 Oct 2024 (2024)
3. Belting, H.: An Anthropology of Images: Picture, Medium, Body (2001). Princeton University Press, Princeton (2011)
4. Bettini, M.: Death and its double. Imagines, ridiculum and honos in the Roman aristocratic funeral. In: Mustakallio, Katariina, et al. (eds.) Hoping for Continuity: Childhood, Education and Death in Antiquity and in the Middle Ages, pp. 191–202. Institutum Romanum Finlandiae, Rome (2005)
5. Bolter, J.D., Grusin, R.: Remediation: Understanding New Media. The MIT Press, Cambridge, MA (2000)
6. Butler, J.: Precarious Life. The Powers of Mourning and Violence. Verso, London–New York (2004)
7. Catoni, M.L.: La comunicazione non verbale nella Grecia antica. Bollati Boringhieri, Torino (2008)
8. Ciammitti, L.: L'Arca di Niccolò: riflessioni e documenti. Marsilio, Venezia (2022)
9. Cometa, M.: Cultura visuale. Raffaello Cortina, Milano (2020)
10. Curtius, E.R.: European Literature and the Latin Middle Ages (1953). Routledge & Kegan Paul, London (1979)
11. De Martino, E.: Morte e pianto rituale. Dal lamento funebre antico al pianto di Maria (1958). Einaudi, Torino (2021)
12. Didi-Huberman, G.: Ninfa Dolorosa. Essays sur la memoire d'un geste. Gallimard, Paris (2019)
13. Didi-Huberman, G.: The Surviving Image (2002). The Pennsylvania State University Press (2017)
14. Diodorus Siculus.: The Library of History. Harvard University Press, London-Cambridge, MA (1939)
15. Doka, K.J., Martin, T.L.: Men Don't Cry, Women Do: Transcending Gender Stereotypes of Grief. Routledge, London (2000)
16. Fisher, M.: Ghosts of My Life: Writings on Depression. Hauntology and Lost Futures. Zero Books, Alresdorf (2014)
17. Formis, B.: Gestes à l'œuvre. De l'incidence, Cherbourg-en-Cotentin (2015)
18. Harvey, J.: Photography and Spirit. Reaktion Books, London (2007)
19. Holst-Warhadt, G.: Dangerous Voices: Women's Laments and Greek Literature. Routledge, London (1992)
20. Huhtamo, E., Parikka, J.: Media Archaeology: Approaches, Applications, and Implications. University of California Press, Los Angeles (2011)
21. Joler, V.: New Extractivism. extractivism.online (2020). Accessed 28 Apr 2025
22. Kohn, T., Nansen, B., Arnold, M., Gibbs, M.: The restless dead in the digital cemetery. In: Moreman, C.M., Lewis, A.D. (eds.) Digital Death: Mortality and Beyond in the Online Age, pp. 111–124. Praeger, Westport (2014)
23. Lagerkvist, A.: Existential Media. Oxford University Press, Oxford (2023)
24. Le, T. T., Jin, R.: The beginning of cyber-necromancy: replacing deceased loved ones with AI agents. SocArXiv (2 Feb 2024). https://doi.org/10.31235/osf.io/8eb76
25. Longhi, R.: Caravaggio. Editori Riuniti, Rome (1982)
26. Makhortykh, M.: AI and the Holocaust: rewriting history? The impact of artificial intelligence on understanding the Holocaust (2024). https://doi.org/10.54675/ZHJC6844
27. Manovich, L., Arielli, E.: Artificial aesthetics. Generative AI, art and visual media. https://manovich.net/index.php/projects/artificial-aesthetics (2024)

28. Natale, S.: Supernatural Entertainments: Victorian Spiritualism and the Rise of Modern Media Culture. The Pennsylvania State University Press, University Park Pennsylvania (2016)
29. Paglen, T., Robin Jones, Ch.: The artist Trevor Paglen, the surveillance state, and your life. SSENSE. https://www.ssense.com/en-us/editorial/art/this-artist-the-military-industrial-complex-and-your-life?utm_source=twitter&utm_medium=social&utm_campaign=portrait&utm_term=trevor_paglen_22_03_2018. Accessed 28 Apr 2025
30. Pasquinelli, M., Joler, V.: The Nooscope manifested: artificial intelligence as instrument of knowledge extractivism. nooscope.ai (2020). Accessed 28 Apr 2025
31. Plautus: Amphitryon. Harvard University Press, Cambridge, MA–London (2011)
32. Polybius: The Histories, vol. 3. Harvard University Press, Cambridge, MA (2010)
33. Pussetti, C. G.: Le donne, la morte, il dolore: pratiche funebri ed elaborazione del lutto tra i Bijagó dell'isola di Bubaque (Guinea Bissau). La Ricerca Folklorica 49, 45–54 (2004)
34. Savin-Baden, M., Mason-Robbie, V. (eds.): Digital Afterlife: Death Matters in a Digital Age. Chapman and Hall/CRC, New York (2020)
35. Sisto, D.: Online afterlives. Immortality, memory, and grief in digital culture (2018). The MIT Press, Cambridge, Massachusetts-London (2020)
36. von Schlosser, J.: History of Portraiture in Wax (1911). In: Panzanelli, R. (ed.), Ephemeral Bodies: *Wax* Sculpture and the Human Figure, pp. 172–303. Getty Research Institute, Los Angeles (2008)
37. Vernant, J.-P.: The figuration of the invisible and the psychological category of the double: the *kolossos* (1965). In Myth and Thought among the Greeks, pp. 321–332. Zone Books, New York (2006)
38. Volli, U.: False icone. Per un'analisi semiotica del fotogiornalismo. In: Del Marco, V., Pezzini, I. (eds.) La fotografia. Oggetto teorico e pratica sociale, pp. 356–379. Edizioni Nuova Cultura, Rome (2011)
39. Wu, Y., Nakashima, Y., Garcia, N.: Gender bias evaluation in text-to-image generation: A survey. arXiv preprint arXiv:2408.11358. https://doi.org/10.48550/arXiv.2408.11358 (2024)
40. Zhong, L., et al.: Animated 3D human avatars from a single image with GAN-based texture inference. Comput. Graph. **95**, 81–91 (2021). https://doi.org/10.1016/j.cag.2021.01.002

Pietro Conte is associate professor of Aesthetics in the Department of Philosophy "Piero Martinetti" at the University of Milan, and a member of the research centers "EXT. Extended Realities" and "HCAI. Human-Centered Artificial Intelligence." His research focuses on illusion, hyperrealism, and the multifarious practices of un-framing, especially in relation to immersive (virtual) environments. He has explored this thematic cluster in the monographs *Unframing Aesthetics* (2020) and *In carne e cera. Estetica e fenomenologia dell'iperrealismo* (2014).

Maria Serafini is a PhD candidate in Philosophy and Human Sciences at the University of Milan and a former visiting PhD student at the Warburg Institute in London. She was a member of the ERC project "AN-ICON" (*An-Iconology: History, Theory, and Practices of Environmental Images*). She holds a BA in Philosophy from the University of Rome "La Sapienza" and MAs from the University of Roma Tre, in Cognitive Sciences and in Philosophical Sciences. Her research combines aesthetics, visual culture, and anthropology, investigating emerging digital practices related to death and mourning, as well as the digital remediation of pre-digital rituals and iconographies of mourning. She focused on the relationship between presence, digital afterlife, and virtual environments. She has presented at international conferences and co-organised an international symposium on digital death and immortality.

GPSR Compliance
The European Union's (EU) General Product Safety Regulation (GPSR) is a set of rules that requires consumer products to be safe and our obligations to ensure this.

If you have any concerns about our products, you can contact us on

ProductSafety@springernature.com

In case Publisher is established outside the EU, the EU authorized representative is:

Springer Nature Customer Service Center GmbH
Europaplatz 3
69115 Heidelberg, Germany

www.ingramcontent.com/pod-product-compliance
Ingram Content Group UK Ltd.
Pitfield, Milton Keynes, MK11 3LW, UK
UKHW020631260926
475829UK00003BA/20
* 9 7 8 3 0 3 2 0 8 7 2 5 6 *